Policies for Sustainable Governance of Global Ecosystem Services

Environment

MIND

Society Economy

Munasinghe Institute for Development
'making development more sustainable-MDMS'

WORLD
RESOURCES
INSTITUTE

Policies for Sustainable Governance of Global Ecosystem Services

Edited by

Janet Ranganathan
Director, People and Ecosystems Program, World Resources Institute, Washington, DC, USA

Mohan Munasinghe
Chairman, Munasinghe Institute for Development (MIND), Colombo, Sri Lanka, and Vice Chair, Intergovernmental Panel on Climate Change (IPCC), Geneva, Switzerland

Frances Irwin
Fellow, World Resources Institute, Washington, DC, USA

Edward Elgar
Cheltenham, UK • Northampton, MA, USA

Published by
Edward Elgar Publishing Limited
Glensanda House
Montpellier Parade
Cheltenham
Glos GL50 1UA
UK

Edward Elgar Publishing, Inc.
William Pratt House
9 Dewey Court
Northampton
Massachusetts 01060
USA

A catalogue record for this book
is available from the British Library

Library of Congress Cataloguing in Publication Data

Policies for sustainable governance of global ecosystem services / edited by Janet Ranganathan ... [et al.].
p. cm.
Includes bibliographical references and index.
1. Ecosystem management. 2. Environmental policy. I. Ranganathan, Janet, 1962–

QH75.P59 2007
333.95'16—dc22

2007039427

ISBN 978 1 84720 244 4

Printed and bound in Great Britain by MPG Books Ltd, Bodmin, Cornwall

Contents

About the editors *vii*
List of contributors *ix*

Editors' introduction 1

Action agenda for sustaining ecosystem services 21
Frances Irwin and Janet Ranganathan

PART I INCORPORATING ECOSYSTEM SERVICES INTO DEVELOPMENT PLANNING AND PRACTICE

1. Mainstreaming and implementing the Millennium Ecosystem Assessment results by integrating them into sustainable development strategy: Applying the Action Impact Matrix Methodology 73
Mohan Munasinghe

2. Development, ecosystems and governance: What are the priorities for policy makers? 108
Antonio La Vina

3. To change global change: Ecosystem transformation and conflict in the 21st century 125
Nicolás J. Lucas, Iokiñe Rodríguez and Hernán Darío Correa

4. Putting the assessment into action: Six steps for governments 149
Lailai Li

5. From assessment to action: Operationalizing environmental sustainability at the national level 169
Guido Schmidt-Traub and Albert Cho

6. Environmental policy priorities for the World Bank 190
Robert Goodland

PART II USING NEW TOOLS AND SKILLS

7. A scenarios approach to developing ecosystem management strategies 215
Mark E. Bateman

8. The implications of interdisciplinary scientific assessments for environmental governance 235
Richard B. Norgaard

9. Looking back and ahead in ecosystems: Reflections on the lessons of the Millennium Ecosystem Assessment 254
Anthony C. Janetos

10. Policy actions to help move us towards ecosystem security 273
Karin M. Krchnak

11. Framing the Millennium Ecosystem Assessment messages for political resonance 296
Frances Seymour

PART III BUILDING INSTITUTIONS TO SUSTAIN ECOSYSTEM SERVICES

12. Meeting the Millennium Ecosystem Assessment challenge: The case for biome stewardship councils 315
Sudhir Chella Rajan

13. Smart globalization: Opportunities for business, scientific and political leadership 329
David Jhirad

14. A commission to spur investments in ecosystem services for the poor 346
Frances Irwin

Index 367

About the editors

Janet Ranganathan is the Director of the People and Ecosystems Program at the World Resources Institute (WRI). The Program's goal is to reverse the rapid degradation of ecosystems and assure their capacity to sustain life and prosperity. Janet works extensively with business, government and civil society to catalyze changes in the way ecosystems are valued and managed. Janet founded and directed WRI's Greenhouse Gas Protocol Initiative, an international multi-stakeholder partnership convened by WRI and the World Business Council for Sustainable Development, for the purpose of developing international greenhouse gas accounting and reporting standards. Before joining WRI she worked as a Senior Lecturer at the University of Hertfordshire, UK and in a policy and regulatory capacity at the UK Department of Environment and Hertfordshire Waste Regulatory Authority.

Mohan Munasinghe is Vice Chairman, Intergovernmental Panel on Climate Change, co-winner of the 2007 Nobel Prize for Peace. A Sri Lankan, he has earned post-graduate degrees in engineering, physics and development economics from Cambridge University (UK), Massachusetts Institute of Technology (USA), and McGill University and Concordia University (Canada). He has also received several honorary doctorates (*honoris causa*). Currently, he is Chairman, Munasinghe Institute for Development (MIND), Colombo; Vice Chair, UN Intergovernmental Panel on Climate Change (IPCC), Geneva; Honorary Senior Advisor to the Sri Lanka Government; and Visiting Professor, United Nations University, Tokyo. During 35 years of distinguished public service, he has been Senior Energy Advisor to the President of Sri Lanka, and Advisor to the United States Presidents Council on Environmental Quality. At the World Bank, he was Senior Advisor/Manager, covering sustainable development, environmental economics and policy, energy, water, and urban transport, etc. He has taught as Visiting Professor at several major universities worldwide. He has won many international prizes and medals for his research and its applications, authoring 90 books and over 300 technical papers on economics, sustainable development, climate change, power, energy, water resources, transport, environment, disasters and information technology. He is a Fellow of several internationally recognized Academies of Science, and serves on the editorial boards of a dozen academic journals.

Frances Irwin, a fellow at the World Resources Institute, currently writes about governance and ecosystem services. Much of her work over the past 35 years at non-governmental organizations ranging from The Institute of Ecology and the Vermont Natural Resources Council to The Conservation Foundation and World Wildlife Fund-US has focussed on ensuring that citizens have a voice in decisions that affect their lives and environment. As director of the pollution prevention program at the WWF-US, she organized the research and participation of non-governmental groups in preparing international guidance on public pollutant registers. Her publications also cover areas such as cumulative environmental impacts, public policy for chemicals, integrated pollution prevention and control, and electronics innovation for climate change.

Contributors

Mark E. Bateman, IW Financial, USA
Albert Cho, McKinsey & Co., USA
Hernán Darío Correa, Wildlife Conservation Society Colombia Program, Colombia
Robert Goodland, formerly World Bank, USA
Frances Irwin, World Resources Institute, USA
Anthony C. Janetos, Joint Global Change Research Institute, USA
David Jhirad, World Resources Institute, USA
Karin M. Krchnak, The Nature Conservancy (TNC), USA
Antonio La Vina, Ateneo de Manila University, Philippines
Lailai Li, Institute for Environment and Development/LEAD, China
Nicolás J. Lucas, Centro Fueguino para el Desarrollo Sustentable, Argentina
Mohan Munasinghe, Munasinghe Institute for Development (MIND), Sri Lanka, and Intergovernmental Panel on Climate Change (IPCC), Switzerland
Richard B. Norgaard, University of California, USA
Sudhir Chella Rajan, Indian Institute of Technology, India
Janet Ranganathan, World Resources Institute, USA
Iokiñe Rodríguez, Wildlife Conservation Society Colombia Program, Colombia
Guido Schmidt-Traub, UN Millennium Project, USA
Frances Seymour, Centre for International Forestry Research (CIFOR), Indonesia

Editors' introduction

Nature is sending us an urgent warning signal. If we had a panel with lights monitoring the status of ecosystem services, 15 (nearly two thirds of 24 examined) would be flashing red, five would be yellow, and only four would beam a steady green (Table 1) (MA, 2005c: 41–5).

Table 1 The condition of ecosystem services

Ecosystem Services	Degraded	Mixed	Enhanced
Provisioning	Capture fisheries Wild foods Wood fuel Genetic resources Biochemicals Fresh water	Timber Fiber	Crops Livestock Aquaculture
Regulating	Air quality regulation Regional and local climate regulation Erosion regulation Water purification Pest regulation Pollination Natural hazard regulation	Water regulation (i.e., flood protection) Disease regulation	Carbon sequestration
Cultural	Spiritual and religious values Aesthetic values	Recreation and ecotourism	

The most comprehensive audit ever undertaken of our natural systems and their capacity to support human well-being finds that the crucial natural assets that all people, but especially the poor, depend upon are declining. These assets or services range from fresh water and pollination to storm protection and a source of spiritual values.

These warnings are highlighted in the Millennium Ecosystem

Assessment. This scientific effort provides an evaluation of the condition and trends of the world's ecosystems and their services on which people and all other living creatures rely. The Assessment takes a century-long view – 50 years back and 50 years forward. It assesses separately conditions and trends of ecosystem services at the global scale and at 33 places ranging from local to national and regional (see Figure 1).

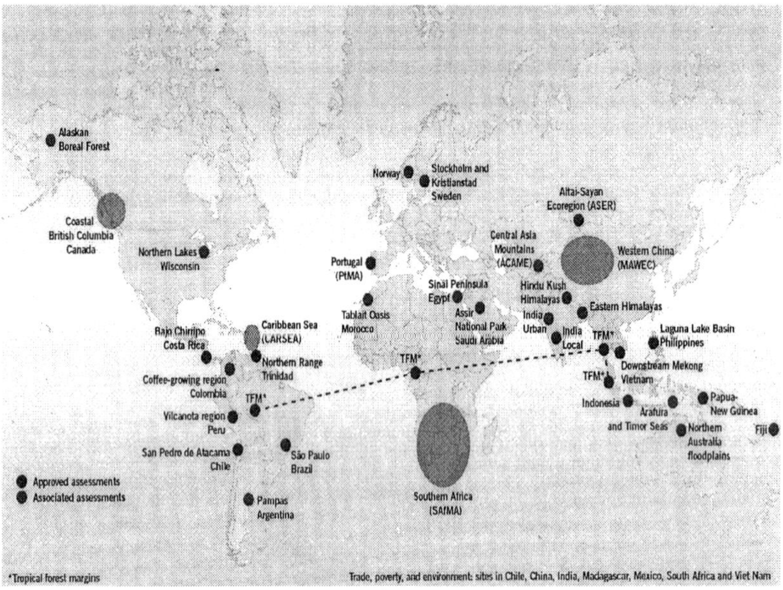

Source: MA (2005c)

Figure 1 Locations of the sub-global assessments of the Millennium Ecosystem Assessment

Released in March 2005, the Assessment is the culmination of a four-year effort involving over 1300 of the world's pre-eminent scientists from 96 countries (see Box 1). By design, the Assessment was not policy prescriptive – proposing policy recommendations was left to others. The World Resources Institute invited 17 experts from diverse backgrounds to review the Assessment's findings, in order to focus the debate on the seeds of a global Action Agenda to turn the now flashing red lights to yellow and green, thereby enabling both people and ecosystems to thrive. The only instructions given to them were that all options should be considered.

DRAWING DOWN NATURAL CAPITAL

The feedback from the Assessment is startling. Every natural system – forest, dryland, inland water, coastal, marine, island, mountain and polar – is experiencing rapid increases in the impact of climate change and pollution.

BOX 1 WHAT WAS UNIQUE ABOUT THE MILLENNIUM ECOSYSTEM ASSESSMENT?

- *It examined ecosystem services – the benefits people obtain from nature.* This approach focuses attention on the often overlooked reality that success in economic development must be underpinned by healthy ecosystems. Environmental assessment usually focuses on the effects of humans on ecosystems rather than the state of natural assets essential for human well-being.
- *It took a multi-scale approach.* To understand the relationship among global, regional and local drivers of ecosystem change and their effects on specific communities and landscapes, the Assessment examined services from a range of temporal and spatial scales. Cutting trees at the local scale may not change the flow of water in a nearby river, but changes at the scale of a river basin can affect the magnitude of downstream flooding. Soil erosion may not affect agriculture productivity over days or weeks, but will have major effects over years or decades.
- *It included a wide range of scientists.* The selection of scientists ensured geographical balance between developed and developing countries, and began to bridge gaps between natural and social scientists.
- *It involved civil society and business.* By including a range of stakeholders in addition to scientists in designing and carrying out the work, the Assessment sought to obtain a broader view of ecosystem services and increase the use of the results.
- *It was designed to provide information requested by governments.* The Conferences of Parties (COPs) to four UN conventions asked for the assessment. The formal request came from resolutions passed by the COPs.

Temperate grasslands (a type of dryland), and inland and coastal water systems already exhibit very high impact from habitat change and pollution (MA, 2005c: 16). These changes have helped meet a growing population's need for food and water, have reduced the proportion of malnourished people, and improved human health (MA, 2005c: 5–6). But these benefits have come at a growing cost to the health of ecosystems, challenging our ability to sustain improvements over time and jeopardizing efforts to improve the lives of the millions of people living in poverty. For the sake of short-term gain, we have undermined nature's capacity to provide the services that vibrant communities need – food, clean water and sanitation, natural hazard protection, and cultural landscapes.

The following are some illustrative findings that highlight the scale of human impacts on natural systems over the past 50 years.

- **Worldwide fish landings peaked in the 1980s and have since remained static**, even though demand has never been greater. Aquaculture (fish farming) has mitigated some of this shortfall, contributing 43 per cent of total fish production in 2004 (FAO, 2007), but sometimes at the expense of extensive mangrove destruction, increasing the vulnerability of coastal communities to natural hazards (MA, 2005c: 43) (see Box 2).
- **Twenty per cent of coral reefs and nearly a quarter of mangroves have been lost since about 1980,** along with their capacity to buffer coastal communities from storms. Nearly half of the world's major cities are within 50 kilometers of a coast (MA, 2005c: 27). People are more vulnerable than ever to extreme events, as demonstrated by the high loss of life and accelerating economic losses from natural disasters such as the Asian tsunami in 2004 (MA, 2005c: 43).
- **More land was converted to crops between 1950 and 1980 than in the 150 years from 1700 to 1850**. Over two thirds of temperate grasslands and Mediterranean forests and over one half of tropical dry forest, temperate broadleaf forest, tropical grassland and flooded grasslands have been converted. Most land suitable for intensive agriculture is already under cultivation (MA, 2005c: 32–3); further increases in agricultural output are likely to result from more intensive management of existing cultivated areas. The economic value of converted land is often far less than that of sustainably managed natural systems, which provide a greater variety of services than intensively cultivated systems.
- **Over half of the synthetic nitrogen fertilizer ever used on the planet has been applied to crops in the past two decades** (MA, 2005c: 2). As much as 50 per cent of this is lost, contributing to rapidly rising nitrate concentrations in rivers, lakes and coastal areas (MA, 2005c: 43, 69). This can lead to 'dead zones', which are areas

where marine life cannot survive because of low oxygen levels. Dead zones can be found around the globe, including in the Kattegat at the mouth of the Baltic Sea and in the Gulf of Mexico, and can reach up to 70 000 square kilometers in size (Diaz, 2001; MA, 2005a: 346).

- **The amount of water impounded by dams has quadrupled since 1960**. Three to six times as much water is now held in reservoirs as in natural rivers (MA, 2005c: 2). As a result of large-scale dam construction during the last half of the twentieth century, tens of millions of people have been displaced and more have suffered a loss of the resources on which their livelihoods depend. Ecosystems have also been adversely affected – freshwater biodiversity is threatened, wetlands are increasingly lost, and the capacity of surrounding ecosystems to provide the services on which society depends is quickly disappearing (World Commission on Dams, 2000).

BOX 2 SHRIMP AQUACULTURE – PROFIT OR LOSS?

Adapted from Sathirathai and Barbier (2001)

Next time you enjoy a shrimp dinner consider for a moment that it may have started its life in a South Asian pond, built in place of a mangrove swamp, and have been fed with wild fish from local fisheries. Two likely consequences of this process are degraded fish nurseries and a weakened natural barrier to the sea for adjacent communities. In the last five years global shrimp production has grown at a rate of 10–20 per cent per annum. Shrimp farms have multiplied along coastal systems in South, East and Southeast Asia, often replacing mangroves (MA, 2005c: 41). Thailand alone accounts for 25 per cent of the world shrimp market, with the number of shrimp farms increasing from 19 700 in 1995 to 35 000 by 2003 (MA, 2005a: 358).

Mangroves are coastal wetlands that are flooded with brackish water, making them ideally suited for aquaculture. From the perspective of a shrimp farmer, converting mangroves to aquaculture makes financial sense, as illustrated by one study in southern Thailand which found that aquacultures had a net present value (using a 12 per cent discount rate) of US$8072 per hectare compared with only US$735 for an intact mangrove. However, from a fuller cost accounting perspective the economics of shrimp aquacultures are not so rosy. Intact mangroves provide a number of important services that are not captured in the marketplace. These include supporting offshore fisheries through the provision of nursery habitat for marine life, as well as storm protection for coastal communities. At the same time shrimp aquaculture has a number of environmental and societal costs not typically borne by the farmer, including water pollution and land degradation. The productive life of a typical commercial shrimp farm in Thailand is only five years, after which yields dramatically decline and disease increases. The site is then often abandoned for another location, leaving behind a wasteland that is unfit for further productive use. Aquaculture also enjoys a number of input subsidies including nominal land rent and taxes. Taking these additional factors into account, a fuller comparison of the economic value of intact mangroves and shrimp farms reveals the net present value of intact mangroves to be in the region of US$31 887 per hectare versus minus US$4918 for shrimp aquaculture. Even if the mangrove restoration costs are excluded, the mangrove benefits are still greater than aquaculture.

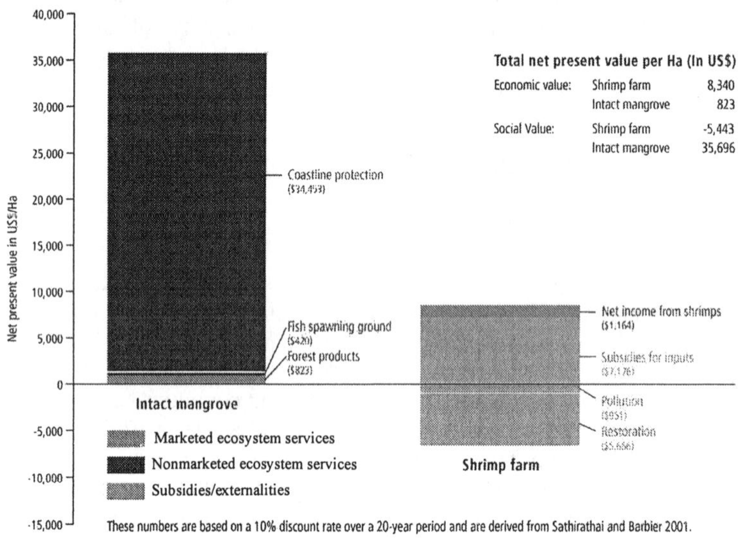

Figure 2 The hidden costs and benefits of mangrove swamps vs. shrimp farms

This example also raises fairness questions in regard to the distribution of costs and benefits from mangrove conversion. The shrimp farmers who gained in this example are mainly people from outside the community who could afford the high initial investment needed. The losers are the poor who depend on mangrove services for their livelihoods and well-being. This example is not unique, but pervasive throughout Southeast Asia and many other coastal tropical regions. In the five Asian countries most affected by the 2004 tsunami, for example, human activities had reduced the area of mangroves by 26 per cent between 1980 and 2000. After the tsunami hit, areas buffered by coastal forests, like mangroves, were found to be less damaged than areas without tree vegetation, highlighting the protective services of mangroves (Danielsen et al., 2005: 643).

RECOGNIZING TWO SIDES OF THE SAME COIN

We tend to take nature's services for granted. Gretchen Daily says that 'the disparity between actual and perceived value is probably nowhere greater than in the case of ecosystem services' (Daily, 1997: 6). Humans seldom think about the many different ways we gain from the natural capital provided by living and non-living resources. As a result, we often overlook the fact that healthy ecosystems underpin long-term economic success. The economy impacts the environment but it also depends on natural assets. Our individual and national security also depends upon services that are provided by nature, though we often take this for granted as well. In our ever more globalized world, threats arising from ecosystem change can easily cross borders. When ecosystem services such as freshwater provision and disease regulation are compromised, the problems that result, such as epidemics, famines, political instability or refugee migration, can affect people around the world.

The Assessment provides a conceptual framework that allows us to see these two sides of the coin (Figure 3) by capturing the circular relationship between the well-being of people and the services that ecosystems offer on one side, and the impacts of human activities on ecosystems on the other side.

This approach goes beyond the familiar process of assessing and mitigating the environmental impact of a decision to build a highway or a mine. It encompasses human impact through both direct drivers such as decisions about what technology to use, and indirect drivers such as patterns of consumption. But it goes beyond this. It extends the focus to how the changes we make to natural systems alter the supply of their services, and how these changes in turn affect our lives. These services are fundamental to the food we eat, the air we breathe, the landscapes we enjoy, the protection we have against storms and disease, and, more broadly, our freedom of choice and action.

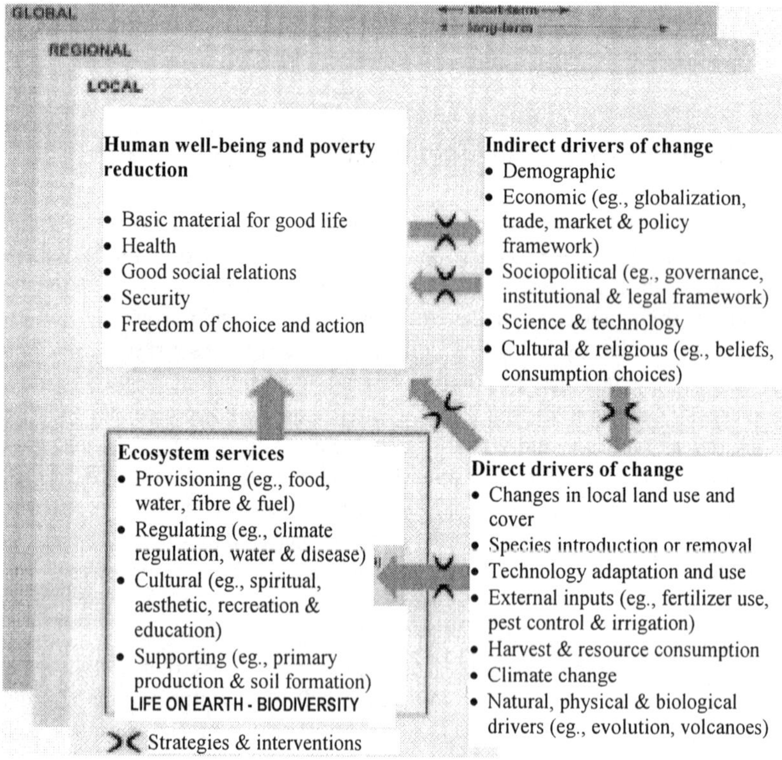

Source: MA (2005e)

Figure 3 The Millennium Ecosystem Assessment conceptual framework

ECOSYSTEM SERVICES: THE BASICS

The term 'ecosystem' emerged in the 1930s. It means a 'dynamic complex of plant, animal and micro-organism communities and their nonliving environment interacting as a functional unit' (CBD 1992). The term 'ecosystem services' as 'the benefits ecosystems provide' was first used in the 1960s.

In the last decade, research on ecosystem services has grown dramatically and the concept has been refined. The Millennium Ecosystem Assessment uses a definition by Costanza and colleagues that includes both natural and human-modified ecosystems as sources of ecosystem services. It follows Daily in using the term 'services' to cover both goods and services (MA, 2005b: 56). Because ecosystem goods and services can be difficult to distinguish, the Assessment defines services to include both the tangible (such as, food) and intangible (that is, aesthetic) benefits that humans obtain from ecosystems. The Assessment categorizes nature's services – which we use interchangeably with ecosystem services – in four groups:

- *Provisioning services:* food, fresh water, fiber and fuel
- *Regulating services:* biophysical processes that control climate, floods, diseases, air and water quality, pollination and erosion
- *Cultural services:* places that provide recreational, aesthetic or spiritual values
- *Supporting services:* the underlying processes such as formation of soil, photosynthesis, and nutrient cycling.

We are still at an early stage of understanding and evaluating the health of ecosystem services and how human alterations to natural systems affect their supply and distribution over time. Several decades of debate about achieving sustainable development have begun to suggest how to incorporate our knowledge into cogent policy and actions. In 1969, the IUCN (now the World Conservation Union) adopted a mandate that recognized 'the perpetuation and enhancement of the living world – man's natural environment – and the natural resources on which all living things depend' and referred to managing 'air, water, soils, minerals and living species including man, so as to achieve the highest sustainable quality of life' (Adams, 2006). The Brundtland Commission's definition of sustainable development – 'development which meets the needs and aspirations of the present generation without compromising the ability of future generations to meet their own needs' – is widely used but not precise enough to guide specific action (WCED, 1987).

The evolution of the sustainable development concept to include social, economic and environmental dimensions is described in Munasinghe (2007). Increasingly, scientists think about sustainability in terms of the ability of socioeconomic and ecological systems to adapt to and benefit from change, rather than as a static balance between people and nature. Societies and the ecosystems on which they depend are said to be resilient if they are able to absorb change while maintaining their basic structure, identity and function. Societies are adaptable if people can manage the links between society and ecosystems to maintain resilience. Information on ecosystem services can help decision makers distinguish between options that are likely to enhance ecological and social resilience (and thus sustainability) and those that may undermine it.

KEY FINDINGS OF THE MILLENNIUM ECOSYSTEM ASSESSMENT

The Assessment looked in detail at 29 of the many services that ecosystems provide people. These services were selected both because they have been significantly affected by recent ecosystem change and because human well-being is likely to be greatly affected as a result. The availability of information and data was also a factor in choosing which services to assess. The 29 ultimately chosen include five supporting services. Since, by definition, supporting services are not directly used by people, these five services were not given a rating of enhanced or degraded. Alterations in these services do, however, greatly affect the other three categories of services and therefore may indirectly be used by people and be enhanced or degraded (MA, 2005c: 45).

Table 1 summarizes how human use has enhanced or degraded the ecosystem services that the Assessment evaluated, namely the provisioning, regulating, and cultural services. Only four of the 24 services rated were enhanced, three of which are provisioning services (crops, livestock and aquaculture), reflecting society's focus on managing ecosystems to produce food for a growing population. In contrast, 15 services were found to be in decline worldwide – six of 11 provisioning services, seven of ten regulating services, and two of three cultural services examined. By using up provisioning services such as timber, groundwater, and capture fisheries faster than they can be replenished, we are depleting the assets available to future generations. Actions to increase provisioning services often lead to the degradation of regulating and cultural services.

Box 3 provides an overview of four of the Assessment's major findings. By converting natural systems into production factories for a handful of provisioning services – crops, livestock and aquaculture – we lose a myriad of other often more valuable services, not captured in the marketplace, but nonetheless critical to our well-being. The costs of these lost services are already being borne by parts of society. The first to suffer are often the poor, who depend upon the lost services for their livelihoods, while the beneficiaries are often wealthier and live in places far away from the degraded services.

BOX 3 FOUR MAJOR FINDINGS OF THE MILLENNIUM ECOSYSTEM ASSESSMENT

The collective findings of the Millennium Ecosystem Assessment document unprecedented ecosystem change in the last 50 years, and their weakened capacity to provide essential services in the future. The findings challenge the soundness of basic premises that underlie the valuation of ecosystems and their services in economies. They also cast doubts on the way we now make decisions to safeguard ecosystems and maintain their services for future generations. The Assessment lays out four main findings.

1. Humans have Radically Altered Ecosystems in just 50 years

In the last half of the twentieth century, humans changed ecosystems more rapidly and extensively than in any comparable period of history, primarily to meet growing needs for food, fresh water, timber, fiber and fuel. This has resulted in a substantial and largely irreversible loss in the variety of life on earth (MA, 2005c: 2). The distribution of remaining species is becoming more similar across regions, mainly as a result of species moving from one region to another, following the increasing movement of people around the world. Almost one third of land is now under cultivation for the purpose of producing provisioning services for people (MA, 2005c: 32). An increase in atmospheric carbon dioxide concentration by one third since 1750, two thirds of which occurred since 1959, has the greatest potential to alter natural systems through changes in global weather systems. Nature's capacity to adapt to such climatic changes is limited by the anticipated speed of climate change and by the changes humans have made to the landscape. Many species are now stranded in small islands of nature, surrounded by urban or intensively farmed land, making it difficult or impossible for them to adapt or move to more suitable areas (MA, 2005d: 13–14).

2. Ecosystem Change has Brought Gains in Human Well-being, but at High Costs to 'Natural' Capital

Taken together, ecosystem changes have resulted in significant benefits to humans, including improvements in health and a reduction in the proportion of malnourished people. However, these gains have come at an increasing cost. The Assessment's findings indicate that increases in the provisioning services that have a market price have inadvertently caused the degradation of other ecosystem services. The degraded services are often 'regulating' services such as those provided by mangroves that limit flooding and provide nurseries for fish. These regulating services are mostly public goods that have no value in the marketplace until they are lost. As the experience of the tsunami demonstrates, the loss of the flood protection services of mangroves becomes strikingly apparent after disaster hits.

The Assessment highlights the growing body of evidence that human-induced ecosystem changes increase the likelihood of abrupt and potentially irreversible ecosystem changes. The increased likelihood of abrupt changes stems from a combination of biodiversity loss and growing pressure from overharvesting, climate change, invasive species, and nutrient loading, weakening the resilience of ecosystems (MA, 2005c: 12). These problems, unless addressed, will substantially diminish the benefits that future generations obtain from ecosystems. Of particular concern are disruptions to the supply of ecosystem services, such as the 1990s collapse of the Atlantic cod stocks off Newfoundland, Canada (see Figure 4). This fishery had supported small-scale, inshore fisheries for centuries. In the late 1950s, offshore bottom trawlers began exploiting the deeper part of the stock, increasing the size of the catch but depleting stocks. After the stocks collapsed in the early 1990s, a commercial fishing moratorium was instigated, but more than a decade later there are few signs of the fish returning (MA, 2005c: 58).

Fish landings in thousands of tons

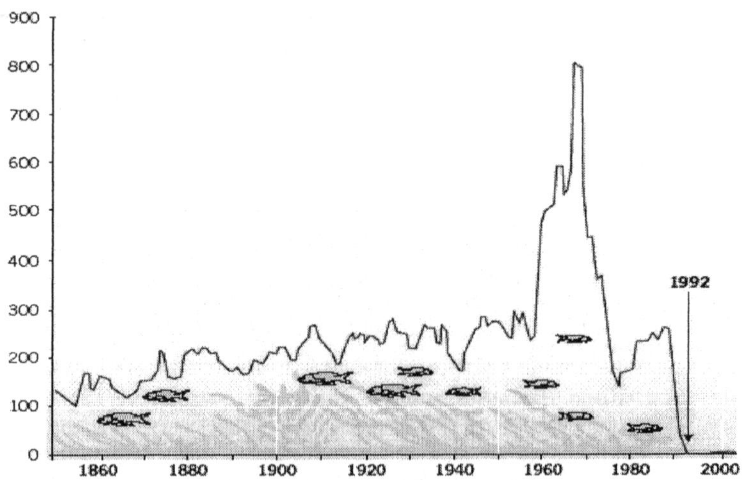

Source: MA (2005e)

Figure 4 Fish landings of Atlantic cod off Newfoundland

3. Further Unsustainable Practices will Threaten Development Goals

Ecosystem degradation, greater risk of ecosystem collapse, and exacerbation of poverty, particularly among the resource-dependent poor, all have their roots in the choice of development paths. If they continue unchecked, these problems will undermine the gains in human well-being. The Assessment concluded that degradation of ecosystems presents a significant barrier to achieving the UN Millennium Development Goals agreed to by world leaders at the United Nations in 2000 to reduce poverty and improve human well-being by 2015. Poverty and ecosystem degradation often go hand in hand. Half the urban population of Africa, Asia, Latin America and the Caribbean suffers from diseases associated with inadequate water and sanitation. Declining fisheries reduce the availability of an inexpensive source of protein in developing countries. Development prospects for drylands, which cover approximately 40 per cent of the Earth's land surface and are home for 2 billion people, were singled out by the Assessment as being especially dependent on actions that reduce or reverse ecosystem degradation. Freshwater availability in drylands is already below the minimum level considered necessary for human well-being and is projected to decrease further. Approximately 10–20 per cent of drylands are already degraded (MA, 2005c: 29).

The Assessment developed four scenarios over a 50-year timeline to explore plausible futures for ecosystems and human well-being (MA, 2005c: 71). The scenarios explored two global development paths, one in which the world becomes more globalized and another in which it becomes more regionalized. In addition, two approaches to ecosystem management were explored, one in which actions are reactive, with problems only addressed when they become obvious, and another in which management is more proactive, with policies designed to manage ecosystem services over the longer term. Under all four scenarios there is significant growth in consumption of ecosystem services, continued loss of biodiversity, and further ecosystem degradation (MA, 2005c: 17). Land conversion remains an important driver of biodiversity loss, although in some regions other factors become increasingly important. Nitrogen build-up in rivers and coastal waters in developing countries, especially in Asia, is damaging human health, fisheries, and habitats such as coral reefs (MA, 2005d: 21).

4. Workable Solutions Require Major Governance Changes

It will be a significant challenge to reverse ecosystem degradation while meeting the demands of a growing population, but options do exist. Three of the four scenarios show that some of the negative effects of rising consumption of ecosystem services can be reduced in both developing and industrial countries by major changes in policies, institutions, and practices on a scale well beyond current practice. Promising interventions identified in the scenarios include investments in environmentally sound technology, active adaptive management, proactive approaches to addressing ecosystem problems, extensive investment in public goods such as education and health, and an emphasis on eliminating socioeconomic disparities (MA, 2005c: 92).

The scenarios diverge in regard to the overall state of services, with the most significant decline in 'futures' where governments favor their own national or regional security over global cooperation and where conservation is given low priority (MA, 2005d: 21). By contrast, improvements occurred across all ecosystem service categories, in 'futures' where regional ecosystems such as watersheds are the focus of political and economic activity and emphasis is given to proactive local ecosystem management strategies. Both provisioning and regulating services improved in a globally connected 'future' with emphasis on the use of environmentally sound technology and a proactive approach to ecosystem management. Improvements occurred only to provisioning services in a globally connected 'future' characterized by global trade and economic liberalization, strong emphasis on poverty reduction and investment in public goods (such as education), but a reactive approach to problems. In all scenarios biodiversity continued to decline, raising questions about the long-term sustainability of improvements in both services and human well-being (MA, 2005c: 92).

The Assessment stopped short of making recommendations to achieve the significant changes in decision making that it had identified as needed. At the same time, it voices optimism that society has the ability 'to ease the strains people are putting on natural services of the planet, while continuing to use them to bring better living standards to all' (MA, 2005d: 23).

OUTLINE OF THIS BOOK: THEMES FOR A GOVERNANCE AGENDA

The chapters in this volume take up the challenge of distilling the implications of the Assessment for what decisions need to be made and how we make them. They focus on policies, governance measures and organizational changes that will foster ecosystem health.

In the next chapter, Frances Irwin and Janet Ranganathan envision what decision making to sustain ecosystem services might look like in 2030 and offer an action agenda for achieving this vision. The authors lay out the potential roles of civil society, business, researchers, local communities, national governments and international organizations. They then describe five barriers to making decisions that recognize the importance of natural assets and propose practices that existing institutions can use to overcome them.

Following this overview, Parts 1 to 3 organize the papers commissioned by the World Resources Institute under three themes: incorporating ecosystem services into the development sector; devising tools and skills to mainstream ecosystem services; and creating new institutions that integrate ecosystem services into decisions.

The authors of chapters in Part I – Mohan Munasinghe; Tony La Vina; Nicolás Lucas, Hernán Correa and Iokiñe Rodríguez; Lailai Li; Guido Schmidt-Traub and Albert Cho; and Robert Goodland – use lenses such as sustainable development, a development planning ministry, the Millennium Development Goals, local communities, and the World Bank to examine ways to incorporate ecosystem services into development planning and practice.

Munasinghe seeks to mainstream the Assessment results by integrating them into sustainable development (SD) strategy at the national and local levels, with localized implementation through communities and individuals. He also argues for explicitly linking the Assessment findings with the Millennium Development Goals. He proposes the Action Impact Matrix (AIM) tool to link ecosystem goals with national development strategy, determine priority issues, and implement development activities to manage ecosystems and restore ecosystem services. The AIM involves a fully participative, consensus-building, multi-stakeholder exercise.

La Vina takes a very hands-on approach. He plays the role of an official in a developing nation exploring the Assessment's implications for his ministry. He discusses the following key questions: How can a process like the Assessment be useful in development planning? How can the specific Assessment findings about the health of ecosystems be addressed in a national development plan? How can the results be integrated in such plans?

Looking more at the specific findings of the Assessment, Lucas, Correa and Rodríguez argue that these findings suggest two fundamental political interpretations, which have consequences for global development. First, there is a direct relationship between the patterns of globalization and the contemporary forms of biophysical transformation of the Earth. Second, global ecosystem change creates socio-environmental conflicts that must be assumed as a major political challenge for the coming decades.

Li focuses on the value of justice and livelihoods for the poor – another important aspect of development practice brought to the fore by the Assessment. She suggests that inter-governmental agencies play an important role influencing national governments to incorporate ecosystem management into their development strategies and urges increased investment in education to disseminate the knowledge generated by the Assessment.

Schmidt-Traub and Cho focus their chapter on the Millennium Development Goals (the world's existing shared framework for addressing poverty and ensuring environmental sustainability) and their links with the Assessment. Drawing on the work of the United Nations Millennium Project, this chapter suggests how developing countries can put the concept of environmental sustainability into practice by developing operational strategies for meeting their environment objectives. This includes carrying out needs assessments to quantify the human and financial resources required to implement the strategies.

Using a global institution as a vehicle for proposing reforms, Goodland argues that the World Bank has incredible opportunities to promote environmental prudence. He offers recommendations on how the Bank can integrate an ecosystems-based approach into its mainstream policy and operations.

In Part II, Mark Bateman, Richard Norgaard, Anthony Janetos, Karin Krchnak, and Frances Seymour explore tools and skills that either arise out of the Assessment process or that might be used to advance its findings.

Bateman offers ten discrete policy ideas that can serve as catalysts to bring about significant changes in ecosystem management and use. He pulls together the positive management aspects of each of the Assessment's scenarios to develop his list of policy recommendations.

In contrast, Norgaard uses the Millennium Assessment conceptual framework to explore the ways in which the structure of the Assessment led to extensive deliberative, democratic and scientific discussion between participating scientists. This gave participating individuals a new process for making sound judgments, which can help elucidate and implement particular policy choices.

Janetos focuses on the learnings from the Assessment process for similar future scientific assessments. He makes four recommendations: ensure that national governments have a strong role and voice in the assessment; ensure that new scientific assessments are part of an ongoing policy process; ensure that there is substantive participation from the private sector; and devote at least some resources to analysis of clearly different policy options.

Similarly focused on the Assessment process, Krchnak determines that the institutional and political issues of coordination, inclusion and participation that are emphasized in the Assessment's approach can help build the needed trust, experiences and confidence across multiple actors and processes to move toward sustainable development.

Seymour examines how environmentalists can relay the complex findings of the Assessment to a general audience in a way that is easily understood and compelling. Her chapter proposes several messaging strategies to mobilize political support around the Assessment's conclusions.

The final cluster of chapters, written by Chella Rajan, David Jhirad and Frances Irwin, suggest new institutions that could help mainstream ecosystem services into everyday decisions at the regional, national and international level.

Rajan argues that action that transcends conventional forms of country-level collaboration is needed to address the enormous challenges to the earth's ecosystems. He proposes a new institutional arrangement in the form of cross-national biome stewardship councils, with strong grassroots representation, to develop biome-specific recommendations that maximize both ecosystem protection and human welfare.

Exploring approaches to 'smart globalization', Jhirad examines incentives for business, financial, scientific and political leaders to participate in this process, and makes several recommendations for these constituencies. He supports reforming global governance through the creation of a Leaders' Forum of 20 nations.

Irwin focuses her chapter from the premise that the same regions that face the greatest ecosystem degradation and most rapid conversion of remaining natural assets also face the greatest development challenges. Her chapter recommends establishing a commission, building on the experience of the Commission for Macroeconomics and Health, to work with national governments to overcome barriers to incorporating into development plans those specific investments in ecosystem services targeted at helping the poor. She stresses the importance of a public process that sets targets and monitors progress

THE WAY FORWARD

Imagine for a moment that the Earth is a business. Planet Earth Ltd is an integrated global conglomerate that provides products and services to customers all over the world. It has been in business a long time, with such huge capital accounts that only recently, in the wake of other business scandals, have its owners decided to audit the books and examine how well it is really doing.

The audit reveals a company in deep trouble. Nearly two thirds of the company's 24 divisions examined are in the red; only four are profitable, while the other five showed mixed results regionally. The auditor's qualified opinion is sobering, pointing to the absence of internal controls: capital accounts undervalued and depleted for short-term gain and extensive use of off-balance sheet assets and liabilities. In addition, business units are pitted against each other with an inefficient transfer pricing mechanism, limited or no strategic planning, and underinvestment in R&D. Markets have distorted values for Planet Earth's assets, and there is no leadership to recognize the synergies between divisions. The company appears to be run without a CEO. The auditor contends that to avoid future bankruptcy, the company's management must change its practices significantly or be replaced.

The Millennium Ecosystem Assessment can be compared to a business audit. The analogy shows that the results of the Assessment merit serious attention. Of the 24 ecosystem services assessed, only four were enhanced, while 15 were being degraded or used unsustainably and five were stable globally, although in trouble in some regions. Overuse of provisioning services – food, fresh water and timber – has run down natural capital assets. Important regulating services, such as water purification, pest regulation and flood control, are off the balance sheet, and their depreciation overlooked. Others such as climate control are undervalued and used inefficiently. Ecosystem stewardship is marginalized in less powerful ministries or relegated to the management of a single service like food, timber and water. Development planners have failed to make the connections between healthy ecosystems and the attainment of social and economic goals, ultimately jeopardizing all three. Investment in knowledge about how ecosystems function and how they relate to society's well-being is inadequate and fragmented in many different disciplines.

The Assessment's framework linking drivers, services and human well-being may well turn out to be its most important contribution. It provides a template for developing a more complete profit and loss statement for Planet Earth. By looking at the contribution of ecosystems through the common lens of services for human well-being, the argument that ecosystem stewardship is a zero sum game is quickly diffused, making clear that healthy ecosystems are integral to healthy economies and healthy societies over the long term. At the same time, it provides a unified approach to assessing and managing services. Such an approach is especially important at a period of growing demand for services to support an expanding population and to help lift millions of people out of poverty, many of whom depend on ecosystems services for their livelihoods.

The Assessment provides the foundation for performing a periodic audit of ecosystem services. The next steps can take advantage of online, interactive, virtual globe tools such as Google Earth to organize and display information in easy-to-use ways. Using these techniques can lead to a sea change in how the public sees and values nature's services and in turn in the way society makes decisions about ecosystems and human well-being.

Everyone needs to be at the table to assist in this transformation: civil society, business, the educational and research communities, local communities, national government and international organizations. The transformation means integrating ecosystem stewardship into all types of governance by increasing the availability of information on ecosystem services, and tipping the balance in favor of local rights to resources and local voices in decision making. It means managing across levels and increasing the use of accountability mechanisms and economic and financial incentives. It means developing individual and societal norms that make ecosystem stewardship an integral consideration in all decisions.

At the same time, new institutions and tools can help put the action agenda into practice: a Forum (discussed by Jhirad) to provide political leadership from a wider range of countries; a Commission (proposed by Irwin) to bridge the gap between macroeconomics and the environment, working toward the common goal of reducing poverty; Biome Stewardship Councils (introduced by Rajan) as focal points for meshing the now disparate data on ecosystem services and experimenting with the most effective ways to disseminate and use the information to make decisions; and the Action Impact Matrix (AIM) (suggested by Munasinghe) to link ecosystem goals with national development strategy, through a fully participative, consensus-building, multi-stakeholder exercise.

The way forward requires rewiring the institutions of governance – making new connections to understand and find solutions to solve the complex interlinked challenges of ecosystem degradation. One thing is abundantly clear: 'business as usual' is no longer an option. The time has come to stop operating Planet Earth Ltd solely for the purpose of making a few shareholders rich in the short term, and instead to manage it as a family trust fund, set up for the benefit of today's and tomorrow's children.

REFERENCES

Adams, W.M. (2006), 'The future of sustainability: Re-thinking environment and development in the twenty-first century', in *Report of the IUCN Renowned Thinkers Meeting*, Gland, Switzerland: The World Conservation Union.

Convention on Biological Diversity (CBD) (1992), *Convention on Biological Diversity Convention Text: Article 2*, Rio de Janeiro, Brazil.

Daily, G. (ed.) (1997), *Nature's Services: Societal Dependence on Natural Ecosystems*, Washington, DC: Island Press.

Danielsen, F., M.K. Sørensen, M.F. Olwig, V. Selvam, F. Parish, N.D. Burgess, T. Hiraishi, V.M. Karunagaran, M.S. Rasmussen, L.B. Hansen, A. Quarto and N. Suryadiputra (2005), 'The Asian Tsunami: A protective role for coastal vegetation', *Science*, **310**(5748), p. 643.

Diaz, R. (2001), 'Overview of hypoxia around the world', *Journal of Environmental Quality*, **30**(2).

FAO (Food and Agricultural Organization of the United Nations) (2007), *State of World Fisheries and Aquaculture: 2006*. FAO: Rome, online at: http://www.fao.org/docrep/009/A0699e/A0699e00.htm.

Millennium Ecosystem Assessment (MA) (2005a), *Ecosystems and Human Well-Being – Current States and Trends: Findings of the Conditions and Trends Working Group*, Washington, DC: Island Press.

Millennium Ecosystem Assessment (MA) (2005b), *Ecosystems and Human Well-Being – A Framework for Assessment*, Washington, DC: Island Press.

Millennium Ecosystem Assessment (MA) (2005c), *Ecosystems and Human Well-Being: Synthesis*, Washington, DC: Island Press.

Millennium Ecosystem Assessment (MA) (2005d), *Living Beyond Our Means: Natural Assets and Human Well-Being: Statement from the Board*, Washington, DC: Island Press.

Munasinghe, M. (2007), *Making Development More Sustainable: Sustainomics Framework and Practical Applications*, Munasinghe Institute for Development, Colombo, Sri Lanka: MIND Press.

Sathirathai, S. and E.B. Barbier (2001), 'Valuing mangrove conservation in Southern Thailand', *Contemporary Economic Policy*, **19**(2), 109–22.

World Commission on Dams (WCD) (2000), *Dams and Development: A New Framework for Decision-Making*, London, UK: Earthscan.

World Commission on Environment and Development (WCED) (1987), *Our Common Future*, Oxford, UK: Oxford University Press, online at: http://www.anped.org/media.php?id=475.

Action agenda for sustaining ecosystem services

Frances Irwin and Janet Ranganathan

The Millennium Ecosystem Assessment makes clear that current approaches to managing our natural assets have not kept ecosystems healthy. Nor are they likely to be adequate for the challenges that lie ahead. Over the next 50 years, an expanding global population and economy will place growing demands on ecosystems at a time when their capacity to provide services has never been lower.

This rising demand for ecosystem services, coupled with dwindling supply, will lead to growing vulnerability and conflicts over who gets the benefits and who pays the costs of disrupted ecosystems in an already highly inequitable world. Eighty per cent of the world's gross domestic product belongs to the one billion people living in the developed world; five billion people in developing countries share the remaining 20 per cent (RWSS, 2005). The regions facing the largest development challenges (south of the Sahara in Africa, Central Asia, parts of South and Southeast Asia, and Latin America) will also confront the largest problems from ecosystem degradation. Overlaid on this fragile world will be the effects of growing pressures such as the build-up of nitrogen in rivers and coastal waters, species extinction, and increased incidence of droughts and flood from climate change.

A VISION

The challenges, while enormous, are not insurmountable. The seeds of change needed to put us on a path to sustaining ecosystem services are already emerging, assembled from existing pockets of best practice around the globe. What might the world of decision making look like in 2030 if these seeds germinate and transform the way we see and value our ecosystem services? Will it become second nature for people to safeguard the capacity of ecosystems to provide the mosaic of services necessary for human well-being? Will ecosystem health and development aspirations be viewed as mutually reinforcing sides of the same coin, instead of a zero-sum game?

Imagine for a moment that the movement to ensure the sustainability of our natural assets and long-term well-being that is already underway in parts of the world grows across economic sectors and political boundaries. Led by participants in the Millennium Ecosystem Assessment, global business leaders with progressive ecosystem policies such as members of the World Business Council for Sustainable Development, coalitions of governments such as the Poverty and Environment Partnership, and multidisciplinary applied scientific groups such as the Resilience Alliance, the movement unleashes a new generation of information and governance approaches.

Civil society performs the role of watchdog, solution provider and trailblazer. It makes information on ecosystem services and the performance of the public and private sectors available, enabling the public to hold business and government accountable. It is an effective advocate at all levels for strengthening the rights of local people, especially the poor, in development decisions. Multilevel partnerships, convened by civil society, that bring together formal and informal actors for the purposes of developing collective solutions provide model governance structures for government and business to emulate. Civil society is at the vanguard of calls for fiscal and policy reform that aligns economic and financial incentives with ecosystem stewardship.

Business views ecosystem stewardship as a source of new markets and competitive advantage. Companies adopt policies on ecosystem services and routinely report to the public on how they are implementing them. Companies in the natural resource sectors operate at the behest of local communities. Local voices are influential in shaping business operations in their locality. Managers routinely have ecosystem stewardship goals incorporated into their performance reviews. New businesses emerge to take advantage of new and growing markets for technologies and products that reduce ecosystem impacts and of payments for ecosystem services.

Research communities embrace ecosystem services as a major research theme. Information from monitoring systems for ecosystem condition and trends is analyzed and broadly disseminated, informing economic and development policies and election priorities. The Assessment conceptual framework becomes the organizing locus for a ballooning interdisciplinary research agenda that builds bridges across the traditionally fragmented natural, social and economic research disciplines.

Empirical research validates the contribution of participatory governance to successful development strategies and decisions.

Local communities have clear rights to resources in their communities and a significant voice in decisions about how they will be used and how the resulting costs and benefits will be distributed. Those in particular who have a major stake in the health of the ecosystem services, such as the vast majority of the world's poor who are concentrated in rural areas, have appropriate opportunities to influence development decisions that affect the services they depend upon. They also have ready access to information on the state and trends of ecosystem services. Locally elected officials keep a close eye on the health of services – whether water, timber, sacred sites or others – because they know they will have to answer to voters at the ballot box. Ecosystem Service Districts are commonplace. They collect information and use it to clarify trade-offs among services. They use tools such as zoning and taxation to protect and restore natural capital.

National governments routinely monitor the health of ecosystem services and include indicators in national accounts, enabling the public to track progress and hold decision makers accountable. They provide the public with access to information, participation and judicial review on decisions affecting ecosystem services. Policies routinely set minimum environmental standards, but respect the rights of individuals and communities over the natural resources where they live and promote fairness in the distribution of costs and benefits associated with ecosystem change, both at home and abroad. Tax and subsidy policies reinforce ecosystem stewardship. Budget systems are in place to identify investment opportunities for restoring ecosystem services, estimate the costs and benefits, and incorporate them into agency budgets. Government officials work regularly with networks at all levels, but particularly regional, to encourage action at the level most likely to be effective.

International organizations play a lead role in establishing the conditions for institutional cooperation on ecosystem stewardship across political levels and geographic scales. They facilitate dialogue among national parties involved in disputes over ecosystem degradation and diminishing services, such as fresh water and fisheries, helping parties to focus on the root causes of ecosystem degradation. They also support national development processes to implement the Millennium Development Goals through investments that restore ecosystem services that the poor in rural areas depend upon for their livelihoods. Local communities and indigenous groups have a voice in the negotiations of international agreements. Information from regional and national monitoring systems and groups like Biome Stewardship Councils (see Rajan's chapter in this collection) informs convention priorities and goal setting and enables the public to track progress by national governments in implementing their commitments under environmental conventions.

Is this vision achievable? Overly optimistic? We believe that it is possible to achieve it and that it is imperative to do so. The next section lays out an action agenda to put society on a path toward this vision. The agenda draws on three sources:

- Technical and synthesis volumes of the Millennium Ecosystem Assessment;
- The Assessment review papers included in this volume;
- The work of the World Resources Institute, much of it reported in the 2002–04 and 2005 editions of the World Resources Report (WRI et al., 2003, 2005)

AN ACTION AGENDA

What would an action agenda that could achieve this vision look like? Despite the immensity of the challenge, we do know where the path to sustaining ecosystem services starts. Table 1 outlines the barriers we face and corresponding actions, including illustrative examples, to address them.

In a generation or two, will we look back and wonder how society could have failed to recognize the importance of natural assets? Why wasn't the health of ecosystems a higher priority? Why was it so difficult to move beyond outmoded ways of making decisions? Most basically, people take the availability of nature's services for granted. As Table 1 shows, five reasons for this stand out. First, people lack relevant information on the connection between ecosystem services and human well-being and often do not use what they have. Second, local people, especially the poor, frequently lack the rights to use services they depend on for their livelihoods and to play a significant role in managing them. Third, our institutions focus narrowly on their mandates and find it difficult to work across boundaries, whether of professional specialties, geographic scale, or political level. Fourth, the measures to hold government and business accountable for decisions affecting natural assets are weak or absent. Fifth, the incentives required to align economic and financial decisions with ecosystem health are not in place.

While much remains to be learned, burgeoning research and experience with a variety of approaches among government, business and civil society groups in the last several decades point the way to making decisions that will sustain ecosystem services.

Table 1 Sustaining ecosystem services: barriers, actions, and examples

Barrier	Action	**Examples**
People fail to make the connection between healthy ecosystems and the attainment of social and economic goals Individuals, business, government, and civil society lack – or don't use – information on how their decisions and goals affect or depend on the availability and quality of food, water, cultural and other services of nature	Develop and use information about ecosystem services	• Perform regular monitoring and assessment • Identify and manage trade-offs • Frame messages that resonate with the public • Tailor information for citizens, producers and purchasers
Local people often lack clear rights to use and make decisions about the ecosystem services they depend on for their livelihoods and well-being People in local communities – particularly indigenous people, women, and the poor – lack secure tenure to land and resources they use and have little or no influence in decisions about how they are used	Strengthen the rights of local people to use and manage ecosystem services	• Ensure that individuals and communities have secure rights to ecosystem services • Decentralize decisions about ecosystem services • Bring local voices to the table to influence development projects and policies
The management of ecosystem services is fragmented among many different agencies and bodies that often work at cross-purposes and fail to coordinate across levels. The complexity and integrated nature of ecosystems necessitates a coordinated management approach. However, decision making and research bodies often lack the mandate and capacity to work across geographical and temporal scales and political levels.	Manage ecosystem services across multiple levels and timeframes	• Establish the conditions for cooperation with communities • Form bridging organizations • Use co-management practices • Raise priority of working across levels in national institutions
Government and business accountability mechanisms for decisions about ecosystem services are frequently absent or weak Corruption flourishes when decisions are not transparent and sanctions are unavailable	Improve accountability for decisions that affect ecosystem services	• Hold elected officials accountable • Publicly decide and track investments in ecosystem services for development • Increase corporate transparency

Responsible management of ecosystem services does not always pay Many ecosystem services have no value until they are lost. The economic and business case for maintaining ecosystem services is weak, obscured or non-existent. The costs of ecosystem degradation are not borne by those responsible.	Align economic and financial incentives with ecosystem stewardship	• Eliminate perverse subsidies and reform taxation policies • Include ecosystem risk in financial evaluations • Support markets and payments for ecosystem services • Incorporate ecosystem stewardship goals in managers' performance objectives

The proposed actions grow out of experience with initiatives undertaken in the last several decades, as conservation, community, business and development leaders, seeing the limits to achieving their goals within narrow focuses, began to explore ways of connecting their work. For example, conservationists have long used protected areas as their primary tool. They have been successful in setting aside more than 10 per cent of the planet's land surface to preserve threatened habitats and species. There is growing recognition, however, of the rights of local people living in or adjacent to such protected areas. As a result, conservation groups have begun to work much more closely with local people. Initial efforts sought to integrate development and conservation goals in projects around protected areas and then moved to community-based natural resource management projects. At the same time, business and development leaders also began to explore more effective ways of achieving their business and economic goals, while minimizing environmental impacts. Business faces resistance from communities in using forest, water and other resources. Civil society is calling for more effective ways to reduce poverty than waiting for the benefits of economic growth to trickle down to the poor.

The projects that emerged out of this ferment tested ways to answer fundamental questions about how to make decisions about natural assets. How can we develop and use information about ecosystem services in everyday individual and institutional decisions, and bring attention to and weigh the trade-offs among immediate ecosystem services of food and the longer-term ones such as flood control and clean water? How can we strengthen the rights of local people to use and manage ecosystem services? What kind of governance structures will facilitate the management of ecosystem services across geographic, political and temporal boundaries? What mechanisms are needed to improve accountability for decisions that affect ecosystem services, and how do we better align financial incentives with ecosystem stewardship? The learnings from these early initiatives inform the actions recommended in Table 1.

On their own, none of the five actions is sufficient to address the enormity of the challenge. But when collectively applied they represent a powerful force for change. Better information can provide political and economic incentives to take ecosystem services into account. Ensuring that individuals and communities have secure access to the ecosystem services they depend on for their livelihoods gives them a financial incentive to invest in the long-term health of those natural assets.

Of course not everything on the action agenda will be implemented at once. Powerful obstacles stand in the way. Elite groups who control land and resources will not want to relinquish control. Corrupt regimes will fight efforts to increase transparency and accountability. Empowered rural communities may be unable to secure the needed investment to restore healthy ecosystems capable of supporting their livelihoods. But the democratization of natural resources has begun, and the winds of change will ultimately disperse today's germinating seeds far and wide.

This chapter lays out how this vision can be achieved, by describing examples of how the action agenda is already being implemented in existing institutions around the globe, that is, by civil society, business, research communities, local communities, national governments and international organizations.

ADVANCING THE ACTION AGENDA IN EXISTING INSTITUTIONS

What steps can existing institutions take to implement the action agenda? This section puts meat on the bones of the agenda outlined above. It explores the five actions listed in Table 1 in more depth by providing promising examples of how civil society groups, business, research communities, local communities, national governmental agencies, and international organizations are already putting them into practice.

The examples are not exhaustive. They are intended to introduce the direction and range of opportunities for carrying out the five actions. They encompass the very specific – such as an effort to make a particular kind of information available about legislative performance – and the much broader – efforts to decentralize governmental decisions about ecosystem services. The actions and examples illustrate how we can bring the health of nature into our everyday decisions by developing and using information, strengthening the rights of local people, managing across multiple levels, improving accountability, and aligning economic and financial incentives. With each example, we note the types of organizations that could scale up these approaches and activities elsewhere.

DEVELOP AND USE INFORMATION ABOUT ECOSYSTEM SERVICES

It is difficult to think of a decision that does not in some way depend on the benefits that nature provides and that in turn does not influence the health of nature's services. Investors, finance and development ministries, companies, local governments, and individuals at home and at work take actions that drive decisions about how resources should be used. These decisions in turn drive ecosystem change. Effective decision making requires relevant and accurate information, but when it comes to ecosystems this information is often unavailable or not used.

Responsibility and power lie with individual organizations and people, yet without sufficient information even the best intentions fail. To make everyday choices in a way that incorporates an understanding of the inevitable trade-offs in utilizing ecosystem services requires regular monitoring and periodic assessment of ecosystem services. It also requires developing and using tools to identify these trade-offs, framing the message of ecosystems' importance in ways that resonate, and tailoring the information so that citizens, producers and purchasers are able to use it. Developing and using information about ecosystem services as illustrated by these four actions will enable individuals and organizations to make more informed decisions about how best to manage and use the services nature provides.

Civil Society, Research Communities, Local Communities, National Governments and International Organizations: Perform Regular Monitoring and Assessment

Regular collection and analysis of data on ecosystem services is fundamental to making informed decisions on issues that affect or are affected by ecosystem health. These data need to become as easily available and commonly used as economic data. The Millennium Ecosystem Assessment provides a methodology for tracking the status and trends of ecosystem services.

While our knowledge of ecosystems has increased dramatically, it simply has not kept pace with our ability to alter them, the Assessment says. Even valid baselines are lacking in many instances, with deforestation estimates, for instance, accurate only to within plus or minus 50 per cent. Sometimes the information is available but the financial resources are not available to process it. For instance, the Assessment points out that this is the case for global monitoring of land cover change using remote sensing. Some types of information are missing almost entirely. The Assessment found little information on the economic consequences of changes in ecosystem services or on the links between ecosystem services and human well-being, except for food and water (MA, 2005d: 101–2). The challenge is to establish regular monitoring and assessment of services at all scales and to fill in the gaps identified in the Assessment.

Several groups are working to spur the process of collecting, interpreting and using data on ecosystem services. To encourage widespread use of the Assessment's methodologies, a consortium of organizations led by the United Nations Environment Programme (UNEP)'s World Conservation Monitoring Centre is preparing an ecosystem assessment technical handbook. WRI is collaborating with the World Business Council for Sustainable Development and Meridian Institute to develop an 'Ecosystem Services Review' methodology to help businesses assess the risks and opportunities associated with their use of and impact on ecosystem services. Another coalition led by WRI is writing a guide to help policy makers and other mainstream ecosystem services in socioeconomic decisions. Development agencies may be appropriate sponsors for regular sub-global assessments of ecosystem services in the future (see La Vina's chapter in this volume).

The experience of the Assessment supports investment in cross-disciplinary and cross-scale learning that involves the policy community and the public as well as scientists (see Norgaard's chapter in this volume). A group called Locally Managed Marine Areas in the Pacific, for example, demonstrates one way to include resource user communities in monitoring an ecosystem service that they depend on for their livelihoods. Finding that their fisheries were deteriorating, in the 1990s coastal communities in Fiji brought back traditional management practices such as closing some zones to fishing, or limiting the amount that could be harvested. Twenty per cent of the inshore fisheries involving 125 communities are now locally managed in this way. To test the effectiveness of these approaches, experts from the University of the Pacific taught local people to monitor their fisheries. Using these skills, the Fiji community of Ucunivanua found that the number and size of clams in the area closed to fishing and adjacent areas increased dramatically (WRI et al., 2005: 144–51).

A new generation of information and communication technology offers opportunities to improve ecosystem monitoring significantly at low cost. In Malaysia, conservationists count elephants by satellite (WWF, 2005). WRI and the University of South Dakota are partnering to develop and apply remote sensing forest monitoring tools to monitor tree cover change on a regular basis using a globally consistent methodology. Civil society groups can encourage local participation monitoring by distributing cell phones and offering a free connection service in return for regular reports on environmental conditions (see Box 1).

BOX 1 USING MOBILE PHONES TO MONITOR LAND DEVELOPMENT PRACTICES IN ARGENTINA

An important advantage of mobile phone technology is that little infrastructure is required for their operation. In addition, new advances in text messaging and Internet access capabilities mean that mobile phones are now a viable and affordable communication platform for the collection of real-time, on-the-ground monitoring data by a variety of groups including non-governmental organizations (NGOs) and local communities (see Bateman's chapter in this volume).

The value of mobile phone technology is well understood by the indigenous Wichi people on the Pizarro Reserve in northwest Argentina. There are no land lines in this remote area, so residents were previously unable to communicate environmental conditions or call for help when developers destroyed land and homes. When the state government threatened to auction off the Pizarro reserve to agribusiness, Greenpeace Argentina distributed mobile phones to various community leaders and taught them how to send text messages to activists for help (Oberman, 2005).

Future ecosystem monitoring and assessment needs to occur at multiple scales and levels, draw on both traditional and scientific knowledge, and emphasize the links between ecosystems and people. New technologies, such as web-based interfaces that display spatially referenced information on a virtual globe like Google Earth, provide ways to share, analyze and disseminate information across different levels. Once more experience has been gained in monitoring and assessing ecosystem services, standards will be needed for factors such as data quality, measurement units and labeling. Such standards can help ensure that the resulting data can be aggregated and compared and that the results are appropriately used.

Civil Society, Research Communities, Local Communities, National Governments and International Organizations: Identify and Manage Trade-offs

Local and national development planners, project developers, natural resource managers, and a myriad of other public and private decision makers regularly make choices that trade one ecosystem service for another and determine who bears the costs and who benefits. However, tools to identify and provide information on these trade-offs are often unavailable or unused (see La Vina's and Goodland's chapters in this volume). Examples of tools that offer promise include Munasinghe's Action Impact Matrix (see his piece in the next section), several types of mapping, and scenarios.

Geospatial mapping offers an approach to identify and consider trade-offs in the distribution of the costs and benefits derived from ecosystem service use. Overlaying high-resolution maps of poverty and ecosystem services, for example, can highlight relationships that are important for policy development and poverty reduction. With the help of detailed poverty maps, analysts can compare, for example, the spatial patterns of poverty with the spatial distribution of investments in piped drinking water. In Kenya the majority of rural communities obtain their drinking water from untreated surface water, groundwater, or a combination of surface and groundwater. Households that rely on surface water are particularly vulnerable to interruptions in water flows and to water contamination. Investments in piped water and water treatment are important means by which people can obtain enhanced benefits from ecosystem services and insulate their livelihoods and well-being from interruptions in the quantity and quality of water supplies. Detailed maps can provide a comparison in Kenya of poverty rates and the degree to which communities rely on ecosystems or on technology for their water supplies. These maps have been created and reveal that the poorest areas in the upper Tana have not benefited yet from investments in piped drinking water (WRI et al. in press).

The Sub-Global Assessment (SGA) for Southern Africa describes another mapping tool. For the Gariep region, it used an irreplaceability concept to assess trade-offs between food services and biodiversity. Food production was divided into two types: calorie production (cereal) and protein (meat). Based on targets for calories, protein and biodiversity, replaceability values were assigned to map grids in the Gariep basin. These ranged from 0 (many options in other locations to achieve goals) to 1 (totally irreplaceable; the goals for services would not be met if this location were not included). While no site was found to be irreplaceable for protein and calorie goals, several sites were irreplaceable for biodiversity. This information supports a land use plan that guides protection of sites with a high degree of irreplaceable biodiversity, while steering grazing or cultivation on other sites (SA/MA, 2004).

Scenarios can inform choices about development in the context of the potential for sudden, non-linear change or the relative merits of different policy interventions. How might deforestation and climate change interact with geological hazards faced by communities in mountainous areas (for more on this scenario, see La Vina's chapter)? How might a new infrastructure development such as a major road system or large dam affect the supply of ecosystem services to different communities? While it is not possible to avoid all trade-offs from ecosystem change, identifying them and taking them into account in decisions will increase the likelihood of creating more winners and fewer losers. Such trade-offs would be facilitated by improving methods of valuing ecosystem services (discussed further in Munasinghe's chapter).

Civil Society and Research Communities: Frame Messages that Resonate with the Public

To resonate with citizens and spur calls by civil society for more effective policies, information on ecosystem services must be framed in ways that resonate. In her chapter in this volume, Seymour suggests several ways that Assessment findings might be framed to create the modifications in people's behavior and policy that will be required for ecosystems to be protected.

- Ecosystem degradation is exposing humans to increasing risk as the ability of ecosystems to provide services decreases. *Ecosystem maintenance can be framed as social insurance against risk* in the same way that social security helps manage the risk of old age. An important element of this approach would be ensuring that the benefits are widely shared and that the equity dimension is addressed. Civil society groups will need to use visionary thinking to prepare to take advantage of events – such as Hurricane Katrina – or other types of political 'trigger points' to build the societal agreement necessary to adopt social insurance. Facing soaring payouts for natural disasters, the insurance industry is a likely ally.

- Costs and benefits of ecosystem degradation are often not equitably distributed. The benefits go to the more concentrated winners and the costs to the more diffuse losers. Thus *ecosystem stewardship can be cast as a fairness or justice issue.* In Indonesia, logging benefited the elite while rural communities suffered the damage of the deforestation. Making the links between ecosystem stewardship and the poor would reverse the caricature of environmental advocates as elitist. Equity of access to ecosystem services such as water and parks may be most successfully applied at the individual and community rather than the country level. Connecting degradation of ecosystems to the exacerbation of poverty may be one way to build a norm in wealthier countries against the exploitation of natural resources to the detriment of the developing world, such as the import of illegal or unsustainably logged timber from tropical rainforests.

To convey these messages, graphic images and simple analogies are likely to be most effective. Ecological clocks might provide graphic images of the levels of degradation. Patterned after the Doomsday Clock used by the *Bulletin of Atomic Scientists* to track the danger of a nuclear catastrophe, ticking timepieces could track trends in the health of fisheries, climate or other ecosystem services.

Civil Society, Research Communities and Local Communities: Tailor Information for Citizens, Producers and Purchasers

There is an urgent need to disseminate the rich body of existing knowledge to the public in order to change how we use ecosystem services (see Li in this volume). Disseminating information in useful ways is as crucial as generating it. People need information about ecosystem services tailored to particular types of decisions ranging from protecting their family and participating in making policy to selecting investments and producing and purchasing goods. Efforts are underway to translate information on ecosystem services into appropriate forms for all of these purposes. The challenge is to improve these practices and to move them into common use.

To protect their families and participate in policy decisions, individuals need many types of information about ecosystem services. One of the most fundamental is information about the quality of the air they breathe and water they drink. An analysis of 16 case studies by a coalition of non-governmental groups called The Access Initiative looked at the quality and accessibility of monitoring data on air and water in nine countries. It found that information on daily air quality and trends was available in the press, on the radio, and/or on the Internet in urban areas in Chile, Hungary, India, Mexico, Thailand, and the US. However, detailed information on surface and drinking water quality was easily available only in the US and South Africa. In Hungary, India, Mexico, Thailand, and Uganda, there was no active dissemination of this information to the public.

The analysis by The Access Initiative also looked at 17 case studies of how information is provided during emergencies ranging from a cholera outbreak and a fish poisoning to industrial fires and transport accidents. Overall the analysis found strong performance in providing information for larger disasters. However, in most cases, both the quality and accessibility of information about emergencies at private facilities was weak. The report concludes that lack of government and civil society capacity constrains performance on providing access, while attention by the media helps improve it (Petkova et al., 2002).

As the demand for sustainably produced goods grows, investment in ecosystem-sensitive practices will increase. The Assessment's Business and Industry report interpreted the Assessment's findings to direct companies to business opportunities offered by investment in such areas as low-input systems for organic farming, including technologies that optimize food yield, nutrient loading and water use (MA, 2005e: 27–9). To create demand for these goods, companies and customers need information assuring them that they are getting goods that meet sustainability criteria. For the wood products industry, satellite imagery offers the promise of providing accurate and regular information for large areas of forests at lower cost than field verification and with greater reliability than self-reporting. WRI's Global Forest Watch initiative has pioneered the use of this technology for mapping the condition of the world's forests, producing maps of forest condition, change and the location of the world' remaining intact forests (WRI, 2006). When linked to leading wood products and financial companies, this information has led to more responsible forest policies. For example, IKEA, the world's leading home furniture retailer, requires that none of its solid wood products be harvested from intact forests as defined by WRI maps.

Pressure both from investors (see the section on paying for improvements in ecosystem service provision) and purchasers combined with business leadership is changing the practices of producers. As at other stages of the chain, geospatial maps and processes involving non-governmental groups, business and government play an important role. Kwa-Zulu Natal (KZN) Wildlife Agency has developed a public planning tool, KZN Biobase, which illustrates the scarcity of ecosystem types relative to current and proposed future land use. Mondi, a paper and packaging company with roots in South Africa, finds the results mapped in KZN Biobase have had a 'profound effect on how we look at things'. The Biobase documents which areas need to be protected and linked and thus helps the company decide where to locate their plantations and environmental corridors. After a multistakeholder process that developed wetland and riparian delineation procedures, the forest industry has agreed to remove most forest plantations from all wetland areas in the next harvesting cycle (10–15 years for short rotation crops) (P. Gardiner, private communication, 2006).

Educational initiatives can help get information to the business community. Coalitions of business schools in China as well as in North America and Latin America are working with WRI to make principles of sustainable enterprise a key part of the education of every business student (Bunch and Powers, 2003: 34–7). WRI's Environmental Enterprise Corps pairs business students with small entrepreneurs who are establishing or expanding environmental companies in Latin America and China (EEC, 2005). These efforts can be replicated in other regions and expanded to other professional training efforts, such as accounting, investment and banking.

STRENGTHEN THE RIGHTS OF LOCAL PEOPLE TO USE AND MANAGE ECOSYSTEM SERVICES

Ecosystem services are a lifeline for the poor in rural communities. Three quarters of the poor worldwide live in rural areas. Yet they often do not have clear rights to the land, fisheries, forests or other resources they use. Nor are they likely to have the ability to influence decisions about managing resources that they depend on for their livelihoods. For example, what rights do indigenous or other local people have when a conservation or protected area is established? This is an important issue given that at least half of all protected areas are on ancestral lands of indigenous or other traditional peoples (Gillis and Southey, 2005: 103; Hutton et al., 2005: 357–61). Similarly, what rights does the local community have when a company or a national government proposes a project that will affect the land or waters it uses? Without rights and the capacity to use them, individuals and communities do not have the incentive and influence to make decisions that will sustain the resources they depend on for their identity and survival.

In most countries, decisions about how to manage and use ecosystem services are dominated by national governments, international donors and multinational companies. Conservation groups have also played a prominent role through their efforts to protect nature by establishing parks and protected areas. However, there is increasing recognition at these levels of the need for more bottom-up approaches as local communities, sometimes with the help of human rights advocates, make the case to shift the balance of power and increase their voice. Similarly, groups working to reduce poverty and promote biodiversity conservation are finding a common cause in pro-poor development policies and projects that also emphasize the central role of the local community.

Of course, devolving more rights to local communities must also be accompanied by appropriate safeguards to ensure that the interests of a range of users are considered. This might include minimum standards for resource use and access set by the national government. It could involve helping communities to establish monitoring systems for ecosystem services such as the earlier example of Locally Managed Marine Areas in Fiji. It is also likely to involve establishing mechanisms to foster cooperation across levels as described in the next section, or mechanisms to hold decision makers accountable such as those described later. Learnings from these movements show the importance of strengthening rights to resources, decentralizing authority to make rules for managing resources, and bringing local voices to the table to influence local development and broader policy decisions.

Civil Society, Local Communities and National Governments: Ensure that Individuals and Communities have Secure Rights to Ecosystem Services

When people have secure rights to the ecosystem services that they rely upon for their livelihoods, it can have a profound effect on how they use and invest in those resources. A farmer who knows he has the right to use a field during his lifetime will be likely to invest in irrigation and soil management, while a farmer with no long-term access over the land will not (Bruce, 1998).

Securing rights over ecosystem services need not involve giving full title or exclusive ownership. An emerging movement for tenure reform calls for legal support for traditional tenure including communal tenure systems (see Box 2). Such systems can include a broad range of rights and obligations such as the right to exclude others from unauthorized use, the right to derive income from the resource, or an obligation to use the land so that it does not harm others. Thailand, for example, has recently provided land registration, deeds and credit to its rural population (WRI et al., 2005: 62). New tenure laws have been implemented in Mozambique, and Tanzania and Uganda now recognize land held in customary tenure (traditional or communal arrangements) as fully legally tendered 'as is' (WRI et al., 2005: 85).

In many countries women are important guardians of ecosystem services provided by water and land, but have few clear rights to them. In many developing countries women produce 60 to 80 per cent of food, yet in those developing countries where data is available, only 2 per cent own their own land, only 5 per cent of those living in rural areas receive technical assistance, and less than 10 per cent receive credit (FAO, 2007). Water and sanitation sector authorities can help empower women and enable them to play a more progressive role in managing ecosystems (see Krchnak's chapter in part II). Partnerships among conservation, environmental, poverty, development and other civil society organizations can advance changes in laws that prohibit women from getting loans or opening bank accounts and support equitable access to land and tenure arrangements by female producers (UN Interagency Taskforce on Gender and Water, 2005).[1]

The Millennium Challenge Corporation, a US fund that supports development based on how well a country performs against governance benchmarks, is adding indicators on access to both water and sanitation and resources. The intent is to encourage countries to make it easier for the poor, often women, to register property or otherwise obtain access to land (MCC, 2006). One challenge is likely to be ensuring that data on these indicators are monitored regularly in poor countries.

Civil Society, Local Communities, National Governments and International Organizations: Decentralize Decisions about Ecosystem Services

Ecosystem services are more likely to be sustained if local communities take a leadership role in their management under broad policies developed at the regional, national and international levels. In the 1980s, international donors began to urge countries to transfer some functions to local governments to improve efficiency and achieve economic growth. By the 1990s, national governments began to grant powers to manage natural resources at the local level. Those given authority ranged from elected local authorities, appointed agents of government resource ministries, and elected or appointed groups using wildlife or agricultural land, to local members of a political party, civil society, or traditional leaders. At the same time, some conservation groups have introduced a community-based management approach to try to move toward collaborative management of protected areas. Under this approach, the community sets the agenda and develops the rules that govern who uses the resources and under what conditions.

Researchers have found that effective decentralization policy meets four criteria. First, it transfers meaningful power, including fiscal power such as collecting fees, taxes, or fines, to a local authority. Second, that institution is representative of local people and has knowledge of the local resources and people's dependence on them. Third, local people can hold the institution accountable through elections or other means. Fourth, fiscal and regulatory incentives are in place to promote sustainable management of resources over the long term (WRI et al., 2003: 91–3).

Moving from top-down natural resource management to incorporate bottom-up approaches is challenging. Analysis of decentralized practices in Senegal, for example, shows that local communities still require prior approval from the forest service for almost any action, often through a complex planning process. Setting minimum standards may be a way to allow the community more flexibility. For example, the community might be given the right to manage dead wood grasses and other products. They might also be given the right to cut for commercial purposes 25 per cent of robust coppicing species, leaving enough to ensure erosion and wind protection. In these cases, no formal permission, planning or replanting would be required. At the same time, cutting of some species above a certain size or of a particular species (such as teak) would be allowed only with permission from the national forest service (J. Ribot, private communication, 2006).

When the community itself sets the rules over natural resources management, systems are most likely to sustain all types of ecosystem services in the long term (Ostrom, 2005: 259). One example comes from Venezuela where managers in Canaima National Park tried for many years to persuade indigenous Pemon to cease their traditional burning of savanna. Recent studies, however, show that far from being a destructive activity in itself, when based on Pemon knowledge of nature and cultural values this practice can actually help control larger, more destructive fires (Rodríguez, 2004). While it is important not to oversimplify the design of rules for managing common pool resources, researchers find that case studies do suggest principles – not blueprints – that can be used as questions in thinking through with local users how to improve the sustainability of a common pool resource (Ostrom, 2005: 271) (see Box 2).

BOX 2 COMMUNAL MANAGEMENT OF NATURAL RESOURCES

Researchers, who examined hundreds of cases, found that beyond providing a legal basis for common property it is also important for national governments to recognize the rights of users to develop and implement their own rules for using the resource. Besides secure tenure, the research shows that other principles, such as defined boundaries and at least minimal rights to devise institutions, are important for communal management of resources to be effective (Ostrom, 2005: 258).

Similarly, the Poverty–Environment Partnership, an informal network of development agencies focused on integrating poverty and environment issues within the framework of international efforts to achieve the Millennium Development Goals, finds the process works best when local communities themselves are empowered to take the lead. In Namibia, conservancies run by elected committees of local people manage wildlife on millions of acres with technical help from the government and NGOs. As a result of their work, formerly endangered wildlife populations are rebounding and the communities are benefiting from jobs, income from tourists, and game meat. Since 1998, more than 95000 Namibians have received benefits of some kind from the conservancies, and the number of conservancies has grown from four to 31. In 2004, the UN Development Programme recognized Namibia's Torra Conservancy by awarding them the Equator Initiative Award for their community-level work using biodiversity sustainably in order to reduce poverty. Ensuring equitable distribution of the benefits of conservancies within communities still presents a challenge. In a positive trend, women currently hold conservancy jobs, serve on committees, and chair three conservancies (WRI et al., 2005: 114–23).

Civil Society, Business, Research Community, Local Communities, National Governments and International Organizations: Bring Local Voices to the Table to Influence Development Projects and Policies

Decisions shaped by broad participation of community members are more likely to be fair and effective (WRI et al., 2003). Local communities bring essential perspectives such as knowledge of local and traditional practices. They know where the sacred sites are located, what cultivation practices work, and where wood and water for the community originate. Without strong local voices, the benefits of an oil well or a park are likely to flow to the country's elite or to other countries. Meanwhile, the local community may become poorer, more fragmented, and lose a sacred site or clean water as well as opportunities for better health and education programs. The high poverty levels of people in undemocratic countries with oil and gas projects starkly illustrate this problem. Half of the combined populations of 34 developing countries that rely heavily on oil and gas resources for exports live on less than $1 a day. Two thirds of the countries with large exports of oil and minerals are not democratic, with the result that local communities are likely to have little say in how projects are planned and how the revenues are used (UNDP, 2005: 124).

Clear rights to participate are essential to ensuring that people can influence decisions that affect their local resources such as fish, forests, water or minerals. The World Commission on Dams recommends that those whose rights are most affected and who face the greatest risk from a development project be recognized as having the greatest stake in the decisions and thus have a corresponding place at the negotiating table. Its report says 'Effective implementation of free, prior and informed consent marks a significant step forward in recognizing the rights of indigenous and tribal peoples' (Dubash et al., 2001). Some countries including Australia and the Philippines are beginning to require that the government obtain the consent of indigenous people for projects within their land (WRI et al., 2005: 72). The Action Impact Matrix method (introduced by Munasinghe in his chapter) is a consensus-building exercise which has been used successfully since the 1990s to mainstream stakeholder environmental concerns into development decisions.

Implementing such provisions and principles raises basic issues about how a community with a range of interests gives consent and how its interests are balanced with those of the country and globe. One analysis of consent related to mining projects suggests that it is the 'right of a community to be informed about mining operations on a full and timely basis and to approve an operation prior to commencement. This includes participation in setting the terms and conditions addressing the economic, social, and environmental impacts of all phases of mining and post-mining operations' (Bass, 2003). Goldman Sachs, a leading investment bank, says: 'We prefer to only finance projects in indigenous areas where free, prior informed consultation results in support of the project by the affected indigenous peoples.' They define this as ensuring that the project sponsor or borrower demonstrates that consultation relies on existing customary institutions; that local, regional or national-level authorities have provided affected communities with ways to be represented, consulted or air grievances; and that applicable laws have been upheld (Goldman Sachs, 2006: 6).

More broadly, citizens need access to information, participation, and judicial review to influence policy as well as project decisions affecting ecosystem resources. These access rights are increasingly recognized in international declarations and laws, such as Principle 10 of the Rio Declaration, but are more slowly implemented in practice.[2] Particularly important for major development projects is the ability to influence the planning and implementation of international financial institutions such as the World Bank and regional banks. Over the past 15 years, these institutions have responded to controversy surrounding development projects by making more information available. Generally, they inform people about decisions that have already been made rather than providing information that would allow people to participate in decisions. The African and Asian Development Banks, however, do require disclosure of social and environmental information at least 120 days before project approval for both public and private sector lending (Bank Information Center and freedominfo.org 2005: 3,7).

Civil society groups are working to improve the performance of national governments in providing access to information, participation, and justice by evaluating current laws and practice (see Box 3).

BOX 3 THE ACCESS INITIATIVE

The Access Initiative, convened by WRI, is a global coalition that aims to 'strengthen the capacity of public interest groups to track progress toward and build a global constituency for national-level implementation of a set of common participation and access standards.' It seeks to raise the awareness and commitment of governments to build participatory systems (Petkova et al., 2002: 12).

An analysis of laws and practices in nine countries around the world carried out by the Initiative found that governments do best in providing citizens with access to information. Laws are frequently strong though implementation is often weaker. Countries rate lowest on providing access to justice (an impartial judiciary or administrative review) when disputes arise. They scored in-between on providing opportunities to take part in making decisions. Participation is now strongest in the environmental sector and weak or non-existent in other sectors such as power production. The analysis urged national governments and the international community to support independent monitoring as a step towards improving performance (Petkova et al., 2002).

Civil society coalitions have now completed access assessments in

27 countries, with assessments underway in another 15. Thailand is now completing its third assessment.

Through the related Partnership for Principle 10, civil society groups and national governments are making measurable, time-bound commitments to improve access. National government partners include Bolivia, Chile, Indonesia, Mexico, Uganda, Ukraine and the UK. Uganda has achieved its commitment to adopt a Freedom of Information Act. The newest member – Indonesia – is incorporating a number of commitments into a revision of its Environmental Management Act. They include increasing the public's involvement in the environmental impact assessment process, incorporating public participation guidelines in new local environmental regulations, responding to public grievances in environmental cases, and publishing more environmental information on the Internet and in regulation booklets (PP10, 2006; TAI, 2006).

MANAGE ECOSYSTEM SERVICES ACROSS MULTIPLE LEVELS AND TIMEFRAMES

Because ecosystems and the services they provide cross political boundaries and operate at a range of timeframes, it is crucial that governance bodies have the authority and capacity to work at multiple timeframes – shorter and longer term – and levels – local, regional, national, continental and global. Similarly, problems arising from ecosystem degradation or potential solutions do not respect political boundaries or occur in single timeframes. For example, in 2004 degradation of drylands in Africa contributed to the formation of dust clouds in the region that traveled thousands of miles, causing toxic algal blooms off the coast of Florida, coral reef erosion in the Caribbean, and respiratory problems throughout North America (MA, 2005e: 11).

Findings from the Millennium Ecosystem SGAs echo this need to work across boundaries; collaboration across several levels was critical to success in 21 ecosystem management cases. In contrast, six of seven ineffective responses to ecosystem problems could be attributed to the lack of such multilevel collaboration. Assessment participants in areas as varied as the Caribbean, Portugal and Sweden identified building the trust needed for interaction between informal networks, usually at the local level, and more formal institutions, often at the national level, as a key challenge to working across levels (MA, 2005c: 214–17). Successful collaborations were associated with local initiative and with a focus on longer term effects, such as the depletion of a fishery or decline in biodiversity (MA, 2005c: 214,186).

Today's institutions, however, usually focus on a single sector such as forestry or finance at a single political level or geographical scale and often on a short timeframe. A community group or social network is concerned about local livelihoods, a national agency about planning national development, and a secretariat of an international convention about improving the state of a specific type of resource such as biodiversity, migratory species, or wetlands globally. At every level, institutions find it challenging to work across geographic and political boundaries or consider longer timeframes because of narrow mandates, limited capacity and the absence of incentives. If ecosystem services are to be sustained to support human well-being, decision makers need to address drivers and effects of change that emerge at different levels and time scales. This will require working outside traditional boundaries in multidisciplinary settings. In the past, such coordination across levels has been costly. Today, modern information and communication technologies make cooperating across political levels and ecosystem scales more feasible.

Nevertheless, thinking and acting across scales and levels remains a challenge. Cultural, technical, institutional and resource barriers abound. Yet by choosing to perform multiscale, place-based sub-global assessments, the Millennium Ecosystem Assessment took an important step towards demonstrating that taking a more coherent approach is possible and leads to more effective management of ecosystem services. WWF's '3 x M' method for promoting poverty reduction and environmental sustainability by linking required changes across the local, subnational and national levels, offers another approach to achieving coordinated action across levels (Reed, 2006). The SGAs, WRI reviews, and other initiatives like WWF's provide insight into four illustrative actions that can improve management across geographic scales and timeframes: establishing the conditions that support cooperation with communities, creating bridging organizations, utilizing comanagement practices, and raising the priority in national governments of working across levels.

Civil Society, Research Community, Local Communities, National Governments and International Organizations: Establish the Conditions for Cooperation with Communities

Much of the conflict over the use of ecosystem services emerges between local communities and government, business or civil society working at the national or international level. Conflicts range from confrontations over use and management of protected areas to disagreements over titling and use of indigenous land. As Table 2 shows, these issues relate not only to environmental concerns such as water pollution but also to changing economic structure and processes such as mega-projects and privatization.

Table 2 Social and Environmental Impacts of Globalization in Latin American Case Studies

Examples	Source of conflict	Form of global market economy	Indicator of environmental change	Indicator of social impact
Salinas de Manaure, Colombia	Large-scale salt extraction and commercia-lization	Salt trans-nationals	Destruction of two coastal marshes, exceptional and unique ecosystems of semiarid Guajira	Exclusion and exploitation of local economy; indigenous social fragmentation
Machu Picchu, Peru	Tourism	Tourists, tourism companies, tour operators	Waste accumulation, anarchic management	Inequity in distribution of benefits from tourism; exclusion and exploitation of local economy
Los Pelambres, Chile	Copper mining	Mining trans-national	Long-term water and land pollution	Erosion of local farming economies; displacement; social fragmentation
Cocha-bamba, Bolivia	Privatiza-tion of water	Water use and management transnational	Change in the use and management of water resources	Lack of respect to local definitions of water use; erosion of local traditions and customs

Source: Rodríguez and Correa (2005)

A basic step in working effectively across levels is to establish the conditions that will support cooperative management. Most fundamentally, these conditions entail designing effective dialogues, looking broadly at the issues and integrating knowledge across levels, and looking broadly at all services – not just those in the marketplace..

To be effective, multilevel dialogues should provide sufficient time for a thorough airing of issues and conflicts, involve all players with a stake in the decision, and balance the power among parties. Effective solutions are elusive when only formal leaders are involved at the local level or when a national or an international body is left out, for example.

As discussed in the previous section, on bringing local voices to the table, effective cooperation requires that local stakeholders have a say in the decision through approaches such as free, prior informed consent by communities. It is also important to address differences in power among the participants of the dialogue in day-to-day interactions, in developing environmental knowledge, and in national legal and political frameworks. This may require investing in community-driven research and training. Scientists who participated in the Assessment and civil society groups, such as those working on access to decision making, can work closely with communities facing pressure on their resources. In some places networks have already been established. For example, a network of the Locally Managed Marine Areas in Fiji provides training to interested communities not only in biological monitoring but in action planning and socioeconomic monitoring as well. Through such networks, local communities can learn the skills needed to play an active role in negotiating decisions and managing resources across levels (WRI et al., 2005: 92).

Successful dialogues across levels also require integration of different forms of knowledge (see the chapter by Lucas, Rodríguez and Correa). Historically, this has not been done. For example, state interventions in agricultural and education practices in rural South Africa have undermined the traditional teachings of local communities which believed certain locations and resources to be sacred, thereby protecting them from overuse. Many of these formerly protected resources which are vital to the rural South African communities for their supply of provisioning, regulating, and cultural services are now degraded (MA, 2005b: 96). The integration of many forms of knowledge provides a basis for broadening the alternatives considered and analyzing trade-offs among provisioning, regulatory and cultural services that might otherwise have been missed.

One of the more challenging conditions thwarting institutional cooperation is the willingness to consider intangible issues as seriously as economic issues. Efforts to resolve conflicts over natural resources often focus on compensation through community development projects or jobs or new management practices, while issues such as different views of 'identity, authority, different cultural constructions of nature and its use and different definitions of development' are ignored (Lucas, Correa and Rodríguez discuss this further in their chapter). If these core intangible issues are not addressed, conflict is likely to continue. The Vilcanota SGA provides one illustration of how traditional beliefs can be accommodated. The Assessment team altered the conceptual framework to encompass the Quechua communities' understanding of ecological and social relationships by adding a category of ecosystem services to reflect the local knowledge of the 'protection' of Mother Earth (MA, 2005b: 110).

The 'water war' in Cochabamba, Bolivia, one of the more widely known examples of recent environmental conflict in Latin America, offers insights into how to establish effective cooperation across levels for the purpose of resolving natural resource conflicts. Cochabamba's 'water war' involved two different views of authority over water and its use (see Box 4 for a definition of the Andean traditional view of water).

BOX 4 THE ANDEAN TRADITIONAL VIEW OF WATER

Water as a living being. Water is a living being; it provides life and gives meaning to the universe. Andean traditional cultures maintain a permanent dialogue with water. Water is treated and nursed like a child.

Water as a divine being. Water comes from Wirakocha, creator of the universe, who fertilizes the Pachamama (Mother Earth) and allows the reproduction of life. It is thus considered a divinity that is present in the lakes, rivers, seas and all sources of water.

Water as a base of social reciprocity and complementarity. Water allows the integration of human beings and nature, and the articulation of nature within human society.

Water as a universal and community right. Water does not belong to anybody. It is a common good. It is distributed equitably in accordance to needs, customs, and community rules and norms.

Water as an expression of flexibility and adaptability. Water acts in accordance to the rules of nature, ecosystems and circumstances, and does not comply with rigid rules.

Water as a life-giving and transformative being. Water follows the rules of nature. It acts in relation to the seasonal cycles and the conditions of the territory. Its sustainable use depends on the application and generation of knowledge and abilities obtained throughout centuries of humans interacting with nature.

Source: Adapted from CONDESAN (2003)

Based on their ancestral view of water, local farmers and indigenous people wanted to maintain their traditional management practices, while the government, viewing water solely as a resource, proposed its privatization along with the sewer systems. Only after a concerted effort by local farming and indigenous organizations, involving a national blockade and lobbying in the National Congress, was the traditional view heard and understood by national decision makers. This new understanding and shift in power relations allowed for the negotiation of a solution to the conflict. A legal change in the Potable Water and Sanitation System Law was made to ensure the protection of traditional uses and customs, and a contract with an international consortium, which was to have been in charge of providing water services, was rescinded.

A legal change in the Potable Water and Sanitation System Law was made to ensure the protection of traditional uses and customs, and a contract with an international consortium, which was to have been in charge of providing water services, was rescinded.

More in-depth dialogue based on an increased understanding of the history of drinking water offers promise of developing an approach to management that recognizes the different natures of water in communities such as Cochabamba. A careful look at practices over centuries shows that drinking water has more often served as a physical, economic, *and* social resource rather than one of these alone (Salzman, 2006).

Civil Society, Business, Research Communities and International Organizations: Form Bridging Organizations

One approach to fostering decision making across multiple levels is the formation of bridging organizations. These groups provide forums for bringing together a range of formal and informal actors, sharing a diversity of knowledge, and developing collective solutions on a specific issue. By identifying and coalescing around common interests, participants develop trust and opportunities for action appropriate to a range of different spatial and time scales and political levels.

The SGAs provide a number of models of bridging organizations. In the Caribbean Sea SGA, a non-governmental group brought together local communities and policy makers in St Lucia to get mangrove forests designated as marine reserves and set up reforestation projects to supply local demand for fuel wood (MA, 2005b: 217). In the Philippines, the Laguna Lake Development Authority created River Rehabilitation Councils that involved people's organizations, environmental groups, industry representatives and local government units. With broader local participation, the Laguna Authority was able to clean up pollution, a goal it had been unable to achieve before the creation of the Councils (MA, 2005b: 212).

A non-governmental group worked with a community in China to help its members present their interests to developers (this example is discussed further by Li in her chapter). The Institute of Environment and Development took part in preparing the social impact assessment for the West to East Gas Pipeline and ensuring that the compensation process for farmers was transparent and fair. At the beginning, the local farmers in a populous province highly dependent on agriculture said: 'The pipeline is like time bombs buried by the government and developers, but the switches are in our hands.' The Institute was able to bridge the differences by improving communication, explaining the compensation process clearly, and enabling the farmers to make their case. It is not standard practice to involve non-governmental groups in working with local communities in UN-managed development projects in China. However, in this case, Shell's previous experience in Nigeria led to their requiring non-governmental participation in the project. Li, in her chapter, urges China to make it standard practice to include NGOs in negotiations around large development projects.

Research Communities, Local Communities, and National Governments: Use Co-management Practices

Under a co-management approach, institutions from different levels jointly manage ecosystem services. Usually this is a partnership between a local community and a regional or national government agency. This approach allows each level to bring its strengths – differing information, policy tools, financial resources and political support – to a joint effort.

National and subnational governments are most likely to have authority to adopt and use regulations and economic incentives. They are also most likely to have scientific knowledge and data, particularly about changes that emerge over longer time scales such as changes in climate and biodiversity. In some cases, they may be able to affect drivers of ecosystem change located outside local communities where the effects of these actions emerge. Thus these higher level institutions can provide the overall view for planning and information and also ensure that local action and national and international rules align.

In contrast, local communities usually have more detailed and longer term understanding of their own land, water and plant and animal life from traditional or indigenous practices. They are the ones who are on the spot to implement changed practices and to monitor activity and the health of ecosystem services, particularly if they have clear legal access to the services. In rural areas they are also likely to be dependent on healthy ecosystems for their livelihoods and thus have a special interest in maintaining a steady supply of services.

An example from Samoa illustrates how co-management can work in managing fisheries in the Pacific. The national government provides legal authority, research and expertise as well as marketing information, credit and transportation. However, communities are given clear tenure rights in resources and authority to manage local resources. The Samoan Fisheries Division has worked with more than 40 villages to adopt management plans. The villages provide on-the-ground implementation and monitoring (WRI et al., 2005: 93, 96–7).

To be effective, co-management requires ensuring that local residents have real authority to make decisions, as noted in the section on decentralizing decisions, and that user groups are accountable to local political authority. If local governance bodies have the ability to experiment with rules, it becomes possible to learn what works while avoiding large-scale failure. Co-management does not require one level being in charge. Rather, specialized districts or private associations at the local level may be nested in general governance structures at regional and/or national levels that provide services such as judicial review (Ostrom, 2005: 283–6).

Civil Society, National Governments and International Organizations: Raise Priority of Working across Levels in National Institutions

At each level from the community to the international level there is a multitude of instruments, organizations and institutions. The Assessment points to the importance of ensuring policy coherence across these levels. 'International conventions must be coherent with national polices and these in turn must be coherent with local policies' (MA, 2005c: 519–20). National institutions are often the pivot between local or regional institutions and international ones. The priority they give this role greatly influences the ability to design and carry out coherent policies. The absence of coherent policies hinders the effectiveness of efforts ranging from scientific assessment to development and conservation initiatives.

High-level officials did not represent national governments in the Millennium Ecosystem Assessment. As a result, national governments are less committed to following up the Assessment's findings either at home or internationally through conventions (see Janetos's chapter in this volume). Similarly, organizations focused on development and poverty reduction were not principal partners. Future assessments need to include high-level participation both from national governments and development groups so that they will incorporate results such as the importance of nutrient loading or increased risk of ecosystem disruptions into their own work (La Vina discusses this in his chapter).

International networks can give national leaders and managers an opportunity to improve skills to do their jobs, as well as help them make the links they need to work with colleagues in other countries as well as at other levels of governance. Implementation of international conventions can be the focus for bringing together representatives from different levels of government to develop policy, whether on preserving wetlands or protecting the climate. Networks can also be formed regionally. The African Ministerial Conference on Water aims to strengthen intergovernmental cooperation on water and sanitation in Africa. As women develop new strategies for managing water, a Women's Association of Water Ministers might be established to encourage information sharing and capacity building across regions of the globe (an idea introduced by Krchnak in this volume).

One of the most developed networks is the International Network for Environmental Compliance and Enforcement (INECE). Founded in 1990, it now links more than 4000 practitioners in 120 countries ranging from inspectors, prosecutors and regulators to parliamentarians, judges and civil society groups. The network provides training and capacity building programs, raises awareness, and enhances enforcement cooperation. Through regular conferences, the network gives its members an opportunity to learn and build long-term relationships to address environmental compliance and enforcement issues (INECE, 2006).

IMPROVE ACCOUNTABILITY FOR DECISIONS THAT AFFECT ECOSYSTEM SERVICES

Public and private decision makers are more likely to make choices that sustain ecosystem services if they are held accountable for them. The public frequently does not have the means to hold governmental or private bodies responsible for their actions affecting ecosystem services. Obstacles include strong resistance from corrupt regimes and lack of public information. Holding decision makers accountable can be particularly difficult because complex chains connect impacts and the actions that cause them. The causes and effects can be separated by decades and by hundreds or even thousands of miles. Monitoring is difficult in remote areas. Consider the degradation of water by fertilizer and pesticide runoff from intensive agriculture. Who besides the environment agency is accountable – farmers, food purchasers, policy makers, the World Trade Organization? Historically, the answer would probably be none of these groups.

Most basically, public officials can be held accountable through elections in countries choosing a democratic government. An informed public, non-government organizations that are free to form, and a media free to report are hallmarks of functioning democracies and provide the building blocks for accountability. On a global scale, democracies are on the rise. In 1950, 22 of the world's 154 nations were democracies, but now the global count has reached 119 democratic states out of 192 countries (Freedom House, 2000).

Civil society, governments and leading firms are engaged in a wide variety of efforts to develop methods of reporting on government and business performance. These include such projects as the 'publish what you pay' initiative, which is a coalition of more than 300 NGOs worldwide that seeks mandatory reporting of payments from oil, gas and mining industries to government officials for natural resource extraction. The group is aimed at helping resource-rich developing countries to use their natural resources for economic development and poverty reduction rather than fueling corruption and social divide (Publish What You Pay, 2006). Buoyed by modern information, monitoring, and communication technologies, transparency – 'deliberately revealing one's actions so that outsiders can scrutinize them' – is fast becoming the basic mode of accountability (Florini, 2003: 32).

Returning to the question of who is accountable for pollution of water by pesticides and fertilizer, in the future the answer will move beyond environmental agencies to include all the other actors – farmers, purchasers, policy makers and world trade officials – if they knowingly took action that contributed to the problem. Tracking votes of elected officials, developing a process to track implementation of the Millennium Development Goals, and increasing corporate transparency illustrate just a few of the measures that can be used to drive greater accountability for managing ecosystem services.

Civil Society, Local Communities and National Governments: Hold Elected Officials Accountable

Elected officials are ultimately responsible for decisions about the ways ecosystem services are used. Information on ecosystem health, the distribution of natural resource benefits, and voting records enables citizens to hold elected officials accountable when they come up for re-election, creating incentives for improved performance. Civil society frequently plays a key role in calling for greater transparency as well as in using information to hold decision makers accountable by bringing it to the attention of voters.

Citizens need information on governmental decision-making processes concerning natural resources. Accountability for management of these services needs to be linked to locally elected officials such as parliamentary representatives. The Ugandan Parliament, for example, passed a motion in 2005 to record the votes of each legislator on bills and motions in parliament. This was in part a response to recommendations from a WRI-sponsored study in Uganda on legislative environmental representation by the Advocates Coalition for Development and Environment. Ugandan citizens, environmentalists and journalists can now hold legislators accountable for their votes on natural resource issues (Tumushabe and Bainomugisha, 2004).

One obstacle to overcome is the short timeframe of politicians who have an eye on the next election and thus little incentive to consider longer range goals and ways to track the condition and value of ecosystem services over decades. One way to address this is to develop and use annual public indicators of ecosystem health at both the national and local levels to provide citizens with a more complete picture of their nation's wealth. The European Union has developed three levels of indicators under its ten headline objectives for sustainable development. One of these objectives is for natural resources management. Level 1, designed to follow advancement of the overarching policy objective, tracks the population of farmland birds and fish catches outside safe biological limits. Level 2 indicators, which give more detailed analysis of core policy arenas, include measurements such as developed area as a percentage of all land area and groundwater extraction as a percentage of total available groundwater. Level 3 indicators are targeted to a more specialized audience, such as the academic community, and cover data such as the portion of land area at risk of soil erosion. All of these indicators are published in annual reports and the data are made accessible to the public online (Eurostat, 2006). With the help of analysis by academics and non-governmental groups, individuals can use annual public indicators such as these to hold elected officials accountable for the state of ecosystem services and their influence on the state of the economy.

While most national governments provide regular reports on economic indicators such as gross domestic product (GDP), unemployment and inflation, few provide parallel accounts for the health of their ecosystems, despite their clear relation to economic performance. The World Bank has developed one such measure of national wealth that includes environmental assets, such as land, soil fertility and forests. When applied to a spectrum of countries, the measure showed that environmental assets are more important, relative to total wealth, in low income countries, amounting to 26 per cent of wealth compared to just 2 per cent in industrialized countries (World Bank, 2005). For some countries, practices such as unsustainable logging or extracting minerals have led to an overall loss in national wealth as high as 50 per cent, as in the case of the Congo (MA, 2005d: 55). Valuations of ecosystem services can also be used to support the development of national accounts that adjust growth in wealth for resource depletion and damage. To facilitate the development of such accounts that integrate with conventional measures of goods and services like GDP, agreement is needed on definitions and the services to be measured (Boyd, 2006). A leadership forum, such as the one proposed by Jhirad in this volume's final section, might push for agreement on common definitions of goods and services in national accounts and encourage use of such accounts.

Civil Society, National Governments and International Organizations: Publicly Decide and Track Investments in Ecosystem Services for Development

The United Nations has adopted eight goals, known as the Millennium Development Goals (MDGs), to be achieved by 2015. These goals call for eradicating extreme poverty and improving health and education, as well as ensuring environmental sustainability and equity. The Assessment shows that 'Any progress achieved in addressing the MDGs of poverty and hunger eradication, improved health, and environmental sustainability is unlikely to be sustained if most of the ecosystem services on which humanity relies continue to be degraded' (MA, 2005d: 2). Yet few, if any, countries are on track to meet the MDG goal on environmental sustainability. A major reason for this is that national governments do not have frameworks for planning and budgeting to reverse environmental degradation as part of development strategies (for more, see Schmidt-Traub and Cho's chapter). Some have further argued that the Assessment findings should be explicitly linked to all the MDGs (not just environmental sustainability) within national development strategies in order to receive more favorable attention from decision makers and donors (this is specifically proposed by Munasinghe in his chapter).

National governments lack a planning and budgeting process to incorporate investments in ecosystem services in their development strategies for several reasons. To begin with, much of the attention to the environment occurs at the international level through conventions or at the local level through individual projects rather than at the national level. In addition, environmental ministries seldom have proposals for specific investments in ecosystem services that document their costs and benefits in a way useful to development and financial planners. This is partly because the outcomes and costs of specific interventions or best practices to improve ecosystem services are only beginning to be identified. In contrast, the health community does have a concrete set of tested investments that reduce disease ready to insert into national budgets.

Practice in scaling up interventions to the national level is similarly at an early stage. Missing data and the site-specific nature of environmental issues also hinder developing proposals for specific interventions. In addition, investment decisions are political because they require trade-offs among regions and sectors. For example, how will water be divided between agriculture and city drinking water? How should use of a forest be shared among timbering, tourism and water conservation and purification?

The challenge for a national government is to convene a *public* process to implement the MDGs. A planning body led by the prime minister and planning or finance ministry would provide overall direction with the participation of other ministries, international institutions, civil society, the private sector and other partners. An MDG Strategy Group made up of specialists from these groups – and including scientists and local practitioners – would identify targets, gather data, choose effective interventions, and translate information into implementation. An Environment Working Group would take the lead in organizing data about ecosystem services and in proposing the interventions needed to maintain or restore the capacity of ecosystems. With other members of the MDG Strategy Group, the Environmental Working Group would identify the financial resources needed to implement the interventions and where they might be obtained. The Strategy Group would prepare a ten-year framework prioritizing and sequencing investments in all sectors, proposing needed policies and reforms, and assigning responsibilities for implementation. To translate this framework into action, the Strategy Group would develop a medium-term (3–5 years) expenditure framework to specify how resources for ecosystem services would be generated and disbursed (discussed further by Schmidt-Traub and Cho).

While planning and budgeting processes can sound arcane, these processes and the resulting documents are in fact basic tools in allowing the public to track where funds are supposed to go and whether they are actually being spent as designated. If data on implementation of specific interventions related to ecosystem services and targets for achieving them are included in a development framework and made public, the media, civil society, and other parts of government can use this information to track progress and hold the government accountable.

One barrier is likely to be the lack of access to information, participation, and judicial review in developing countries. All but one of the 40 countries classified by the World Bank as Heavily Indebted Poor Countries currently lack laws providing freedom of information (Banisar, 2006; World Bank, 2006). A process to provide accountability for achieving investments in ecosystem services may be the trigger for civil society groups to introduce or strengthen access laws and practices guaranteeing public information and participation as they have in Uganda (as discussed earlier in this chapter). Doing so is not easy. In Ghana and Honduras, two countries in which poverty reduction strategies have served as a focus to initiate efforts to adopt freedom of information laws, such bills became mired in legislative bodies. Supporters of Ghana's proposed law wrote 'perhaps due to a desire by some members to perpetuate a culture of secrecy for self interest, the bill has developed cold feet' (McIntosh, 2006).

A Commission on Macroeconomics and Ecosystem Services for Poverty Reduction, proposed by Irwin in her chapter, would help national governments build the links among finance, development and environmental ministries needed to put into practice planning and budgeting processes to incorporate ecosystem investments into development strategies.

Civil Society, Business and National Governments: Increase Corporate Transparency

Although national governments started requiring companies to submit public financial reports early in the twentieth century, standards for reporting on impacts and use of ecosystem services have just begun to emerge in recent decades. Experience with them reaffirms that what gets measured gets managed – especially when the results are available to the public. A facility-based, mandatory reporting program known internationally as Pollutant Release and Transfer Registers has demonstrated that when information is publicly reported in a standardized format and actively disseminated, it acts as a powerful accountability tool, driving companies to reduce their toxic releases dramatically (WRI et al.,

2003: 110–15).

Recent studies have found that approximately 700 multinational companies now use the Global Reporting Initiative (GRI) guidelines, which are developed by an international consortium of stakeholders, to prepare sustainability reports. Although select businesses have voluntarily begun disseminating information, and governments increasingly mandate some reporting by industrial facilities on pollution and wastes released to the environment, stronger incentives are needed to make the emerging reporting norms on the use of water and other natural resources as commonplace as financial reporting (Bazilchuk, 2005: 38–9). Stock exchanges may be in the best position to represent investors' interests. They could mandate the uniform disclosure of corporate interactions with ecosystems (a possibility that is explored by Bateman). The Johannesburg Securities Exchange (JSE) has considered requiring that all companies publish a report based on the guidelines of the GRI. Corporations that fail to follow socially responsible and sustainable policies may have their charters revoked and their listings removed from public stock exchanges (discussed further by Goodland in his chapter).

Another form of accountability comes in the guise of consumers who, if provided with information on a company's ecosystem management practices, can choose whether or not to purchase their products. High profile campaigns and boycotts against a variety of products from fish to timber to coffee have focused consumers' attention on how their purchases affect natural resources. In response, a plethora of product labeling and certification schemes have emerged, providing information and 'seals of approval'. The jury is still out on their effectiveness in changing practices. The Assessment noted that forest certification, for example, has been more successful certifying already well-managed land in the North than trouble spots in the tropics (MA, 2005a). Stepwise approaches, focusing, for example, on basics such as compliance with forest laws in countries where these are not routinely enforced, offer promise of raising standards at the bottom of the performance curve.

Pressure is increasing for companies and governments to make public the amounts of revenues from oil and gas and other extraction projects and where they are going. Huge amounts are expected to go to countries such as Nigeria (US$110 billion) and Angola (US$43 billion) between 2004 and 2010. In the past, much money from extraction projects has disappeared through corruption rather than being used to improve human well-being. Canada is currently the only country that requires companies (through its securities law) to disclose royalty payments at a country level (National Instrument 51-101, 2003), although South Africa does require access to information held by the private sector if knowing that information is required for the defense or use of a different right ('Promotion of Access to Information Act', 2000). The International Accounting Standards Board is reviewing its standard for the mineral sector. This process and changes in securities legislation offer opportunities to spur governments to include strong reporting on revenues in company accounting standards (Save the Children, 2005).

When linked to public education programs and supported by financial incentives, greater transparency and accountability can support more informed choices at the shops and at the ballot box, driving more responsible ecosystem management by the private sector.

ALIGN ECONOMIC AND FINANCIAL INCENTIVES WITH ECOSYSTEM STEWARDSHIP

Economic and financial incentives must be aligned with ecosystem stewardship goals if good practices are to be the rule rather than the exception. The Assessment found that 'most resource management decisions are most strongly influenced by ecosystem services entering markets; as a result the non-marketed benefits are often lost or degraded. These non-marketed benefits are often high and sometimes more valuable than the marketed ones' (MA, 2005d: 6).

The earlier example discussed in this volume's introduction of converting mangroves to shrimp aquacultures demonstrates how incentives can inadvertently support ecosystem degradation. From the aquaculture owner's perspective, the short-term financial benefits of converting a mangrove swamp to an aquaculture make good business sense. The resulting loss of the mangrove's non-marketed services (for example, nursery habitat and storm protection) is not borne by the owner, but accrues to adjacent communities who depend on these for their livelihoods and well-being. A bank which provides the owner with the initial investment to convert the ecosystem was almost certainly oblivious to the risks of ecosystem degradation. With no markets or payments available to financially support the provision of the nursery habitat and storm protection, these services would not have been considered by the owner.

How can incentives be changed in cases like the mangrove conversion discussed above to encourage ecosystem stewardship rather than degradation? Here we discuss four examples of approaches that help align incentives with sound management. The first two focus on how governments, through the use of taxation and subsidy policies, and financial institutions, through discounting ecosystem risk, can send more sustainable signals to markets. The third explores emerging markets and payments for ecosystem services, and the last focuses on how to ensure individual managers' incentives are aligned with sound ecosystem management.

Civil Society, Research Communities, Local Communities and National Governments: Eliminate Perverse Subsides and Reform Taxation Polices

Subsidy and taxation systems strongly influence how individuals and businesses use or affect ecosystem services – whether they look for short-term gains or focus on the longer term. Policies now often create incentives for overuse of provisioning ecosystem services such as fisheries, food and timber rather than for sustaining these services and also regulatory services. Numerous international and domestic subsidies in Argentina have exacerbated biological collapse in fish populations off the Argentine coast. Among other government policies, fishermen in Patagonia are given tax subsidies for purchases of fuel for their fishing vessels and are given access to publicly owned waters at a cost which is disproportionately low compared to the commercial value of catches. According to a UN Environment Programme report, over the past decade these policies have cost the Argentinian economy US$500 million, not including the increasing problem of disappearing fish stocks (UNEP, 2001). Farmers who irrigate in the Great Plains of the US receive a tax deduction for extracting groundwater in volumes that exceed what is naturally replenished each year (US IRS, 2004).

Providing market incentives for sustaining ecosystem services requires reforming subsidy programs and eliminating perverse subsidies. As is clear from the above examples, national governments frequently subsidize activities that degrade ecosystem services rather than sustain them. Organization for Economic Co-operation and Development (OECD) countries, for example, provide about US$300 billion annually in agricultural subsidy payments (Chigunta et al., 2004). These perverse subsidies support unsustainable practices at home and marginalize poor farmers in developing countries.

Subsidies for activities that degrade ecosystem services should either be eliminated or transferred to activities that support or restore ecosystem services. Analysts suggest that national governments review their budgets and eliminate subsidies for activities such as agriculture, capture fisheries, logging virgin forests, building hydro dams, and developing and using fossil fuels. Instead, these funds should be used to restore ecosystem services such as those provided by wetlands or for sustainable agriculture (this practice is discussed further by Bateman and Goodland in this volume).

Opportunities also exist to align taxation policies with ecosystem stewardship. Moving from income-based taxation to consumption-based taxation could provide price signals that better capture the costs of using ecosystem services (Box 5). As noted in the next section, a portion of such consumption taxes might be used to pay for maintenance of ecosystem services, although most of the revenue would be used for other purposes.

BOX 5 A MOVE TO CONSUMPTION-BASED TAXES

The Assessment notes that markets currently do not capture the real value of ecosystem services. Bateman and Goodland, in their chapters included in this volume, propose that taxing consumption offers an opportunity to place a price directly on the use of ecosystem provisioning services, such as food, water and timber. Governments can use the multiplier effect of consumption taxes on costs embedded in products to encourage 'ecosystem friendly consumption'. If only levied on new consumer items a consumption tax would also provide an incentive to use items longer or to purchase used goods.

Businesses would also be encouraged to produce goods that are of higher quality and durability – touting the consumers' lower tax bill over the life cycle of a product. Individuals would benefit by being able to exercise more control over their taxes by regulating their consumption patterns.

Clearly, a number of complex issues need consideration when making such a radical shift, including protecting the buying power of the poor for essential goods and services. However, if these issues are not treated as overly politically divisive, they do have solutions.

Taxation policy can be used to curb the current preference of the majority of investors for short-term gains over longer-term profitability, a practice that inhibits investment in new technologies that could conserve ecosystem services. Governments can move from giving preferential tax status for capital gains (share appreciation) to favor dividends (profitability). At the same time, environmental activists can team up with corporate governance activists to put pressure on companies to find alternatives to stock option plans as compensation for corporate executives. As recent financial scandals demonstrate, managers can manipulate stock prices through short-term actions that undermine long-term profitability. Instead, executive compensation can be linked to long-term profitability measures (proposed and discussed by Bateman in this collection).

Civil Society and Business: Include Ecosystem Risk in Financial Evaluations

Ecosystem degradation creates new financial risks and opportunities for the private sector. Among the risks are scarcity and rising costs of inputs like fresh water, new policies and regulations to address degradation, and changing consumer preferences in response to concerns about ecosystem health. Opportunities include technologies and products that serve as substitutes for ecosystem services, markets for scarce resources such as water quality and wetlands, and new revenue streams from payments for ecosystem services. If financial institutions evaluated these risks and opportunities when making decisions such as whether to take an equity stake in a company or to extend credit to a development project, it could have an enormous influence on how companies use and impact natural resources. Conversely, the profitability of such decisions can be significantly impacted if key ecosystem services (for example, availability of fresh water, flood protection) that support the investment are being degraded.

So far, the socially responsible investment community has taken the lead in investing in 'green' or 'clean technology'. But these investments are relatively small compared to the trillions of dollars that comprise the mainstream investment market. Some inroads have been made with the mainstream financial market. WRI, for example, has worked with a variety of financial institutions, including multilateral development banks, export credit agencies and private banks, to quantify the financial implications of climate change. In a first of its kind partnership between a US brokerage house and an environmental NGO, WRI collaborated with Merrill Lynch to evaluate the interrelated risks of energy security and climate change on the automotive sector (Merrill Lynch and WRI, 2005).

There has been less progress engaging the financial sector on ecosystem degradation, although some financial institutions are now developing policies. Goldman Sachs, a leading global financial institution, has adopted an environmental policy committing to reporting on the impact of its own operations ranging from greenhouse gas emissions to construction of facilities and procurement of goods. It will also influence investors by incorporating environmental, social and governance criteria into analysis of companies. The policy explicitly recognizes the relationship between ecosystem protection and poverty elimination. With academic and NGO partners, Goldman Sachs is undertaking to provide and disseminate independent research on how markets can contribute to solving environmental and related social problems (Goldman Sachs, 2006).

To be effective, policies like those that Goldman Sachs is undertaking need to be supported by information and tools that identify ecosystem risks. Increased recognition of risks will stimulate calls for greater disclosure of data by regulators and financiers. Better data and information will in turn enable financial institutions to routinely discount ecosystem risk, thereby steering investments toward actions that sustain ecosystem services.

Civil Society, Local Communities and National Governments: Support Markets and Payments for Ecosystem Services

The Assessment shows the importance of taking into account the value of both marketed and non-marketed ecosystem services in natural resource management decisions. Studies reveal that the benefits of managing landscapes to sustain their regulatory services such as water purification and flood control can exceed those of converting the ecosystem to intensive production.

For example, the net economic benefit per hectare from intact wetlands that can maintain water quality is almost three times the value after it has been converted to intensive agriculture (MA, 2005e: 57). While correct from a broader societal perspective, such valuations are irrelevant to individual landowners unless there are mechanisms to translate the non-marketed services into economic value.

Programs that pay communities, organizations or individuals to maintain or restore ecosystem services such as water purification, flood control or carbon sequestration are one way to provide a direct economic incentive (see Box 6). These programs are at an early stage of development, but an increasing number of examples are emerging. In Ecuador, a Water Conservation Fund (FONAG) collects user fees from those who benefit from the water in the Condor Bioreserves. It uses these funds to support watershed management projects (this example is discussed by Krchnak in her chapter). In Brazil, states allocate some revenues from their largest source of funds – a tax on goods, services, energy and communications – to municipalities to help them support protected areas for forests or other resources. Analysts of this Brazilian approach to use of taxes urge providing the public with regular information about the amounts of funds transferred and how they are used in order to provide accountability for the program (May et al., 2002).

BOX 6 COMPANIES PAY FOR REFORESTATION TO KEEP THE PANAMA CANAL OPEN

For decades the 50-mile Panama Canal has been the preferred shipping route between the Atlantic and Pacific Oceans, avoiding the longer 8000 nautical mile passage around the tip of South America. Unfortunately, deforestation of the surrounding land has threatened the canal's operations in two ways. First, deforested land has a lower absorption capacity during heavy rains, increasing the risk of floods and erratic water supply. This is a problem because each ship that passes through the canal requires millions of gallons of fresh water to be pumped into the locks to raise and lower it as it passes through the system. Second, deforestation increases soil erosion leading to an increase in silt flowing into the canal. Removing this silt and the associated growth of aquatic vegetation that thrives in thc nutrients in the soil is very expensive.

Concerned by the growing risks to the canal's operations, insurance companies have been requiring shipping companies to pay increasingly high premiums for cover in the event that the canal has to close. Escalating insurance costs and risks of closure seemed inescapable until John Forgach, chairman of ForestRe, a specialist insurance entity focused on forest risks, convinced insurance and major shipping companies to finance a 25-year bond to restore the region's forest ecosystem. Reforestation will lower erosion rates and provide a more controlled flow of fresh water to the canal. Insurance companies benefit by bearing a smaller risk of paying out damages, while shippers enjoy lower insurance premiums (*Economist*, 2005a; BSR, 2006).

One concern with payment schemes is whether they reach the poor. A project in Columbia's Cauca Valley showed some success in using water fees from downstream sugarcane growers hurt by flooding to help mainly poor upland communities change their land management practices to protect the watershed (WRI et al., 2005: 107). The poor are more likely to benefit from payment schemes if they have secure tenure over the services that are being marketed (as noted at the start of this chapter).

A detailed analysis of experience with creating markets for ecosystem services notes critiques of payments such as violation of the polluter-pays principle and undermining a land ethic. The analyst says: 'Service markets clearly pose potential concerns but, in an imperfect world, may well provide the most effective and desirable means of providing services in many more settings than we currently assume.' The vision is that farmers in the future might earn a living slowing floodwaters or purifying water as well as by growing crops and raising livestock (Salzman, 2005). Box 7 illustrates how a forest landowner's revenue streams might change in the future to include payments for services not currently captured in the balance sheet.

BOX 7 FROM GOODS TO SERVICES – IMAGINING THE REVENUE POTENTIAL OF TOMORROW'S FORESTS

The Millennium Ecosystem Assessment highlights the fact that in the future the demand for ecosystem services will grow at a time when the capacity for ecosystems to provide these services will decline. In the not too distant future it is possible to imagine a transformation of the way landowners manage natural assets from managing for a single provisioning service to capturing the value of multiple services.

Consider a community-owned and harvested 3000 hectare forest landscape in Indonesia:

Customer	Ecosystem service	Share of Revenue	
		2006	2026
Global timber market	Timber sales		
Eco-tours Indonesia Ltd	Ecotourism/hunting		
NextPower, Indiana, US	Carbon sequestration		
Biodiversity Offset Exchange	Biodiversity credit		
Municipal Water Treatment Agency	Watershed protection		
Provincial government	Flood protection credit		
Local markets	Nontimber forest products		

Civil Society, Business, Research Community, Local Communities, National Governments and International Organizations: Incorporate Ecosystem Stewardship Goals in Manager's Performance Objectives

Business and other organizations routinely evaluate managers' abilities to meet specific objectives within budget. A basic step that all businesses –

and other organizations – can take to help ensure ecosystem services are on their managers' radar is to incorporate them into performance objectives. If these goals are met, annual bonuses and other rewards may follow.

In the past, most businesses have considered environment a cost, and relegated these issues to an environment department without much clout. Increasingly, companies are seeing that these performance objectives can apply across the company. Financial advisors can improve their understanding of ecosystem services and educate their clients. Facilities managers can save money by finding ways to use less energy, water and materials, while product designers and marketers can look for opportunities to introduce products that place less demand on ecosystems.

Responsible companies with environmental policies and commitments should consider including environmental goals for managers in every part of the organization, not as stand-alone performance goals, but rather connected to the attainment of core social and economic goals. Goldman Sachs's policy says that 'Each of our major business areas has an important role to play in integrating this policy into our operation' (Goldman Sachs, 2006). One way to help ensure that Goldman Sachs's businesses play this role is to include the attainment of healthy ecosystems in managers' performance goals. Novozymes, a biotechnology company with headquarters in Denmark and operations worldwide, links bonuses of top managers to achieving social and environmental targets as well as to reaching financial targets. At the executive management level, one person is responsible for specific targets and that person's bonus is linked to these targets (C. Frier, private communication, 2006).

The action agenda outlined above seeks to ensure that ecosystem health is an integral consideration in the decision-making processes of organizations. Its implementation will challenge existing wealth and power structures and face strong resistance from those who stand to lose. To be successful it must be accompanied by a broader societal norm shift that changes the way people see and value ecosystems. This shift will take time. In today's increasingly urbanized world, people take the availability of nature's services for granted – clean water at the turn of a tap, milk from a carton, and a regular supply of fresh vegetables and fruit from the local supermarket. They forget that a steady supply of nature's services and their own well-being depend on sustaining healthy ecosystems. They are often unaware that nearly two thirds of ecosystem services examined by the Assessment are degraded. A concerted effort is needed to educate people about their dependence on nature and that its fraying services can no longer be assumed. Like organizations, societal norms must shift to make ecosystem stewardship second nature in everyday actions.

NOTES

1 Land tenure is also a broader issue that needs to be addressed for sustainable development to be achieved. As the Assessment notes: 'rural communities in the formal tribal "homelands" in South Africa had no rights of permanent residence outside those areas' (MA, 2005a: 4). Land tenure starts a process of building wealth, one that allows for inheritance and taxation systems that provide funding to the state, but allow families to build wealth as well. When the fall of communism occurred in 1989 in Central and Eastern Europe and the former Soviet Union, one of the first steps in many countries was implementation of land tenure programs. The rapid economic growth occurring, for example, in Central and Eastern Europe is due to a number of factors but property ownership as a wealth-building process should not be underestimated (Krchnak expands on this point).

2 Principle 10 of the Rio Declaration, adopted by 178 nations at the UN Conference on Environment and Development in 1992, says 'Environmental issues are best handled with the participation of all concerned citizens. At the national level, each individual shall have appropriate access to information concerning the environment that is held by public authorities, including information on hazardous materials and activities in their communities, and the opportunity to participate in decision-making processes. States shall facilitate and encourage public awareness and participation by making information widely available. Effective access to judicial and administrative proceedings, including redress and remedy, shall be provided.' The UN Economic Commission for Europe's Convention on Access to Information, Public Participation in Decision-making, and Access to Justice in Environmental Matters, known as the Aarhus Convention, provides a framework of key elements that have been elaborated by several protocols. Ratified by nearly 40 countries, it is open to participation by all interested countries.

REFERENCES

Banisar, D. (2006), 'Freedom of information around the world 2006: A global survey of access to government records and laws', freedominfo.org, online at: http://www.freedominfo.org/documents/global_survey2006.doc.

Bank Information Center and freedominfo.org. (2005), 'Opening the international financial institutions: A comparative analysis and introduction to the IFI transparency resource', online at: http://www.ifitransparencyresource.org/documentupload/Comparative.Analysis.pdf.

Bass, S. (2003), *Prior Informed Consent and Mining: Promoting the Sustainable Development of Local Communities*, Washington, DC: Environmental Law Institute, online at: http://www.elistore.org/reports_detail.asp?ID=10965.

Bazilchuk, Nancy (2005), 'Business come clean', *Conservation in Practice*, **6**(4).

Boyd, James (2006), 'The nonmarket benefits of nature: What should be counted in green GDP?', Discussion Paper, Washington, DC: Resources for the Future.

Bruce, J. (1998), 'Review of tenure terminology', *Tenure Brief: No.1*, Madison, WI: University of Wisconsin at Madison, Land Tenure Center, online at: http://agecon.lib.umn.edu/cgi-bin/pdf_view.pl?paperid=964&ftype=.pdf.

BSR (Business for Social Responsibility) (2006), 'Environmental markets: opportunities and risks for business', online at: http://www.bsr.org/meta/BSR-Trends-Report_Enviro-Markets.pdf#search=%22BSR%20panama%20canal%22.

Bunch, R. and N. Powers (2003), 'The greening of China's B-Schools', *BizEd Magazine* (May/June), online at: http://www.greenbiz.org/news/reviews_third.cfm?NewsID=24819.

Chigunta, F., R. Herbert, M. Johnson and R. Mkandawire (2004), 'The changing policy environment facing african agriculture', in S. Haggblade (ed.), *Building on Successes in African Agriculture*, Washington, DC: International Food Policy Research Institute.

Consorcio para el Desarrollo Sostenible de la Ecorregion Andina (CONDESAN) (2003), *La Visión Andina del Agua. Perspectivas y Propuestas de los Indígenas Campesinos de los Andes*, La Paz: CONDESAN.

Dubash, Navroz, Mairi Dupar, Smitu Kothari and Tundu Lissu (2001), *A Watershed in Global Governance?: An Independent Assessment of the World Commission on Dams*', Washington, DC: WRI. Citing quotations from World Commission on Dams (2000), *Dams and Development: A New Framework for Decision-Making*, London: Earthscan, online at: http://www.dams.org/report/contents.htm.

Earthwatch Institute, The World Conservation Union (IUCN), World Business Council on Sustainable Development (WBCSD), and the World Resources Institute (WRI) (2006), *Ecosystem Challenges and Business Implications*, Switzerland: Atar Roto Presse SA.

Economist (2005), 'Are you being served?' (April 21), online at: http://www.economist.com/science/displaystory.cfm?story_id=E1_PRRGRQJ.

EEC (Environmental Enterprise Corps) (2005), 'Linking environmental entrepreneurs with investors', Washington, DC: WRI, online at: http://www.new-ventures.org/eec.html.

Eurostat (2006), 'Sustainable development indicators', online at: http://epp.eurostat.ec.europa.eu/portal/page?_pageid=1998,47433161,1998_47437045&_dad=portal&_schema=PORTAL.

FAO (Food and Agriculture Organization of the United Nations) (2007), 'Women and food security', online at: http://www.fao.org/FOCUS/E/Women/Sustin-e.htm.

Florini, Ann (2003), *The Coming Democracy: New Rules for Running a New World*, Washington, DC: Island Press.

Freedom House (2000), 'Democracy's century: A survey of global political change in the 20th century', New York: Freedom House.

Gillis, N. and S. Southey (eds) (2005), *A Community Dialogue for Meeting the Millennium Development Goals*, New York: Fordham University Press, online at: http://www.energyandenvironment.undp.org/undp/index.cfm?module=Library&page=Document&DocumentID=5519.

Goldman Sachs (2006), 'Goldman Sachs environmental policy framework', online at: http://www.gs.com/our_firm/our_culture/corporate_citizenship/environmental_policy_framework/docs/EnvironmentalPolicyFramework.pdf.

Hutton, J., W.M. Adams and J.C. Murombedzi (2005), 'Back to the barriers?

Changing narratives in biodiversity conservation', *Forum for Development Studies*, 2, online at: http://www.frameweb.org/file_download.php/Back+to+Barriers.pdf?URL_ID=14432&filename=11413077711Back_to_Barriers.pdf&filetype=application%2Fpdf&filesize=387052&name=Back+to+Barriers.pdf&location=user-S/.

International Network for Environmental Compliance and Enforcement (INECE) (2006), online at: http://www.inece.org.

King Committee on Corporate Governance (2002), *The King Report on Corporate Governance for South Africa – 2002 (King II Report)*, Johannesburg: Institute of Directors in Southern Africa.

May, P., F. Neto, V. Denardin and W. Loureiro (2002), 'Using fiscal instruments to encourage conservation: Municipal responses to the "ecological" value-added tax in Paraná and Minas Gerais, Brazil', in S. Pagiola, J. Bishop and N. Landell-Mills (eds), *Selling Forest Environmental Services: Market-Based Mechanisms for Conservation and Development*, London: Earthscan.

MCC (Millennium Challenge Corporation) (2006), 'Millennium Challenge Corporation Adopts Environmental and Land Access Criteria for Selection of Eligible Countries', online at: http://www.mca.gov/press/releases/2006/release-091106-natlresourcesindicators.php.

McIntosh, T. (2006), 'Freedom of information laws added to the development agenda', (March 22), online at: http://www.freedominfo.org/features/20060322.htm.

Merrill Lynch and WRI (2005), *Energy Security and Climate Change: Investing in the Clean Car Revolution*, online at: http://www.asria.org/ref/library/csrguidelines/lib/050616WRI_Report.pdf.

Millennium Ecosystem Assessment (MA) (2005a), *Ecosystems and Human Well-Being – Current States and Trends: Findings of the Conditions and Trends Working Group*, Washington, DC: Island Press.

Millennium Ecosystem Assessment (MA) (2005b), *Ecosystems and Human Well-Being – Multiscale Assessments: Findings of the Sub-Global Assessments Working Group of the Millennium Ecosystem Assessment*, Washington, DC: Island Press.

Millennium Ecosystem Assessment (MA) (2005c), *Ecosystems and Human Well-Being – Policy Responses: Findings of the Responses Working Group*, Washington, DC: Island Press.

Millennium Ecosystem Assessment (MA) (2005d), *Ecosystems and Human Well-Being: Synthesis*, Washington, DC: Island Press, online at: http://www.maweb.org/proxy/Document.356.aspx.

Millennium Ecosystem Assessment (MA) (2005e), *Living Beyond Our Means: Natural Assets and Human Well-Being: Statement from the Board*, online at http://www.maweb.org/en/ Products.BoardStatement.aspx.

Munasinghe, M. (2007), *Making Development More Sustainable: Sustainomics Framework and Practical Applications*, Munasinghe Institute for Development, Colombo, Sri Lanka: MIND Press.

Munasinghe, M. and R. Swart (2005), *Primer on Climate Change and Sustainable Development*, Cambridge, UK: Cambridge University Press.

National Instrument 51-101 (2003), 'Standards for disclosure for oil and gas activities', online at http://www.albertasecurities.com/dms/1144/

2232/13038__2105907_v1_-_CONSOLIDATED_VERSION_OF_NI_51-101_AS_AT_MARC.pdf.

Oberman, J. (2005), 'SMS mobilizes to demobilize rainforest destruction', online at: http://www.personaldemocracy.com/node/756.

Ostrom, Elinor (2005), *Understanding Institutional Diversity*, Princeton, NJ: Princeton University Press.

Petkova, Elena, Crescencia Maurer, Norbert Henninger and Fran Irwin (2002), *Closing the Gap: Information, Participation, and Justice in Decision-making for the Environment*, Washington, DC: World Resources Institute.

PP10 (Partnership for Principle 10) (2006), online at: http://www.pp10.org.

'Promotion of access to information' (2000), online at: http://www.info.gov.za/gazette/acts/2000/a2-00.pdf.

Publish What You Pay (2006), online at: http://www.publishwhatyoupay.org.

Reed, D. (2006), *The 3×M Approach: Bringing Change Across Micro, Meso and Macro Levels*, Washington, DC: WWF.

Rodríguez, I. (2004),. 'Conocimiento indígena vs. científico: el conflicto por el uso del fuego en el Parque Nacional Canima, Venezuela', *Interciencia*, **29**(3): 121–9.

Rodríguez, I. and H. Correa (2005), 'Lecciones, enfoques y retos en la transformación de conflictos socioambientales. La experiencia del Programa "CyC"', in H. Correa and I. Rodríguez (eds), *Encrucijadas Ambientales en America Latina. Entre el Manejo y la Transformación de Conflictos por Recursos Naturales*, San Jose: Universidad Para la Paz, online at: http://www.upeace.org/cyc/libro/index.htm.

RWSS (Report on the World Social Situation) (2005), *The Inequality Predicament*, New York: United Nations, online at: http://www.un.org/esa/socdev/rwss/rwss.htm.

Salzman, J. (2005), 'Creating markets for ecosystem services: Notes from the field', *New York University Law Review*, **80**: 870–961, online at: http://www.law.nyu.edu/journals/lawreview/issues/ vol80/no3/NYU302.pdf.

Salzman, J. (2006), 'Thirst: A short history of drinking water', *Yale Journal of Law & the Humanities* (forthcoming), online at: http://eprints.law.duke.edu/archive/00001261/01/17_Yale_J.L._&_Human._ (2006).pdf.

Save the Children, UK (2005), 'Beyond the Rhetoric: Measuring revenue transparency in the oil and gas industries', online at: http://www.savethechildren.org.uk/mt/Linkage.pdf.

The South African Millennium Ecosystem Assessment (SA/MA) (2004), *Ecosystem Services in the Gariep Basin*, Stellenbosch, South Africa: Sun Press, online at: http://www.maweb.org/proxy/document.71.aspx.

TAI (The Access Initiative) (2006), online at: http://www.accessinitiative.org.

Tumushabe, Godber W. and Arthur Bainomugisha (2004), *Constitutional Reform and Environmental Legislative Representation in Uganda: A Case Study of Butamira Forest Reserve in Uganda*, Kampala, Uganda: Advocates Coalition for Development and Environment (ACODE) Policy Research Series.

UN (United Nations) Interagency Taskforce on Gender and Water (2005), *A Gender Perspective on Water and Sanitation*, United Nations Department of Economic and Social Affairs, Commission on Sustainable Development, online at: http://www.un.org/esa/sustdev/csd/csd13/documents/bground_2.pdf

UNDP (United Nations Development Programme) (2005), *Human Development Report 2005*, New York: Oxford University Press, online at: http://hdr.undp.org/reports/global/2005/.

UNEP (United Nations Environment Programme) (2001), *Fisheries Subsidies and Marine Resources Managements: Lessons learned from Studies in Argentina and Senegal*, online at: http://www.unep.ch/etu/etp/acts/capbld/rdtwo/FE_vol_2.pdf.

USIRS (United States Internal Revenue Service) (2004), *Farmer's Tax Guide*, Washington, DC: US Department of Treasury, online at: http://www.unclefed.com/TaxHelpArchives/2004/HTML/p225/index.html.

World Bank (2005), *Where is the Wealth of Nations? Measuring Capital for the 21st Century*, Washington, DC: World Bank Publications, online at: http://siteresources.worldbank.org/INTEEI/214578-1110886258964/20748 034/All.pdf.

World Bank (2006), 'Heavily Indebted Poor Countries (40 Countries)', online at: http://web.worldbank.org/WBSITE/EXTERNAL/TOPICS/EXTDEBTDEPT/0,,contentMDK:20260049~menuPK:528655~pagePK:64166689~piPK:64166646~theSitePK:469043,00.html.

World Resources Institute (WRI) (2006), 'Global forest watch', online at: http://www.globalforestwatch.org.

World Resources Institute (WRI), International Livestock Research Institute (ILRI), Department of Resource Surveys and Remote Sensing, Ministry of Environment and Natural Resources, Kenya (DRSRS), and Central Bureau of Statistics, Ministry of Planning and National Development, Kenya (CBS) (in Press), *Nature's Benefits: An Atlas of Ecosystems and Human Well-Being in Kenya*, Washington, DC and Nairobi.

World Resources Institute (WRI), United Nations Development Programme, United Nations Environmental Programme, and the World Bank (2003), *World Resources 2002–2004: Decisions for the Earth – Balance, Voice, and Power*, Washington, DC: WRI.

World Resources Institute (WRI), United Nations Development Programme, United Nations Environmental Programme and the World Bank (2005), *World Resources 2005: The Wealth of the Poor – Managing Ecosystems to Fight Poverty*, Washington, DC: WRI, online at: http://multimedia.wri.org/worldresources2005.cfm.

WWF (2005), 'AREAS Project: Technology', online at: http://www.worldwildlifefund.org/action/areasproject/technology.cfm.

PART I

Incorporating ecosystem services into development planning and practice

1. Mainstreaming and implementing the Millennium Ecosystem Assessment results by integrating them into sustainable development strategy: Applying the Action Impact Matrix Methodology

Mohan Munasinghe

1. OVERVIEW AND FRAMEWORK

This chapter starts from the fundamental premise that measures to protect ecosystems and ecosystem services ultimately must be mainstreamed into conventional development strategy. The Millennium Ecosystem Assessment has assessed the state of ecosystems and the services they provide, the impacts on human well-being, and the strengths and weaknesses of response options. However, the comprehensive and far-reaching recommendations proposed by the Assessment are rather general and mainly at the global and regional levels. Recognizing the multi-scale dimensions of protecting ecosystem services (ranging from the local to the global levels) and the need to strengthen in-country activities, we should seek to mainstream the broad Millennium Assessment results by integrating them into sustainable development (SD) strategy at the national and local levels and implementing them on a more disaggregate basis through local communities and individuals. As a parallel second step, the Assessment findings will also receive more favourable attention from the international development community and donors if an explicit analysis is carried out of the findings' linkages with global development concerns – for example, the millennium development goals (MDG).

One method of achieving these mainstreaming objectives would be to use the Action Impact Matrix (AIM), which is a tool devised to assist environment–development integration. The AIM analytical process will help to identify and prioritize:

(a) key policy linkages between the Millennium Assessment and sustainable development at the national level – upward to the national macro-economic level, and downward to the micro-project level; and
(b) key Assessment-MDG interactions at the global level.

The basic principle of the AIM methodology focuses on '*making development more sustainable*' (MDMS) globally, and within countries and communities (see Appendix 1).

1.1 National Level Mainstreaming through Millennium Assessment–SD Linkage

National decision makers normally pay attention to conventional development strategies that seek to achieve objectives like economic growth, poverty alleviation, food security, improved human health and employment. As shown in the Figure 1.1, sustainable development (SD) is considered a special (and rather obscure) aspect of conventional development.

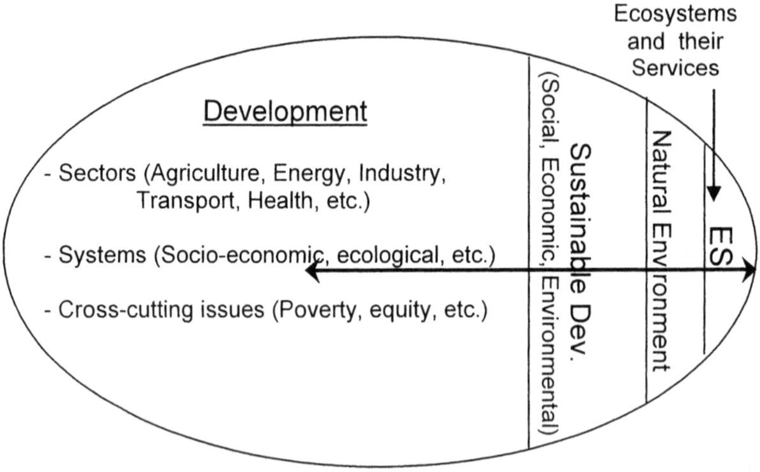

Source: Adapted from Munasinghe (1992)

Figure 1.1 National level ecosystems–sustainable development nexus: decision makers see ecosystems as a minor element in the national development strategy

In turn, the environment is one component of SD, and finally ecosystems are a single element of the environment. The arrow in the figure shows that the essential step in practical implementation of the Millennium Assessment results on a country-by-country basis is to show the interaction between conventional development activities and key ecosystem services, identify priority issues, and determine appropriate policy remedies and responses in the development domain.

1.2 Global Level Mainstreaming by Linking the Millennium Ecosystem Assessment and MDGs

Figure 1.2 indicates the preferred method of linking the key ecosystems identified in the Millennium Ecosystem Assessment report with the MDGs. There are multiple, two-way interactions. The common pitfall is that the development community tends to identify ecosystems and their services with just one MDG – that is, MDG 7 dealing with the environment.

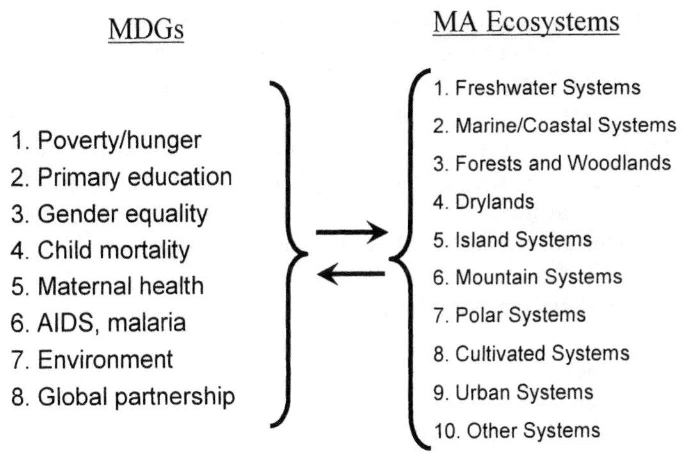

Figure 1.2 Global level mainstreaming – desired viewpoint is that key MA ecosystems and services should be linked to all MDGs, not only MDG 7 on environment

For example, a recent United Nations Development Programme (UNDP, 2005) review of 100 country-level MDG reports concludes that

> an assessment of the extent to which environmental considerations are integrated into MDG reports other than Goal 7 shows that environmental issues

> do not receive much attention outside of MDG7. While referred to in over 80% of the cases, the causal link between environment and other goals is not well recognized or articulated and response systems are not developed.

The above misperception is unduly restrictive. The Millennium Assessment has correctly emphasized that ecosystem services will affect human activities widely, and therefore should be linked with all MDGs (see Schmidt-Traub and Cho's chapter in this volume). Furthermore, establishing the MDG–MA integration at the global level will facilitate efforts to make the same linkage at the national level.

2. IMPLEMENTING THE MILLENNIUM ECOSYSTEM ASSESSMENT RESULTS

Overall, the Millennium Assessment has achieved a key objective of bringing to the world's attention the most credible and up-to-date scientific information on ecosystems and their services. The only comparable scientific exercise on a similar scale is the work of the Intergovernmental Panel on Climate Change (IPCC) (IPCC, 2001a; 2001b; 2001c). The Millennium Assessment has judiciously adapted and even improved on many of the IPCC procedures to carry out their work. Thus, the Assessment's results constitute a good first step and may be compared favourably with the IPCC First Assessment Report (1990) despite the uneven quality of some of the chapters.

2.1 Key Policy-relevant Issues

As the next stage of interpreting and ultimately implementing the Millennium Assessment findings is contemplated, it is useful to pose two wider policy-relevant questions which form a background to the technical material in the reports. They further reinforce the main theme of this chapter – that is, the Assessment results need to be interpreted and incorporated within a broader sustainable development (SD) framework.

1. First, is the balance between humans and nature appropriate?
   The Millennium Assessment correctly favours the anthropocentric viewpoint, and gives prominence to the relevance of ecosystems for human well-being. This will be helpful in convincing decision makers and stakeholders to undertake some of the difficult policy remedies that are needed.

2. Second, is the south–north balance appropriate?
   The Millennium Assessment process devoted considerable effort to incorporating the southern perspective fairly (for example, in the recruitment of managers, lead authors and reviewers, case study material, and so on). Nevertheless, more effort is required in the dissemination and implementation phase to reassure the developing countries. The broad results suggest that:

   (a) ecosystems and their services are under threat globally, and the poor are most vulnerable to loss of ecosystem services;
   (b) policy options and measures are available to manage ecosystems, but they will require radical changes; and
   (c) such management will yield significant benefits (both locally and globally), for human well-being.

Two statements of the Millennium Assessment Board of Directors sum up the arguments clearly.

> A striking part of this assessment is that the people lacking these minimum standards of human well-being are generally those most vulnerable to the deterioration of natural systems (MA, 2005d, 23).

> The over-riding conclusion of this assessment is that it lies within the power of human societies to ease the strains we are putting on the natural services of the planet, while continuing to use them to bring better living standards to all. Achieving this, however, will require radical changes in the way nature is treated at every level of decision-making (MA, 2005d, 31).

The Millennium Assessment also highlights an additional fact, relevant for the billions of poor people in the developing world who are barely able to survive: the bulk of the world's biodiversity resides in the developing countries.

In this context, two key poverty and equity-related questions arise.

(i) Who bears the costs and who reaps the benefits of exploiting ecosystems and their services?
(ii) Who will bear the burden of (and benefit from) preserving ecosystems and their services?

The first issue revolves around the fact that 'in many cases, it is the poor who suffer the loss of services caused directly by the pressure put on natural systems to bring benefits to other communities, often in different parts of the world' (MA, 2005d). If the rich in distant places exploit ecosystem services and the local poor have to pay the costs, there are obvious equity implications – both internationally (South–North) and within countries (and indirectly inter-generational, as well).

On the second issue, the Millennium Assessment recommends many policy prescriptions, but is rather reticent about one central fact – poor countries hardly have resources to address priorities they consider to be most urgent, such as poverty and hunger. If preserving ecosystems and biodiversity have global benefits, both efficiency and equity considerations argue in favour of resource transfers to poor countries that will help them to manage their ecosystem better. Existing mechanisms (for example, the Global Environmental Facility) are inadequate for this purpose.

These questions need to be addressed in order to help the international development community and developing country decision makers respond to the urgent Millennium Assessment challenge of better ecosystem management, while simultaneously addressing existing priority issues (poverty alleviation, human health, food security, water resources, energy, and so on) against a general backdrop of resource scarcity.

2.2 Role of the Millennium Ecosystem Assessment Conceptual Framework

Starting from the global framework provided by the Assessment, strategies, policies and measures for implementation need to be generated at the level of individual communities and sovereign nations so that the responses are case specific. Thus, assessing the advantages and disadvantages of ecosystem management (that is, with social, economic, and environmental costs and benefits, broadly defined) within a national decision-making framework will be an essential prerequisite for establishing the balance between ecosystem management and other urgent development priorities and for identifying and mainstreaming key policies for implementation.

The Assessment's conceptual framework (Figure 1.3) clearly shows the circular linkages between human well-being and ecosystem services. At the global, national and community levels, it needs to be determined more specifically whether these interactions make development more or less sustainable. Then specific remedial policy options may be determined and implemented to minimize negative effects and enhance positive ones.

In Figure 1.3, the impacts of ecosystem services on development (upper horizontal arrow) may be better understood in terms of effects on the three major elements of sustainable development – economic, social and environmental. Figure 1.4 shows an expanded version of this sustainable development triangle which will help to assess the specific impacts of ecosystem services on the key social, economic and environmental indicators of sustainable development (see Appendix 1).

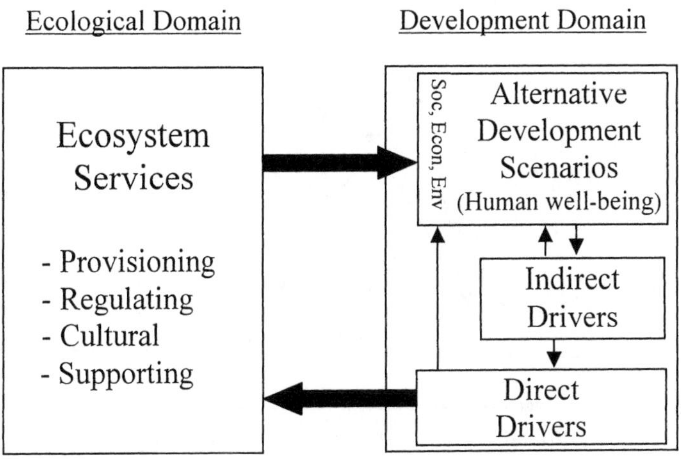

Source: Adapted from the Millennium Ecosystem Assessment (2005a, 37)

Figure 1.3 MA conceptual framework: circular interaction between ecosystem and development domains

Win–win policy options that simultaneously yield ecosystem and development (social, economic and environmental) benefits are unambiguously desirable. In other cases, scarce resources may need to be shared between ecosystem management measures and other sustainable development needs through a careful assessment of trade-offs. The Millennium Ecosystem Assessment recognizes several such areas which involve trade-offs between the present and future, among ecosystem services, among constituents of human well-being and among stakeholders (MA, 2005b, 6).

Those who are called upon to bear the burden of ecosystem management will need to be reassured that the benefits they reap (often over the long term) will exceed the costs (invariably short-term). Equitable burden-sharing will be required, especially to ensure that disproportionate additional costs are not imposed on already impoverished groups, globally and within nations. Such a process will also require the engagement of a broad range of stakeholders, through multilevel mechanisms, to build both ownership and consensus for following up on the Assessment results within specific countries and communities; the Action Impact Matrix (AIM) is a useful instrument for this purpose.

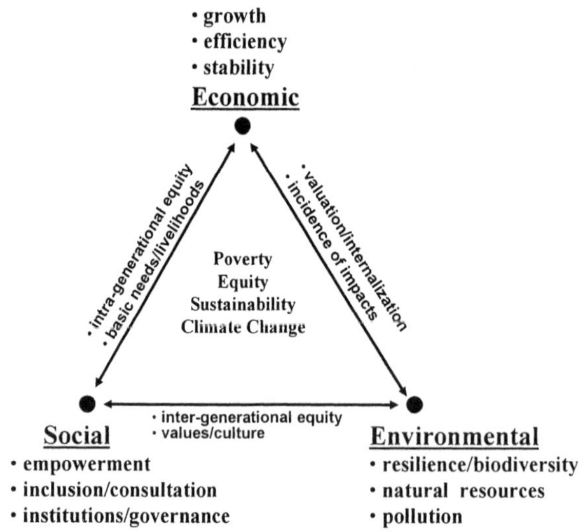

Source: Munasinghe (1992) Rio Earth Summit

Figure 1.4 Sustainable development triangle – key elements and interconnections (corners, sides, centre). Ecosystem services affect all dimensions of sustainable development.

3. ACTION IMPACT MATRIX (AIM) METHODOLOGY AND APPLICATIONS

The AIM is an integrating and linking tool which has several desirable properties identified in the Millennium Assessment – for example, the deliberative, information-gathering, and planning elements (MA, 2005b, 8). It may be used to mainstream ecosystem issues into initiatives which make development more sustainable, at both global and national levels. Globally, MDGs can be linked with key ecosystem services identified in the Assessment. At the national level, the linkage may be made in two main ways, which are complementary and interlinked:

(a) an upward link, where ecosystem concerns are embedded in the macro-level national development strategy of a country via the medium- to long-term sustainable development path, including the building-up of adaptive capacity; and
(b) a downward link, where ecosystem concerns are integrated into the subnational-level development strategy in the short to medium term by carrying out sustainable development assessments aimed at making specific ecosystem projects and policies more sustainable.

The importance of mainstreaming is recognized in the Millennium Assessment: 'the quantity and quality of ecosystem services available are often determined to a greater extent by macroeconomic, trade and other policies than by policies within the environmental sector itself' (MA, 2005b, 7). Furthermore, by linking the global, macroeconomic and micro-local aspects of decision making, the AIM approach helps us to coordinate implementation across scales (also a theme echoed in the Assessment); 'effective management of ecosystems requires coordinated responses at multiple scales' (MA, 2005b, 9).

The AIM has been widely used since the early 1990s, and originally was presented as part of the Sustainomics methodological framework at the 1992 Rio Earth Summit (see Annex 1) (Munasinghe, 1992; 2007). Initially, AIM was used in Brazil, Chile, Ghana, the Philippines and Sri Lanka to integrate a range of environmental concerns into development planning (Munasinghe, 1997; 2002a; 2005; Munasinghe and Cruz, 1994). Subsequently, expanding and 'adapting' the AIM approach to address ecosystem–SD interactions was a logical next step (Munasinghe, 2002b; Munasinghe, 2007; MIND, 2005).

The AIM approach may be used to better understand two-way interactions between (a) development policies and goals and (b) key ecosystem areas relevant to sustainable development (SD). In the methodology, the effects of development policies and goals on ecosystems are first explored, and then the reverse, effects of ecosystems on sustainable development prospects, are identified. The AIM approach analyses key economic–environmental–social interactions to identify potential barriers to making development more sustainable (MDMS), including the progressive degradation of ecosystem services. This tool also helps to determine the priority strategies, policies and projects in the economic, environmental and social spheres that facilitate implementation of measures to manage ecosystems and restore damaged ecosystem services. After completing a national-level AIM exercise, it is possible to apply the process at a subnational or community level to fine-tune the analysis.

A national-level AIM is generated through a fully participative, consensus-building, multi-stakeholder exercise involving 10–50 experts drawn from government, academia, civil society and the private sector representing various disciplines and sectors relevant to both sustainable development and ecosystems. In the initial exercise, the experts usually interact intensively over a period of about two days to build a preliminary AIM. This participative process is as important as the product (that is, the AIM), since important synergies and cooperative team-building activities emerge (for more on the participative process of the Millennium Assessment, see Norgaard's chapter in this volume). The teambuilding is a key element of the investment in human and institutional capacity required to implement the Assessment findings (MA, 2005b: 2). The transparent stakeholder collaboration helps participants to better understand opposing viewpoints and resolve conflicts. It promotes cooperation and ownership across decision-making agencies and ultimately facilitates implementation of the agreed upon policy remedies. Subsequently, the AIM exercise may be repeated quite easily to assess progress toward goals, update the knowledge base, address new challenges, and renew the stakeholder consensus.

The AIM process addresses several implementation process issues that are clearly recognized by the Millennium Assessment's responses working group. They state:

(a) 'insufficient participation and transparency in planning and decision-making have been major barriers to the design and implementation of effective responses' (MA, 2005b: 4);
(b) 'at the national and sub-national levels successful response interventions often involve situation-driven integration across decision-making agencies' (MA, 2005b: 4); and
(c) 'for participatory approaches to succeed, the stakeholders involved need access to information both on the resources being managed and the decision-making process' (MA, 2005b: 5).

For maximum effectiveness, the AIM meeting needs trained instructors to conduct the exercise, documentation (for example, AIM Manual), screening and pre-selection of a balanced group of participants, and advance gathering of relevant background data.

4. SAMPLE APPLICATION TO SRI LANKA AT THE NATIONAL LEVEL

The AIM approach may be illustrated by sketching out a very preliminary application for Sri Lanka at the national level based on the author's experience. The examples provided below are only indicative, and need to be re-checked and validated by a full-blown AIM stakeholder exercise involving a variety of experts.

The AIM process involves several key practical steps:

4.1 Step (a): Determine the Most Important National Goals and Policies

The Millennium Development Goals (UNDP, 2003) provide a generic starting list of possible goals and priorities, but the MDGs need to be tailored to be country-specific. In order to focus attention on key priorities, the list of goals needs to be restricted to less than ten items. In the case of Sri Lanka, a brief perusal of past and recent national development documents helps to identify the following dominant national goals and policies (GOSL, 2002; 2004):

- economic growth,
- poverty alleviation,
- food security,
- employment,
- trade and globalization,
- budget deficit reduction and
- privatization.

4.2 Step (b): Determine Critical Ecosystems and Services Areas Relevant to SD

Similar to the MDGs in the first step, the broad areas identified by the Millennium Ecosystem Assessment under the supporting, provisioning, regulating and cultural categories of ecosystem services provide a useful generic starting point for the second step (MA, 2005a: 56). However, they too need to be fine-tuned on a country-specific basis. Again, in order to concentrate on the key issues, the list of ecosystems and their services needs to be restricted to less than ten items. In the case of Sri Lanka, a review of past and recent environmental reports and documents indicate the following critical ecosystem-related areas (Ministry of Environment and Natural Resources, 2002a; 2002b; 2003):

- Forests
- Managed ecosystems 1 (grain)
- Managed ecosystems 2 (tree crops)
- Coastal & Marine systems
- Wetlands
- Water resources

4.3 Step (c): Identify how Development Goals/Policies might affect Ecosystems (DE–AIM)

The two lists determined by consensus in steps (a) and (b) above are put together to establish the basic AIM framework (see Figure 1.5 for Sri Lanka sample). For convenience of presentation, the row and column headings are given concisely, but the full spreadsheet will contain detailed notes describing each category in further detail. Thus, within each column, all relevant ecosystem services are considered sub-categories (not shown here for the sake of brevity).

The row (S) marked 'status' indicates the current condition of the ecosystem under consideration. This is a useful benchmark. For example, cell S1 has a value of −2 (moderately negative status) because deforestation has reduced forest cover significantly in Sri Lanka and decreased all associated services.

Next, the impacts of national goals and policies on vulnerable ecosystems are identified, as indicated by the direction of the arrows at the top left-hand corner of the figure. As an example, a few shaded cells have been selected to illustrate some key effects. The numbers are qualitative, and represent net outcomes due to a variety of impacts, including trade-offs among different ecosystem services. Brief explanations of the relevant linkages are provided for each cell. Typical examples include:

Cell B1 = +2: Moderate positive effects of poverty alleviation on forests.

The poor often exploit forests unsustainably for their survival (for example, slash and burn agriculture). Poverty alleviation programmes that promote sustainable livelihoods would encourage poor people who depend on natural resources such as forests, to move away from marginal lands into other areas and become less dependent on the natural resource

		Critical Ecosystems and Services					
		(1)	(2)	(3)	(4)	(5)	(6)
		Forests	**Managed Ecosyst. 1 (grain)**	**Managed Ecosyst. 2 (tree crops)**	**Coastal & Marine Systems**	**Wetlands**	**Water Resources**
(S) Status		**−2**	**−1**	**0**	**−1**	**−1**	**−1**
Development Goals/Policies							
(A)	Growth				-2		
(B)	Poverty alleviation	+2					
(C)	Food Security						
(D)	Employment					+1	
(E)	Trade & Globalization						
(F)	Budget Deficit Reduction						
(G)	Privatization						

KEY: 3 = High; 2 = Moderate; 1= Low; Minus (−) = negative impact/status; Plus (+) = positive impact/status

Figure 1.5 Sri Lanka sample AIM: Development effects on ecosystems (DE–AIM)

base (for example, see Upper Watershed Management Project, Ministry of Environment and Natural Resources; and Participatory Forestry Project, Forest Department). Helping the rural poor through community forestry projects would also restore some forest ecosystem functions.

Cell A4 = −2: Moderate negative effects of economic growth on coastal and marine systems.
Growth often leads to destruction of coastal ecosystems for industry and urbanization, intensive construction on the coastal belt, coral and sand mining, depletion of fish stocks and so on. On the positive side, growth may rid some of the population of their dependence on vulnerable coastal resources (for example, increased employment from tourism decreases coral mining), while higher incomes and awareness could encourage conservation.

Cell D5 = +1: Low positive effects of employment promotion on wetlands.
Greater employment opportunities in sectors that do not depend directly on natural resources and biodiversity, especially for the rural poor, would decrease their dependence on marginal natural resources, thus helping to preserve wetlands.

Such summaries would usually be supplemented by longer text descriptions (several pages), giving details of mechanisms involved and citing relevant reports and research studies.

4.4 Step (d): Identify how Ecosystems might affect Development Goals/Policies (ED-AIM)

Next, we seek to identify the impacts of ecosystem vulnerability on national goals and policies, as indicated by the direction of the arrows at the top left-hand corner of Figure 1.6. Once again, a few shaded cells provide examples of key issues, and brief summaries of the relevant linkages are provided for each cell. Examples include:

Cell B1 = −2: Moderate negative effects of forest degradation on poverty alleviation.
Deforestation leads to land degradation, causing further negative impacts on poorer groups who depend on natural resources for their livelihoods. Thus, a vicious cycle might be established.

		Critical Ecosystems and Services					
		(1)	(2)	(3)	(4)	(5)	(6)
		Forests	**Managed Ecosyst. 1 (grain)**	**Managed Ecosyst. 2 (tree crops)**	**Coastal & Marine Systems**	**Wetlands**	**Water Resources**
(S) Status		**−2**	**−1**	**0**	**−1**	**−1**	**−1**
Development Goals/Policies							
(A)	Growth						
(B)	Poverty alleviation	−2					
(C)	Food Security						−3
(D)	Employment				−2		
(E)	Trade & Globalization						
(F)	Budget Deficit Reduction						
(G)	Privatization						

KEY: 3 = High ; 2 = Moderate; 1= Low; Minus (−) = negative impact/status; Plus (+) = positive impact/status

Figure 1.6 Sri Lanka sample AIM: Ecosystems effects on development (ED–AIM)

Cell D4 = −2: Moderate negative effects of coastal and marine habitat destruction on employment: Employment provided through activities like fishing, prawn cultivation and tourism would decrease due to coastal and marine system degradation.

Cell C6 = −3: High negative effects of water resource degradation on food security.
Many crops are highly dependent on the availability of water. Decreases in both quantity and quality of water (for example, due to over-extraction of groundwater or pollution) will undermine food production.

Such summaries may be supplemented by longer text descriptions which give information on the tools involved and note relevant research (see Appendix 2 for an example involving Cell C6).

4.5. Step (e): Prioritize the Most Important Interactions and Determine Appropriate Remedial Policies and Measures (Preliminary AIM)

The Millennium Assessment states that 'an effective set of responses to ensure the sustainable management of ecosystems would address the indirect and direct drivers that lead to the degradation of ecosystem services' (MA, 2005b: 10). The Assessment identified 78 response options. This list is useful as a general base for ideas, but needs to be tailored during the AIM exercise to suit country-specific needs. Other in-country documents such as Poverty Reduction Strategy Papers (PRSPs), National Environmental Action Plans (NEAPs), and relevant Multilateral Environmental Agreements (MEAs), could be used to develop donor consensus and provide national and subnational entry points into policy, thereby facilitating the implementation of measures that address both development and ecosystem issues. Practical and political constraints that limit the range of policy options need to be factored in at this stage.

As an illustrative example, one priority linkage (among many) is selected for further analysis – in this case the key nexus of water and food security is discussed below, and a range of appropriate policy options are identified.

Water resources and food security
A cross-sectoral focus is urgently needed from policy and decision makers that emphasizes securing water resources in the context of achieving food security to make development more sustainable and

improve human well-being. Governments and water managers are faced with the need to increase water supply to meet a still expanding population's increasing demand for food and water while at the same time ensuring that the water supply is sustainable and that ecosystems contributing to that sustainability are protected. Management of water resources and associated ecosystems are most successfully addressed through integrated management at the river basin (or lake or aquifer) scale.

In regions where water is already stressed or scarce, meeting increasing demand for all water uses including ecosystem management becomes increasingly difficult and expensive. However, there are many options for providing the necessary water, many of which may be better and cheaper than constructing new projects. Some options are:

Supply-side management options:

- constructing additional water storage and distribution systems;
- making better use of natural systems, such as wetlands and ground cover, to reduce erosion, store and filter water and recharge aquifers;
- improving the efficiency of existing storage and distribution systems (especially O&M activities); and
- improving water management techniques and institutions.

Demand-side management options:

- increasing water productivity (e.g., recycling and multiple-pass through);
- improving water pricing;
- applying water quotas;
- using economic incentives to reduce withdrawals and pollution;
- improving water quality regulations;
- introducing a pollution permits market;
- importing more food rather than growing it; and
- encouraging less water-intensive crops.

Water pricing in rural areas Water pricing is one of the most important elements of recent water management frameworks because it is the basis for achieving efficient allocation of water resources. Conversely, inappropriate water prices could encourage inefficient use of water, contributing to shortages or depletion of water resources, and the long-run degradation of ecosystems.

Typically, water prices in agriculture, when they exist, more or less cover the variable cost of water supply (like purification costs), while public authorities cover the fixed costs (investment costs). Sometimes prices are set according to farmers' 'ability to pay'. The structure of water pricing systems usually takes one of the following forms: standard volumetric and fixed tariffs, area-pricing, tiered or block-rate pricing, land betterment levy pricing or passive trading, volumetric pricing with bonus or water markets.

Water markets and tradable water rights could help to give water a value separate from land and provide incentives to use water more efficiently and flexibly, since water savings can be sold for extra revenue or used to increase production. Water markets are being promoted by international organizations such as the World Bank. Meanwhile, in developing countries, the practicality and true ecological and livelihood impacts of water pricing and markets are under scrutiny. Higher water prices may deprive the rural poor of their basic water needs, and ration water for the benefit of the rich (who could afford it). Because of this, a broader use of the term economic incentives will certainly be important. For example, positive incentives could be given to farmers for saving water, rather than penalizing the rural poor. Often it is the urban wealthy who ultimately benefit from low food prices (produced with subsidized water), although they could better afford the cost of dealing with negative externalities.

Water pollution Agriculture is the single largest user of water resources. Except for water lost through evapo-transpiration, agricultural water is recycled back to surface water and groundwater. However, agriculture is both a cause and a victim of water pollution. Agriculture causes water pollution through its discharge of pollutants and sediment into surface and groundwater, net loss of soil from poor agricultural practices, salinization and water logging of irrigated land and salt water intrusion in coastal aquifers due to over pumping. Agriculture also suffers from water pollution, because wastewater and polluted surface and groundwater could contaminate crops and transmit diseases to consumers and farm workers. Furthermore, pollution of both fresh and brackish water also harms natural ecosystems, which provide key services to agriculture.

The inadequacy of the standard instruments environmental policy uses to deal with pollution has led, in recent years, to the development of policy schemes which can be divided into two broad categories: (1) ambient taxes where the scheme is based on the observed ambient pollution and (2) input-based schemes, where the policy scheme consists of taxes applied to observable polluting inputs. Common policies against water pollution include user charges for sewerage and sewage treatment, water effluent charges and charges in agriculture, along with a number of more specific policies.

Charges used for agricultural practices are often more profound cases of input-based schemes. Charges on fertilizers, for example, are applied in many countries based on the fertilizer's nitrogen and phosphorus content (the main contributors to pollution in surface water). More specific farm policies aimed at addressing pollution problems have been effective in many countries. A number of off-farm management methods also exist for reducing phosphorus levels in bodies of water such as vegetation buffer stripes, riparian zones and dredging lake sediment.

4.6 Step (f): Perform More Detailed Studies and Analyses of Key Interactions and Policies Options Identified in Step (e) above

The critical issues, policy options and measures identified in step (e) may be subject to further analysis and research to provide greater confidence to decision makers before they begin to implement remedial measures. A variety of macro-, regional or local models are usually applied to possible policies that focus on the specific questions at hand (Munasinghe 1994; 1997; 2002a). This procedure evaluates appropriate country-specific ecosystem management options, selected from the broad response categories listed in the Millennium Assessment – institutions and governance, economic measures and incentives, social and behavioural responses, technological response and knowledge and cognitive responses (MA, 2005b: 11–17).

In our water–food example for Sri Lanka, further multi-scale exploration is needed. At the national level, integrated spatial water basin models linked with a multi-sector macroeconomic model (with disaggregated water uses including irrigation, drinking water, hydropower, ecosystem use, and so on) would be used to test some of the ecosystem management policy options. At the micro-level, more detailed models of local areas and communities would also be helpful to assess specific agricultural and irrigation projects.

4.7 Step (g): Update and Refine Steps (c) to (e) based on Step (f): Revised AIM

The results of the detailed research and analysis in step (f) are introduced into the AIM process in order to update and refine the information in the cells and begin the next stage of implementing remedial actions. A package of policies and projects would be assembled, based on modelling and detailed studies, to address the most important linkages in the AIM. When testing and validating policy alternatives, those that give rise to win–win outcomes (for example, simultaneously addressing economic, social and environmental objectives) would have the highest priority. In other cases, trade-offs among conflicting objectives need to be carefully assessed. Practical and political constraints also need to be kept in mind.

5. SAMPLE APPLICATION TO GLOBAL AND SUB-GLOBAL LEVELS

The procedure outlined in the Sri Lanka example in section 4 could be applied to provide a convenient summary of the most important two-way interactions between global-level policy goals and ecosystem services.

5.1 Steps (a) and (b): Determine the Most Important Global Goals and Critical Ecosystems

As shown in Figure 1.7, the set of global policy objectives (that is, the Millennium Development Goals) are selected and set as the rows of the AIM, and the set of key ecosystems from the Millennium Assessment are used as the AIM columns. As before, within each column, all relevant ecosystem services should be explicitly considered as sub-categories.

5.2 Step (c): Effects of Millennium Development Goals on Critical Millennium Assessment Ecosystems and Services (DE–AIM)

Figure 1.7 shows a few examples of how the MDGs might affect critical ecosystems and services identified in the Assessment, as a part of step (c).

The status row (S) shows the condition of the relevant ecosystem. For example, the −2 in Cell S3 indicates a moderately bad status, based on information from the Millennium Assessment. 'The global area of naturally-regenerating forest has declined throughout human history, and has halved over the past three centuries. Forests have effectively disappeared in 25 countries, and more than 90% of the former forest cover has been lost in a further 29 countries' (Condition Summary – Board, page 26). Two further examples are:

		Critical MA Ecosystems and Services									
		(1)	(2)	(3)	(4)	(5)	(6)	(7)	(8)	(9)	(10)
		Freshwater systems	Marine & coastal systems	Forests & woodlands	Drylands	Island systems	Mountain systems	Polar systems	Cultivated systems	Urban Systems	Others
(S)	Status	−2	−2	−2	−1	−1	−1	−1	−1	0	
Millennium Dev. Goals (MDG)											
(A)	Eradicate poverty & hunger			+1							
(B)	Universal primary education			+2							
(C)	Gender equality & women empowerment										
(D)	Reduce child mortality								+1		
(E)	Improve maternal health										
(F)	Combat HIV/AIDS, malaria & other diseases								+1		

		Critical MA Ecosystems and Services									
		(1)	(2)	(3)	(4)	(5)	(6)	(7)	(8)	(9)	(10)
		Freshwater systems	**Marine & coastal systems**	**Forests & woodlands**	**Drylands**	**Island systems**	**Mountain systems**	**Polar systems**	**Cultivated systems**	**Urban Systems**	**Others**
(S)	**Status**	**–2**	**–2**	**–2**	**–1**	**–1**	**–1**	**–1**	**–1**	**0**	
Millennium Dev. Goals (MDG)											
(G)	Ensure environmental sustainability	+2									
(H)	Develop a global partnership for development			–1							

KEY: 3 = High; 2 = Moderate; 1= Low; Minus (–) = negative impact/status; Plus (+) = positive impact/status

Figure 1.7 Global level: effects of Millennium Development Goals on critical millennium assessment ecosystems and services (DE–AIM)

Cell A3 = +1 (net effect): There will be fewer poor people who exploit forests unsustainably for their survival (for example, slash and burn agriculture).
+2 impact: Poverty alleviation programmes that promote sustainable livelihoods would encourage poor people who depend on natural resources such as forests, to move away from marginal lands into other sectors that are less dependent on the natural resource base. This will lessen pressure on ecosystems and enable them to recover or be restored.
−1 impact: Alleviating poverty can result in some negative consequences on the environment if not carefully managed. Because poverty alleviation can only occur in tandem with economic development, there may be increased demand for high value products. Increased preferences for meat and other livestock products take place with income growth. Effective policies such as those that discourage unsustainable land clearing and deforestation while improving food distribution channels to increase availability and accessibility could help mitigate this effect. Changes in land use would also be triggered by the need to expand commercial, industrial and social services as demand for such services increases with continued improvement in the well-being and economic status of the poor (MA, 2005b: Section 19.2.1.4).

Cell H3 = −1: Expansion of international trade may be harmful to forest ecosystems.
Unrestricted trading in timber and forest products, coupled with aggressive trade expansion policies, would most likely further degrade forest and woodland ecosystems.

5.3 Step (d): Effects of Critical Millennium Assessment Ecosystems and Services on Millennium Development Goals (ED–AIM)

Figure 1.8 summarizes a few examples of how critical ecosystems and services might affect the Millennium Development Goals at the global level. Two typical examples are:

Cell A3 = −1 (net effect): Forest degradation will undermine food security and livelihoods of the poor.
−2 impact: Degraded forest ecosystems cannot provide a wide range of essential services for food production. The depletion of forest and woodland areas would deprive the poor of the resources on which they are dependent, thus aggravating poverty and hunger by decreasing their livelihood options, preventing livelihood diversification and decreasing

food security (Policy responses, Ch. 19, pp. 8–11).
+1 impact: Agriculture and forests compete for land. Some forest depletion may contribute to greater agricultural output. Currently about 11 per cent of the terrestrial biosphere has been converted to cropland (some 36 per cent of the world's land base has crop production potential to some degree), and another 30 per cent is used as grazing land. Some of the world's food needs may be met by expanding agricultural areas, but this comes at the expense of the services provided by natural landscapes such as forests. (Policy options, Ch. 19, Section 19.2.2.3).

Cell D1 = −2: Polluted fresh water systems sharply increase child mortality.
Unhygienic and unsafe environments place children at high risk. Ingestion of unsafe water, inadequate water for hygiene and lack of access to sanitation contribute to millions of deaths among children. A significant portion of deaths occur due to diarrhoea, malaria and dengue outbreaks, particularly in developing countries. Underweight status and micronutrient deficiencies among children that decrease their immune and non-immune host defences have made many of them more vulnerable to infectious diseases. Degraded ecosystems, especially those that allow considerable amounts of standing water, are sources of waterborne diseases (like malaria, dengue, and diarrhoea) to which children are highly susceptible (MA, 2005b: Section 19.2.5.2).

5.4 Steps (e), (f), (g): Prioritise Issues, Determine Policy Options, Conduct Further Research and Update AIM

These steps may be carried out, as appropriate, to mainstream the Millennium Ecosystem Assessment findings into the Millennium Development Goals' implementation process.

Note that the status row (S) may be modified to take other scenarios into account. For example, the effects of further stresses imposed by a variety of global-level drivers like globalization, trade expansion, and climate change could be introduced to help identify appropriate international policy responses.

Furthermore, similar matrices may be constructed to mesh the MDGs with other multilateral environmental agreements (MEAs), like those on climate change (UNFCCC), desertification (UNCCD), and biodiversity (CBD). A recent UN Environment Programme (UNEP)-sponsored 'brainstorming' meeting concluded that 'modalities for greater cooperation should include combining Millennium Project Programmes

		Critical MA Ecosystems and Services									
		(1)	(2)	(3)	(4)	(5)	(6)	(7)	(8)	(9)	(10)
		Freshwater systems	**Marine & coastal systems**	**Forests & woodlands**	**Dry lands**	**Island systems**	**Mountain Systems**	**Polar systems**	**Cultivated systems**	**Urban Systems**	**Others**
(S)	**Status**	**−2**	**−2**	**−2**	**−1**	**−1**	**−1**	**−1**	**−1**	**0**	
Millennium Dev. Goals (MDG)											
(A)	Eradicate poverty & hunger		−2	−1					−2		
(B)	Universal primary education										
(C)	Gender equality & women empowerment										
(D)	Reduce child mortality	−2									
(E)	Improve maternal health										
(F)	Combat HIV/AIDS, malaria & other diseases			−1							

		Critical MA Ecosystems and Services									
		(1) Freshwater systems	(2) Marine & coastal systems	(3) Forests & woodlands	(4) Dry lands	(5) Island systems	(6) Mountain Systems	(7) Polar systems	(8) Cultivated systems	(9) Urban Systems	(10) Others
(S)	Status	−2	−2	−2	−1	−1	−1	−1	−1	0	
Millennium Dev. Goals (MDG)											
(G)	Ensure environmental sustainability										
(H)	Develop a global partnership for development										

KEY: 3 = High; 2 = Moderate; 1= Low; Minus (−) = negative impact/status; Plus (+) = positive impact/status

Figure 1.8 Global Level – Effects of critical millennium assessment ecosystems and services on Millennium Development Goals (ED–AIM)

and MEA programmes to bear upon specific ecosystems identified as country priorities. ....A matrix should be created in order to identify where environmental concerns intersect with initiatives to reduce poverty' (UNEP, 2005).

Analogous exercises can also be conducted at the regional and continental levels as appropriate. For trans-boundary ecosystems (for example, the Amazon, multi-country rivers.), the AIM application needs to be tailored to fit the relevant resource basin and stakeholder objectives which may span many countries and communities.

6. CONCLUSIONS AND RECOMMENDATIONS

1. The Millennium Ecosystem Assessment has achieved a major objective by bringing to the world's attention the *serious global threats faced by ecosystems and their services, and the potential adverse consequences for human well being – especially the world's poor.*
2. Further dissemination and implementation of the Assessment results need to focus on *interpreting the generic global and regional level findings at more geographically disaggregate levels*, especially the assessment of the advantages and disadvantages to specific stakeholders resulting from ecosystem management measures.
3. One key step in this process would be *the incorporation and mainstreaming of ecosystem management measures into sustainable development strategy* at the global, national, subnational and community levels.
4. *The Action Impact Matrix (AIM) methodology is a useful tool* for decision makers and analysts to better integrate policy-relevant Millennium Assessment findings within sustainable development strategy and implement them at multiple scales (global to local), especially within specific countries. The AIM may also be applied to integrate issues arising from other MEAs. The AIM participative process will help to build human resource capacity that facilitates practical implementation of the Assessment's results.
5. *A number of case studies should be carried out to test and validate the AIM approach* in a range of countries.
6. Based on the above case studies, the existing Millennium Assessment multi-scale sub-global assessments, and other recent environment and development literature, *a Special Report should be prepared on an Action Programme to Implement the Millennium Ecosystem*

Assessment results, which will provide practical guidance to decision makers, policy analysts and other stakeholders in interpreting, mainstreaming and ultimately implementing the far-reaching Assessment findings both globally and on a country-specific basis.

APPENDIX 1: UNDERLYING PRINCIPLES

The AIM methodology draws on the following basic principles of the sustainomics framework (Munasinghe 1992; 2002b; 2007):

(a) **MDMS approach**
The step-by-step approach of 'making development more sustainable' (MDMS) becomes the prime objective, while sustainable development is defined as a process (rather than an end point).[1] Although MDMS is incremental, it does not imply any limitation in scope (for example, restricted time horizon or geographic area).

(b) **Sustainable development triangle**
SD requires integration of three main perspectives: social, economic and environmental (Figure 1.4). Each viewpoint corresponds to a domain (and system) that has its own distinct driving forces, objectives and indicators. The economy is geared mainly towards improving human welfare, primarily through increases in the consumption of goods and services. The environmental domain focuses on maintaining the integrity and resilience of ecological systems. The social domain emphasizes the enrichment of human relationships and achievement of individual and group aspirations. The balance among the three domains will be location-specific depending on the stakeholders involved. The sides of the triangle also represent key interactions. For example, the economy–environment interface leads to valuation and internalization of environmental externalities and trade-offs between economic and environmental goals.

(c) **Trans-boundary approach**
The analysis transcends conventional boundaries imposed by discipline, space, time, stakeholder viewpoints and operationality. The scope is broadened and extended in all domains to ensure a comprehensive view.

(d) **Full cycle application of integrative tools**
Two complementary approaches based on 'optimality' and 'durability' may bc used to integrate across economic, social and environmental sub-models within an integrated assessment modelling framework.

A policy mapping method would help to translate actions in the environmental and social domains, into the conventional national economic planning and implementing mechanisms within line ministries and departments.

The Action Impact Matrix (AIM) process is the key link from initial data gathering to practical policy application and feedback. Ecosystems and their services are integrated into SD strategy in two main ways: an upward link where ecosystem concerns are embedded in the macro-strategy of a country via the medium- to long-term sustainable development path and a downward link where ecosystem concerns are integrated into the national development strategy in the short to medium term by carrying out sustainable development assessments (SDA) of micro-level ecosystem-related projects and policies.

APPENDIX 2: SAMPLE LONGER TEXT DESCRIPTION OF CRITICAL CELL LINKAGES

Detailed Descriptive Text of Effects of Water Resources Degradation on Food Security (Cell C6, Figure 1.6)

Water scarcity and access to fresh water is a globally significant and accelerating problem for 1–2 billion people worldwide, leading to reductions in food production, human health and economic development. With population growth and the overexploitation and contamination of water resources, the gap between water demand and supply is growing. Water scarcity, both in physical and economic terms, along with limited or reduced access to water are major challenges facing human society and are key factors limiting economic development in Sri Lanka.

The principal supply of renewable fresh water for humans comes from an array of wetland types, including lakes, rivers, swamps and groundwater aquifers.

It is well established that the degradation of water resources is often a consequence of population growth and demands from increasing economic development, such as infrastructure development, land conversion, water abstraction, eutrophication, pollution, over-harvesting and over-exploitation. Degradation tends to be more rapid where populations are increasing most, leading to demands for increased economic development. There are a number of broad, interrelated economic reasons, including perverse subsidies, why water resources continue to be degraded. Global climate change is expected to further exacerbate the depletion and degradation of water resources. Clearing and drainage, often for agricultural expansion, and increased abstraction of fresh water are the main reasons for the loss and degradation of inland wetlands such as swamps, marshes, rivers and associated floodplain water bodies.

It is clear that there are alterations in the hydrological cycle due to increases in surface temperature, rates of evaporation and, in some regions, increases in precipitation. Changes in the total amount of precipitation and its frequency and intensity directly affect the magnitude and timing of runoff and the intensity of floods and droughts. Such changes will have significant impacts on water resources (Dharmasena, 2004).

Increasing concentrations of greenhouse gases (GHGs) in the atmosphere cause rises in temperature, which in turn enhance potential and actual evapotranspiration. It is not certain how individual water catchment areas will respond to changing evaporation rates and precipitation. It is likely, however, that currently dry regions will be more sensitive to changes in climate. Relatively small changes in temperature and precipitation could cause relatively large changes in runoff. Arid and semi-arid regions will therefore be particularly sensitive to reduced rainfall and to increased evaporation.

An increase in the duration of dry spells will not necessarily lead to an increased likelihood of low river flows and groundwater levels, since increases in precipitation may be experienced during other seasons. More probably, increased rainfall will lead to an increased likelihood of river flooding. Changes in seasonal patterns of rainfall may affect the regional distribution of both ground and surface water supplies. A study by Bandara and Wickramagamage (2004) found that annual rainfall at Nuwara Eliya and the western slopes of the highlands decreased significantly. On the western slopes, the decline was observed mainly during the southwest monsoon period which accounts for a major portion of the rain in this area. This area receives the highest annual rainfall of the country, often exceeding 5000mm, and also contributes to the largest volume of water for hydro power generation and irrigation. There is also a clear indication of the expansion of the dry zone and shrinking of the wet zone (Bandara and Wickramagamage, 2004).

The area under tea in the highlands has declined rapidly as these crops are significantly affected by dry spells (Bandara and Wickramagamage, 2004). However, in some areas such as Watawala and Hatton, tea production is benefited by decreased rainfall through a decline in pests and diseases associated with high rainfall.

Primary food security comes from locally grown foods such as rice. Seventy to eighty per cent of total rice production comes from major and minor irrigation schemes. Non staples such as yams and legumes are grown in rain-fed highlands. Changing patterns of rainfall distribution negatively affect food security.

Crop growth is often limited by temperature. Moisture and water availability will be affected by a temperature increase, regardless of any change in rainfall. Higher temperatures increase the evaporation rate, thus reducing the level of moisture available for plant growth, although other climatic elements are involved. A very large decrease in moisture availability in the dryer regions would be of great concern to the subsistence farmers of these lands. Reduced moisture availability would only exacerbate the existing problems of infertile soils, soil erosion and poor crop yields. In the extreme case, a reduction in moisture could lead to desertification. The southwest monsoon, northeast monsoon and annual rainfall are projected to increase in the future (Basnayake and Vithanage, 2004). Lower rainfall increments on the leeward side of the central hills may lead to water scarcity problems due to population growth and increasing demand for water in the future (Basnayake and Vithanage, 2004).

ACKNOWLEDGEMENTS

The author is grateful to Yvani Deraniyagala, Sudarshana Perera of MIND and Karen Bennet of WRI for assistance in preparing this document. Helpful comments were provided by Kanchan Chopra, Tony Janetos, Tony La Vina, Janet Ranganathan and Walter Reid.

NOTES

1 Sustainable development is defined as a 'process for improving the range of opportunities that will enable individual human beings and communities to achieve their aspirations and full potential over a sustained period of time, while maintaining the resilience of economic, social and environmental systems' (Munasinghe, 1992).

REFERENCES

Bandara, C.M.M. and P. Wickramagamage (2004), 'Climatic change and its impact on upper watersheds of the hill country of Sri Lanka', in S. Herath, A. Pathirana and S.B. Weerakoon (eds), *Proceedings of the International Conference on Sustainable Water Resources Management in the Changing Environment of the Monsoon Region, Vol. 1*, Tokyo: United Nations University.

Basnayake, B.R.S.B. and J.C. Vithanage (2004), 'Rainfall change scenarios for Sri Lanka under anticipated climate change', in S. Herath, A. Pathirana and S.B. Weerakoon (eds), *Proceedings of the International Conference on Sustainable Water Resources Management in the Changing Environment of the Monsoon Region, Vol. 1*, Tokyo: United Nations University.

Dharmasena, G.T. (2004), 'Primary and secondary processing of hydrological data', in S. Herath, A. Pathirana and S.B. Weerakoon (eds), *Proceedings of the International Conference on Sustainable Water Resources Management in the Changing Environment of the Monsoon Region, Vol. 1*, Tokyo: United Nations University.

Government of Sri Lanka (GOSL) (2002), *Regaining Sri Lanka: Vision and Strategy for Accelerated Development*, Colombo, Sri Lanka: Government Press.

Government of Sri Lanka (GOSL) (2004), *Rata Perata: Five Fold Vision for the*

Future of our Motherland, Colombo, Sri Lanka: Government Press.
IPCC (2001a), *Climate Change 2001: The Scientific Basis: Third Assessment Report*, London: Cambridge University Press.
IPCC (2001b), *Climate Change 2001: Impacts Adaptation and Vulnerability, Third Assessment Report*, London: Cambridge University Press.
IPCC (2001c), *Climate Change 2001: Mitigation: Third Assessment Report*, London: Cambridge University Press.
Millennium Ecosystem Assessment (MA) (2005a), *Ecosystems and Human Well-Being: A Framework for the Assessment*, Washington, DC: Island Press.
Millennium Ecosystem Assessment (MA) (2005b), *Ecosystems and Human Well-Being: Volume 3: Policy*, Washington, DC: Island Press.
Millennium Ecosystem Assessment (MA) (2005c), *Ecosystems and Human Well-Being: Synthesis*, Washington, DC: Island Press.
Millennium Ecosystem Assessment (MA) (2005d), *Living Beyond Our Means: Natural Assets and Human Well-Being: Statement from the Board*, Washington, DC: WRI.
Ministry of Environment and Natural Resources (2002a), *State of the Environment in Sri Lanka: A National Report Prepared for the South Asian Association for Regional Cooperation*, Colombo, Sri Lanka.
Ministry of Environment and Natural Resources (2002b), *Sri Lanka's Middle Path to SustainableDevelopment in the 21st Century: National Report of Sri Lanka to the World Summit on Sustainable Development*, Colombo, Sri Lanka.
Ministry of Environment and Natural Resources (2003), *Caring for the Environment 2003–2007: Path to Sustainable Development*, Colombo, Sri Lanka.
Munasinghe, M. (1992), *Environmental Economics and Sustainable Development*, presented at the UNCED Earth Summit, Rio de Janeiro, Brazil, Washington, DC: World Bank.
Munasinghe, M. (ed.) (1997), *Environmental Impacts of Macroeconomic and Sectoral Policies*, Solomons, MD and Washington, DC, USA: International Society for Ecological Economics and the World Bank.
Munasinghe, M. (ed.) (2002a), *Macroeconomics and the Environment*, Cheltenham, UK and Northampton, MA, USA: Edward Elgar Publishing.
Munasinghe, M. (2002b), 'The sustainomics trans-disciplinary meta-framework for making development more sustainable', *International Journal of Sustainable Development*, **5** (1–2), 125–82.
Munasinghe, M. (2007), *Making Development More Sustainable: Sustainomics Framework and Practical Applications*, Colombo, Sri Lanka: MIND Press, Munasinghe Institute for Development.
Munasinghe, M. and W. Cruz (1994), *Economywide Policies and the Environment*, Washington, DC: World Bank.
Munasinghe, M., R., O'Ryan, R., Seroa da Motta, C., de Miguel, C., Young, S. Miller and C. Ferraz (2005), *Macroeconomic Policies for Sustainable Growth – Analytical Framework and Policy Studies of Brazil and Chile*, Cheltenham, UK and Northampton, MA, USA: Edward Elgar Publishing.
Munasinghe Institute for Development (MIND) (2005), *Action Impact Matrix Manual*, Colombo, Sri Lanka: MIND.
UNDP (2003), *Human Development Report 2003 Millennium Development Goals: A Compact among Nations to End Human Poverty*, New York: UNDP.

UNDP (2005), *Environmental Sustainability in 100 Millennium Development Goal Country Reports*, New York: UNDP.
UNEP (2005), *High-Level Brainstorming Workshop for MEAs on Mainstreaming Environment Beyond MDG 7: Conclusions*, Nairobi, Kenya: UNEP.

2. Development, ecosystems and governance: What are the priorities for policy makers?

Antonio La Vina

INTRODUCTION

In this chapter, the results of the recently concluded Millennium Ecosystem Assessment[1] are reviewed using a political lens. Instead of reviewing the Assessment report in a dispassionate and objective manner, the best way to identify lessons, priorities and imperatives from the Assessment is to take it on – understand and analyze it – from a very specific and concrete standpoint: a national development perspective.

In particular, in reading the three main volumes of the Millennium Assessment, the relevant summaries for decision makers, and the statement of its Board, I imagined myself as a senior technocrat of a developing country. This is a country of intermediate size, one with a significant coastline and moderately forested (although lately the rate of deforestation has accelerated), and with a population of 70–80 million. Agriculture continues to be a major economic activity, especially for the rural poor, but the country is rapidly industrializing and urbanizing.

I imagined myself not as an environmental official of that country, but as an official of its national development agency tasked with developing national development plans for the country. In that agency, I imagined myself being assigned by the Minster of National Development Planning to lead the drafting of the 2005–2015 medium term development plan (MTDP) for the national government. Because of pressure from donors as well as from a strong and vocal domestic environmental constituency, the Minister has asked me to review the results of the Millennium Ecosystem Assessment. The Minister is not happy with the pressure put on him by donors and environmentalists so he gives me a very strict mandate, one I must follow rigorously:

- The MTDP must above all be about poverty alleviation. This is its principal objective. The national priority is enhanced food production, accelerated job creation and improved livelihood opportunities.

- Integrate into the MTDP whatever relevant information, analysis, approaches, results and findings of the Millennium Assessment so long as these are supportive of the principal objective of the MTDP – reducing poverty in the country.

In this exercise of imagination, I found myself asking four questions that I decide to address through a memorandum I will write for the Minister. They are:

1. Is the Millennium Assessment a legitimate and credible source of information and analysis for my imagined country to consider in making decisions about its national development plans and strategies?
2. What is in the Assessment's framework and approach that is useful for my imagined country's national development planning process? How can we adopt these so that the regular planning process is in fact permanently transformed and improved?
3. What substantive findings about the health of ecosystems, as found in the Assessment, should we address in a national development plan? What is new in the Assessment that we have not considered before?
4. What concrete steps should we take so that we can integrate the results of the Assessment into the national development plan? In particular, the Assessment should inform the development of specific policies on such areas as agriculture, forestry, protected areas and tourism.

My answers to these four questions are in sections 1 to 4, respectively. These sections are written and should be seen as a 'memo' within the working paper, one directed to decision makers in my imagined country. In this 'memo', I articulate what I think is relevant and important in the Assessment for these decision makers. I also highlight what I think are gaps in the Assessment from a development standpoint. In some cases, where appropriate, I give concrete examples of what could be incorporated into the national development plan.

Finally, in the conclusion of this chapter, I revert back to my lens as a researcher, and identify lessons from these reflections using the political lens of a developing country government official. The point of the exercise of imagination is to identify the priorities and imperatives for action that come after this first experience of a global ecosystem assessment. The point is to identify pathways so that people, particularly those in the community that believes that human well-being is directly linked to the health of ecosystems, can maximize the outcomes from the Millennium Ecosystem Assessment.

1. IS THE MILLENNIUM ECOSYSTEM ASSESSMENT A LEGITIMATE AND CREDIBLE SOURCE OF INFORMATION AND ANALYSIS?

This is an important preliminary question that we should answer before we incorporate its findings and results into our national planning processes. If this process was not legitimate, that is, not authorized and governed by some kind of process where stakeholders, including our government, had a say, we should not even consider it in our own planning processes. For us to consider the Millennium Assessment, it must also have been undertaken credibly by the right institutions and individuals and following best practices in scientific assessments. We cannot just include or interject into our planning process any kind of report issued by organizations that may have their own agenda.

To address the questions of legitimacy and credibility, we ask the following questions:

- Who authorized this assessment process?
- Who made the decisions about how the process was governed and undertaken?
- Who conducted the assessments?
- What process was used to ensure that the information and analysis was based on good science?

The cumulative result of my answers to all these questions is the conclusion that the Millennium Assessment is a legitimate and credible source of information that we should consider, indeed with close attention, in developing our national development plans. The Millennium Assessment was authorized by many governments, including our own, through various United Nations (UN) processes and conventions. As pointed out in the synthesis report, the Assessment responds to government requests for information received through four international conventions – the Convention on Biological Diversity, the UN Convention to Combat Desertification, the Ramsar Convention on Wetlands, and the Convention on Migratory Species. Governments are not the only audience of the Assessment, however; it is also designed to meet needs of other stakeholders, including the business community, the health sector, non-governmental organizations, and indigenous peoples (MA, 2005a). Did the fact that the Assessment was officially authorized ultimately strengthen or weaken the credibility of its findings? Given its intergovernmental character, the Assessment's process followed the normal procedures of similar processes (for example, asking governments for comments on drafts of the report) and the usual constraints (on language, for example, or on respecting jurisdictional boundaries of various conventions and UN bodies and agencies). But the question remains, did these affect the integrity and independence of the findings? The authorization by governments has made the Assessment more valuable for government agencies, and therefore this weakening did not happen.

The governance structure of the Millennium Assessment was characterized by inclusive participation and multi-stakeholder engagement. There was geographical, scientific, gender and political diversity at all levels of governance of the Assessment. The work was conducted by well-established and well-known institutions and individuals. Internationally recognized principles of peer review were also clearly followed in the Millennium Assessment.

I do have one comment on the type of institutions and individuals who participated in the Millennium Assessment – its governance and conduct. The institutions involved were mostly scientific, academic and environmental organizations, and the individuals who participated in the Assessment were mainly natural science and environmental experts.[2] More engagement by development and poverty-focused organizations and individuals would have made the Assessment more legitimate and credible from a development standpoint. To some extent, this gap in the Millennium Assessment was dealt with through reaching out to development and other such audiences (for example, indigenous peoples), but more direct participation by development experts would have been better. The development community should not be just an audience of the Millennium Assessment, but should be a principal partner and collaborator so that it can claim ownership of its results. In this context, as an experiment in multi-stakeholder authorization and engagement, the Assessment (as recommended later in this chapter) should be reviewed so that lessons can be learned on how to initiate, organize and implement such an ambitious and important global effort.

There should be no doubt about the formal legitimacy and the scientific credibility of the Millennium Assessment. However, the

Assessment, because of the preponderance of participation by environmental and scientific organizations and individuals, clearly has an environmental starting-point and it is the environmental community which 'owns' the Assessment. Whether it is useful and relevant to development decision-making is dependent on how it links the health of ecosystems to human well-being in its framework and approach.

2. IS THE MILLENNIUM ECOSYSTEM ASSESSMENT FRAMEWORK AND APPROACH USEFUL FOR OUR NATIONAL PLANNING PROCESS?

Legitimacy and credibility are one thing; usefulness is a different issue. While the Assessment process is clearly legitimate and credible, is the Millennium Assessment useful to us development planners? More specifically, what is in the Assessment's framework and approach that we can actually incorporate into the way we analyze the present situation and project the future of our country? Does the Assessment provide concrete ideas and tools that we can use in making development decisions?

After reviewing the various reports of the Millennium Assessment, including its first product, *Ecosystems and Human Well-being: A Framework for Assessment*, we can conclude that the MA framework and approach can be useful in developing the medium term development plan (MTDP) (MA, 2003). Indeed, some concepts elaborated and implemented in the course of the Assessment have a direct relevance to the planning process that we are undertaking. These include:

- how human well-being and ecosystems are linked;
- the relationship between poverty and ecosystems;
- the concept of trade-offs;
- the diversity of ecosystem services;
- the implications of multi-scale analysis; and
- the role of integrated responses.

The Relationship between the Health of Ecosystems and Human Well-being

While this is not a new concept, the Millennium Assessment succeeds in a very detailed way in linking human well-being, from food security to the protection of livelihoods to cultural values, to the fate of ecosystems such as forests and coastal systems. As we make choices in our national development strategy, for example in making decisions about land use (for example, conversion of forests into lands for agricultural/industrial use), we need to continually make this link and quantify them as much as possible. The finding in the Assessment that gains achieved in human well-being came at a high price not only for ecosystems, but ultimately for human well-being as well, is an important insight that we should remember as we create our development plan. The direct linkage between ecosystem health and poverty is also a consequence of this overall relationship between ecosystems and human well-being. For purposes of our national development planning process, Finding #2 of the Millennium Assessment is the most important:

> The changes that have been made to ecosystems have contributed to substantial net gains in human well-being and economic development, but these gains have been achieved at growing costs in the form of degradation of many ecosystem services, increased risks of non-linear changes and the exacerbation of poverty for some groups of people. These problems unless addressed, will substantially diminish the benefits that future generations obtain from ecosystems (MA, 2005a: 1).

The Linkage between Poverty and the Degradation of Ecosystems

The Assessment can be instrumental to our development of the MTDP because it highlights very well the linkage between poverty and the degradation of ecosystems. The results of the sub-global assessments have been particularly useful in this regard. If the principal objective of the MTDP is poverty alleviation, it is important to note that the deterioration of ecosystems, as found by the Assessment, is a significant driver of poverty (and vice versa). When we adopt poverty alleviation goals and consider the means we will use to reach those goals, we must be mindful of what the consequences are to ecosystems as we exert efforts to meet the goals. The same is true for the reverse. The analysis in the Assessment of the implications of achieving the Millennium Development Goals (MDGs) for ecosystems is the type of analysis that we should do for our own development goals. This would enable us to understand trade-offs and identify mitigation or alternative strategies (for more on the Assessment's implications for the MDGs, see Schmidt-Traub and Cho's chapter in this volume).

The Need to Consider Trade-offs more Carefully

This is one clear recommendation from the Millennium Assessment. We need to be more systematic about this in developing the MTDP and in adopting decisions that favor one set of goods over others. In making decisions about public infrastructure, agriculture, energy, tourism, and so on, we should automatically undertake a standard trade-off analysis so that we are better informed about the implications of our decisions. Actions taken to promote development and poverty alleviation, but that result in degradation of ecosystems which leads to worsening poverty, clearly do not make sense. Actions which increase or expand some ecosystem services at the expense of others must be carefully considered.

Distinguishing among Ecosystem Services

One of the major contributions of the Millennium Assessment to any trade-off analysis we will need to make in the MTDP is the way it distinguishes the various services that ecosystems perform. We should be aware of the complex role ecosystems play in enhancing (or pulling down) human well-being. We need to avoid over-emphasizing the provisioning services at the expense of other services, such as the regulating and cultural services. For example, when we define our food production goals or our wood supply needs, it is important for us to know the effect of this on the other functions of ecosystems. Many of these other services, when lost, become the responsibility of government and government budgets (for example, the provision of water purification, flood control, disease regulation). Along with these important regulating services, the findings in sub-global assessments that emphasize the importance of cultural services provided by ecosystems (for example, the importance of sacred groves in India as well as in many developing country communities) are important insights that should be considered in the drafting of the national development plan.

The Usefulness of Multi-Scale Analysis to Development Decisions

An important lesson from the Millennium Assessment is that scale matters in ecosystem assessments and is relevant in policy-making that affects ecosystems and human well-being. This is a lesson equally applicable to development decisions. Frequently, we make decisions from a national and macroeconomic scale alone; it is rare that we consider the global, regional and local scales. It is essential for us to develop the capacity to do multi-scale analyses to inform our development planning. The MTDP would certainly be much stronger if it resulted from such an analysis.

The Necessity of Integrated Responses

It is very clear from the results of the Millennium Assessment that integrated responses, solutions which are based on an understanding and appreciation of the complexity of the linkages between natural systems and their interaction with human activities, are necessary if ecosystems are to play a positive role for human well-being. Unfortunately, as the Assessment shows, we have few good examples of responses that are truly integrated. Indeed, even the Assessment volume on responses is organized according to ecosystems/sectors and does not provide good examples of integrated responses. At the same time, the assessment of responses conducted in the Assessment is critical information that can be used as we develop our responses to the conditions of our own ecosystems. Certainly, the methodology adopted in the Assessment to evaluate responses can be useful as we make our own choices regarding response strategies. The MTDP should be evaluated on whether or not the strategies and responses it adopts are in fact 'integrated'.

The MA framework and approach is useful for development planning. The challenge is to apply specific concepts and principles such as ecosystem services, trade-offs, multi-scale analysis, and integrated responses to concrete decisions that must be made in developing national development plans and strategies.

3. SUBSTANTIVE ISSUES IN THE MILLENNIUM ECOSYSTEM ASSESSMENT

In one sense, there is really nothing substantively new in the Millennium Assessment. Because the Assessment is principally a review of the state of existing knowledge about ecosystems, those familiar with the literature on this subject should not be surprised by any of the individual findings. However, for development decision-makers, much of this information is new, and if communicated properly, can be influential in decision-making. The major limitation, of course, is that much of the assessment in terms of evaluating the conditions of ecosystems is at a global scale. This is useful background information, but not very significant from the point of view of those actually making development decisions at a national level. Obviously, to the extent that they are relevant to a specific country, the sub-global assessments can be more useful as information tools for decision-making. Indeed, it is clear that integrating the Assessment's approach into development decision-making requires information like that found in the sub-global assessments, so these should be further refined and promoted.

It must be said, however, that although the global assessments are not that useful for national development decisions, these assessments are still valuable with respect to the global issues that they highlight, issues which are critical to the achievement of national development goals. For example, the Assessment does highlight critical global issues that we must consider seriously and integrate earnestly in the MTDP. Examples of these are climate change and nutrient loading. The Assessment also identifies how certain global decisions and systems, such as those relevant to global trade, play a significant role in stabilizing or disrupting the ecosystem health crucial to human well-being.

Climate change is already a familiar issue to most of us as it is something we have been looking at for many years now. The Millennium Assessment simply affirms what we already know about the impacts of climate change on ecosystems and the challenge of dealing with such impacts. What comes out clearly in the Assessment's discussion of climate change is the need, indeed the urgency, for development planners to incorporate adaptation strategies into development plans.

While climate change is not new for us in development circles, the Assessment's similar findings regarding eutrophication (the oversupply of nutrients, mainly nitrogen) stands out as an important new problem area that needs attention. Although technical in its approach and language, the Millennium Assessment should be congratulated for its treatment of the nutrient challenge. Like climate change, dealing with this problem requires modifying a variety of human economic activities. Because of the highly technical nature of the problem, however, to date it has been difficult to galvanize the appropriate response for this challenge. The Assessment sounds a clear alarm for the coming problem of large-scale eutrophication; we should seek to understand it better, so we can integrate ways to respond to it in our MTDP. Realistically, we can only do this if we are able to frame the problem in a way that links it to food security and the livelihoods of the poor.

In addition to these global environmental concerns, the global assessment emphasizes the important role that global trade, and the rules that accompany it, play in determining the health of ecosystems and subsequently human well-being. An example of this is the discussion in the Millennium Assessment of the role of agricultural subsidies and their implication for poverty and the environment.

The findings and results from the global assessment are not specifically useful for national development decisions. For these decisions, sub-global assessments are more relevant and applicable. However, the Millennium Assessment findings on the challenge of climate change and, more significantly because it is not well known, nitrogen overloading, and on the role of global trade are relevant to national decisions. Responses to them should be integrated into national development plans and strategies.

4. INTEGRATING THE FINDINGS OF THE MILLENNIUM ECOSYSTEM ASSESSMENT INTO NATIONAL DEVELOPMENT PLANS

Integrating the Millennium Assessment's approach and its findings into a national development plan is difficult and challenging, but it can be done through a number of approaches that can be followed exclusively or in combination. Some of these possible ways are:

- using as a starting point the indirect drivers of ecosystem change;
- integrating the Assessment's scenarios into national development plans;
- identifying economic sectors critical to ecosystems;
- adopting a restoration and rehabilitation strategy for ecosystems;
- identifying implications to development cooperation;
- adopting specific policies that truly integrate ecosystem health and human well-being goals; and
- recognizing the role of good governance.

Many of the Critical Factors causing Ecosystem Changes are also Central Drivers for Development

Development pathways and ecosystem change share the same demographic, economic, socio-political, cultural and technical drivers. One way of integrating the Assessment's approach into national development plans is to carry out a comparative analysis of how these drivers affect ecosystem health and development options. These could easily be done for such drivers as population, taxes and subsidies, property rights such as land tenure, and technologies. Indeed, an analytical tool can be developed that allows planners to compare the interaction between ecosystem and development issues (for a specific example, see Munasinghe's chapter in this volume).

The Implications of the Assessment's Scenarios on Development Choices

Another approach to integrate the Assessment's findings into national development planning is to consider the implications of the scenarios elaborated in the Assessment on the development options faced by our country. The potential consequences of a particular development pathway – for example, implanting intensification or extensification in agriculture to achieve maximum agricultural export crop production – could be examined using the Assessment scenarios as the frame of reference. This includes taking into account the Assessment's finding that ecosystem change is likely to be highly discontinuous, characterized by sudden, non-linear events and even catastrophic collapse. Concretely, this means asking specific questions such as:

- How should projections of the impacts of climate change affect decisions the country needs to make about public infrastructure, including the building of coastal highways and reclamation projects?
- What are the potential and specific impacts of nutrient loading on the country's water systems and supply? What measures must be taken to avoid what could be very serious consequences?
- As a consequence of deforestation and occupation of mountain areas, coupled with changes in climate patterns, what are the geological hazards faced by human settlements? Can rehabilitation and natural regeneration strategies still work? Should evacuation plans be drawn up at all?

Integrating the Millennium Assessment through a Sectoral Approach

The ultimate test on whether the Assessment's findings, including the approach and framework it applies, are useful for national development decision-making is whether or not they can inform sectoral planning and decisions. In particular, the integration of the Assessment into the plans and strategies for such sectors as agriculture, energy, forestry, and tourism is crucial. For example:

- For agriculture, technologies should be identified, from biotechnology to organic approaches, which limit the environmental footprint of agricultural expansion while increasing food production. More equitable distribution approaches should also be part of the plan for agriculture.
- In the energy sector, the challenge is to promote renewable sources of energy to ensure sustainable energy supply for both national development and for poor communities.
- For forestry, guaranteeing wood supply for construction and other public purposes has to be done in the context of degraded forest systems that provide other important services, not just timber.
- In the tourism sector, the challenge is to put into place policies and mechanisms that protect cultural services, such as aesthetic, historical and religious services, provided by ecosystems while building the capacity of the country to welcome more visitors willing to spend considerable sums of money for tourism purposes.

Over the long term, a new approach is needed to development decision-making that promotes more integration in assessing impacts and responding to both developmental and environmental challenges. The sectoral approach currently used will probably survive as a kind of specialization, and division of labor will still be needed to address specific issues and concerns. This said, more interagency and multi-stakeholder collaboration in assessing options and in implementing strategies is necessary if integrated approaches are to be put into place.

Restoring or Rehabilitating Ecosystems is a Development Goal

Recognizing that there is a need to restore or rehabilitate specific ecosystems so that they are regenerated or maintained to provide provisioning, regulating and cultural services is another way of integrating the Millennium Assessment into national development decisions. Recognition should be made concrete by such steps as providing a mandate and budget for government agencies or other institutions to lead the necessary restoration or rehabilitation work. The rural poor could also be mobilized through financial incentives and direct compensation to participate in this effort.

The Role of Development Cooperation

One practical way of integrating the Assessment's findings into the national development plan is to design a development cooperation strategy that invites donors such as bilateral and multilateral agencies to invest in programs and projects that emphasize the promotion of human well-being in a manner that protects and enhances ecosystem health. Concrete examples include:

- programs that strengthen environmental enforcement by government agencies;
- programs that build capacity for implementing sustainable development and for effective management of ecosystems;
- programs that support integrated responses;
- restoration or rehabilitation projects;
- projects that support poor communities in managing ecosystems; and
- projects that support the development of local responses to global issues, such as climate change and the nitrogen challenge.

To integrate the Millennium Assessment into the national development plan ultimately requires the development, adoption and implementation of new policies. Specific policies that should be considered include:

- *Policy requiring continuing assessment of the impacts on ecosystem and human well-being of development options*: Making such an assessment mandatory for development decision-making is the best way to transform development planning so that decision-makers will truly begin integrating development and environment goals in an effective and operational way.
- *Policies on sustainable agriculture*: Because agriculture is important to the rural poor, who are still the majority in most of the world, and to the environment, since agriculture is a major driver of ecosystem well-being, it is critical that national and international policies on agriculture are designed to benefit the rural poor and to protect the environment. Thus, for example, agricultural subsidies in developed countries need to be reduced while developing countries must enforce environmental norms in making decisions that accelerate land conversion or otherwise intensify agriculture.
- *Policies on sustainable forestry*: National policies on forests must be examined to ensure that the capacity of forests to perform their diverse ecosystem functions and serve various human needs is protected. A broad range of strategies, from market-based

approaches to community-based natural resources management, should be employed to achieve this goal. Strong regulation and enforcement by government agencies (in partnership with communities and other stakeholders) to prevent illegal logging as well as monopolization by a few of the benefits of forest utilization must be central in policies to ensure sustainable forestry.

- *Policies on protected areas*: Integrating the Assessment's approach into national development planning requires an appreciation of the critical role of protected areas in conserving and enhancing ecosystems that benefit human beings and societies. Countries must adopt and implement policies that define, create and sustain protected areas. These policies must include, among others, mechanisms that recognize the role of communities occupying these areas, promote sustainable use of resources in these areas, and enhance the financial sustainability of conservation projects and activities within the protected areas.
- *Policies on sustainable tourism*: Because of the potential of conserved ecosystems to generate revenue from recreational use, national policies that promote sustainable tourism are a major step in integrating the Millennium Assessment into national development decisions. For example, consistently applying the 'carrying capacity' concept to tourism infrastructure and development decisions is consistent with the Assessment's approach and should be systematically required for all related decisions.

Good Governance is the Key to both Ecosystem Health and Human Well-being

A clear finding from the Millennium Assessment is the importance of good governance for both ecosystem health and human well-being. The combining of substantive policies which favor the poor and protect the environment with procedures that are transparent and participatory, bringing different stakeholders together to achieve a common purpose, is needed if ecosystems are to be managed well and poverty truly reduced. A development plan, to be effective, must recognize this, and specific strategies to ensure the adoption and implementation of such norms, policies and procedures must be part of that plan.

In sum, integrating the Millennium Ecosystem Assessment approach and findings into a national development plan, though difficult and challenging, is possible and doable. A combination of the approaches described above is the best way for this integration to be done.

CONCLUSION: LESSONS, PRIORITIES AND IMPERATIVES

When I first began the exercise of imagining myself to be a development technocrat reviewing the Millennium Assessment with a view to integrating what is relevant in it into an imaginary national development plan, what stood out immediately was the temptation to ignore the Assessment completely. On its face, the Assessment stands out as just another environmental report which does not say anything new and warns us that things are bad, but the worse is yet to come. Read and understood this way, the Assessment is not very useful in development decision-making.

It is for this reason that I began the exercise of imagination by examining first the legitimacy and credibility of the Millennium Assessment. For government officials, how an assessment process was authorized and the subsequent governance of that process is critical for any report to be taken seriously. In this respect, the Assessment stands out as having clearly done its political homework, but this is not enough for governments to actually integrate the Assessment's findings, including the approach and framework it applied, into development decisions. The substance of the approach and framework and its findings matters.

As indicated in the 'memo', the Millennium Assessment is valuable in a general way because it introduces an integrated approach to ecosystem health and human well-being. The relevance of its specific findings to national development decisions is not as clear-cut given that most of the actual ecosystem assessments are at a global scale. However, as also indicated in the 'memo', its highlighting of such issues as climate change and nutrient loading is significant and should be taken into account by development decision-makers.

In doing this exercise of looking at the results of the Millennium Assessment through the political lens of a development decision-maker, I was able to identify priorities and imperatives that could inform future action after this first experience with a global ecosystem assessment. These are:

- Review and learn from the Millennium Assessment experience;
- Catalyze more sub-global assessments, especially at the national level;
- Respond to the nitrogen challenge; and
- Change the discourse on sustainable development.

Reviewing the Assessment Experience

Given the costs of doing the Millennium Assessment, should we do this again? If so, how frequently should a global ecosystem assessment be conducted? Is there a cheaper way of doing this? How can future ecosystem assessments be designed so that the development community can have a greater engagement, stake and ownership over it? *The World Resources Institute (WRI), because of its leadership role in initiating the Millennium Assessment, should take the lead in evaluating the Assessment experience.*

Catalyzing more Sub-global Assessments

The specific findings of the Assessment on the state of particular ecosystems are not directly relevant to many national-level development decisions. The global scale of the Assessment dictates this limitation. Doing the sub-global assessments, although a sobering experience for many, was a good step forward. Ultimately, as we learn to do sub-global assessments better, including gaining more experience in conducting national-level assessments, this type of assessment could prove more useful and relevant to development decision-makers. *Leadership for getting more sub-global assessments done should fall on a UN organization such as UNEP or UNDP, both of which have existing national networks that can be tapped for this work.*

Responding to the Nitrogen Challenge

Is it time to develop and implement a global initiative that takes on the challenge of nutrient loading? Is it time to think about an international agreement to deal with the nitrogen challenge? The Millennium Assessment clearly emphasizes the urgency and importance of this issue, and for the first time, the nitrogen issue is framed as a global concern. But the Assessment still describes the problem in very technical terms. There is clearly a need to 'popularize' the issue so that people and countries see it for what it is: together with climate change, it is the most serious environmental challenge we are facing.

Changing the Discourse on Sustainable Development

The most striking insight I have obtained from reviewing the Millennium Assessment reports (both in the context of this chapter and as peer reviewer of a numbers of chapters in the *Responses* and *Scenarios* volumes) is how far we still are from integrating environmental approaches to mainstream development planning and decision-making. This non-integration, from a governance point of view, continues to be a major obstacle in doing things right for ecosystems and people. The Millennium Assessment, by linking ecosystem health to human well-being, is an important step forward in changing this discourse. Moving this even further along requires that we think not necessarily less of ecosystems, but certainly that we think more of, indeed care more for, poverty and development.

NOTES

1 The principal document reviewed is the synthesis report of the Millennium Ecosystem Assessment (MA, 2005a), but other related documents were consulted as well, including the Statement from the Board (MA, 2005b). The author also looked at the three main volumes of the Assessment and the other summaries for decision makers.

2 See Appendix C of *Ecosystems and Human Well-Being: Synthesis* (MA, 2005a) which lists the authors, coordinators and review editors of the MA.

REFERENCES

Millennium Ecosystem Assessment (MA) (2003), *Ecosystems and Human Well-Being: A Framework for Assessment*, Washington DC: Island Press.

Millennium Ecosystem Assessment (MA) (2005a), *Ecosystems and Human Well-Being: Synthesis*, Washington DC: Island Press.

Millennium Ecosystem Assessment (MA) (2005b), *Living Beyond Our Means: Statement from the Board*, Washington DC: Island Press.

3. To change global change: Ecosystem transformation and conflict in the 21st century

Nicolás J. Lucas, Iokiñe Rodríguez and Hernán Darío Correa

1. INTRODUCTION

Never before has the world undergone such a massive process of transformation in such a short period of time. In the words of British sociologist Anthony Giddens, we seem to be living in a runaway world (Giddens, 1999). To many, globalization, commonly known as the process of denationalization of culture, markets, politics and legal systems, has resulted in the intensification of risk, uncertainty and social inequalities in today's world.

The human map that globalization has configured is reflected on the new biophysical map. One of the most important political interpretations of the Millennium Ecosystem Assessment (Assessment) findings is that ecosystem change over the last 50 years is not only driven by socioeconomic and political change, but that it is its biophysical print. Global ecosystem change reflects the regressive distribution of costs, benefits, opportunities and vulnerabilities, the greater interconnection between and within social and natural systems, and the large scale and accelerated rate of change that characterizes the globalization of the last few decades. Consequently, conflict is a central political problem of globalization – and of ecosystem change, too.

Based primarily on the Assessment findings, this chapter explores two propositions:

i. Global ecosystem change is the biophysical reflection of globalization. The social, political and economic drivers that shape human relations are inscribing themselves on the biosphere. Decision-making processes generally fail to recognize how influential human systems have become on the biosphere.

ii. The central political challenge that global ecosystem change presents for the next decades is to bring socioenvironmental conflict to the center of the political arena in a constructive way. Global ecosystem change inherently creates conflict. This change occurs at multiple levels and in increasingly complex social and political conditions. Decision-makers seldom recognize this and responses to conflict are often inappropriate and unsustainable.

For conflicts to be resolved in ways that both acknowledge their complexities and address their structural causes, global ecosystem change demands a response that is capable of transforming current patterns of ecosystem service use towards more equitable ones. This will entail significant changes in institutions, policies and human relations.

Section II explores the two propositions outlined above to draw conclusions and make recommendations in Section III.

2. A POLITICAL INTERPRETATION OF THE FINDINGS OF THE MILLENNIUM ASSESSMENT

1. Global Ecosystem Change is the Biophysical Reflection of Globalization

There is a direct relationship between the global socioeconomic patterns that evolved over the last 50 years and the associated biophysical transformation of the Earth as documented by the Assessment. This relationship is particularly visible in terms of (a) the concentration of ecosystem and associated economic benefits among certain communities; (b) the regressive distribution among poorer communities of vulnerabilities and costs associated with ecosystem change; and (c) the large scale and accelerated rate of socioeconomic and biophysical change.

The main institutional, economic, technological and cultural factors that have driven ecosystem change during the last 50 years are not random, but components of a broadly coherent process: globalization. 'Population change, change in economic activity, sociopolitical factors, cultural factors, and technological change' are not independent of each other, but interact 'in complex ways in different locations to change pressures on ecosystems and uses of ecosystem services' (Millennium Ecosystem Assessment, 2005: 64). More than discrete elements interacting in complex ways, these drivers are the expression of an underlying movement that results in a 'de-localization' or a denationalization of culture, markets, politics and institutions.

The Assessment findings stopped short of explaining the complex interactions among drivers of ecosystem change, let alone describing any rationale in globalization.

> Driving forces are almost always multiple and interactive, so that a one-to-one linkage between particular driving forces and particular changes in ecosystems rarely exists. ... The causal linkage is almost always highly mediated by other factors, thereby complicating statements of causality or attempts to establish the proportionality of various contributors to changes (Millennium Ecosystem Assessment, 2005: 64).

Explaining such statements of causality would be difficult, not only because it would require interdisciplinary research, but also because such statements would call for an assessment of globalization which would have strong political implications.

However, the Assessment does show four important parallel trends that, while not proving 'causality', strongly hint at significant relations:

a. Ecosystem transformation has generated a pattern of cost and benefit accumulation that corresponds with the political and economic patterns created by globalization

In order to understand the systemic relation between globalization and global ecosystem change, one needs to move away from the abstract notion of 'aggregation' used in the Millennium Ecosystem Assessment to discuss indirect drivers of change and to look instead at the more concrete notion of 'accumulation'.

The pattern of social, political and economic predominance of OECD countries and, lately, China, over the rest of the world during the last 50 years of globalization is visible on the map of the biophysical transformations that have taken place during those same years. Large-scale ecosystem transformation has resulted in a certain accumulation of ecosystem benefits among and within nation-states. The shape of this accumulation (the what, who, where and how) is determined by the prevailing social, political and economic conditions in which ecosystem transformation occurred. Over the last 50 years these conditions have supported a certain concentration of power among particular regions and social sectors; internationally among former colonial powers and, lately, China, and nationally among urban populations, especially sectors that are integrated and have adjusted to globalization.

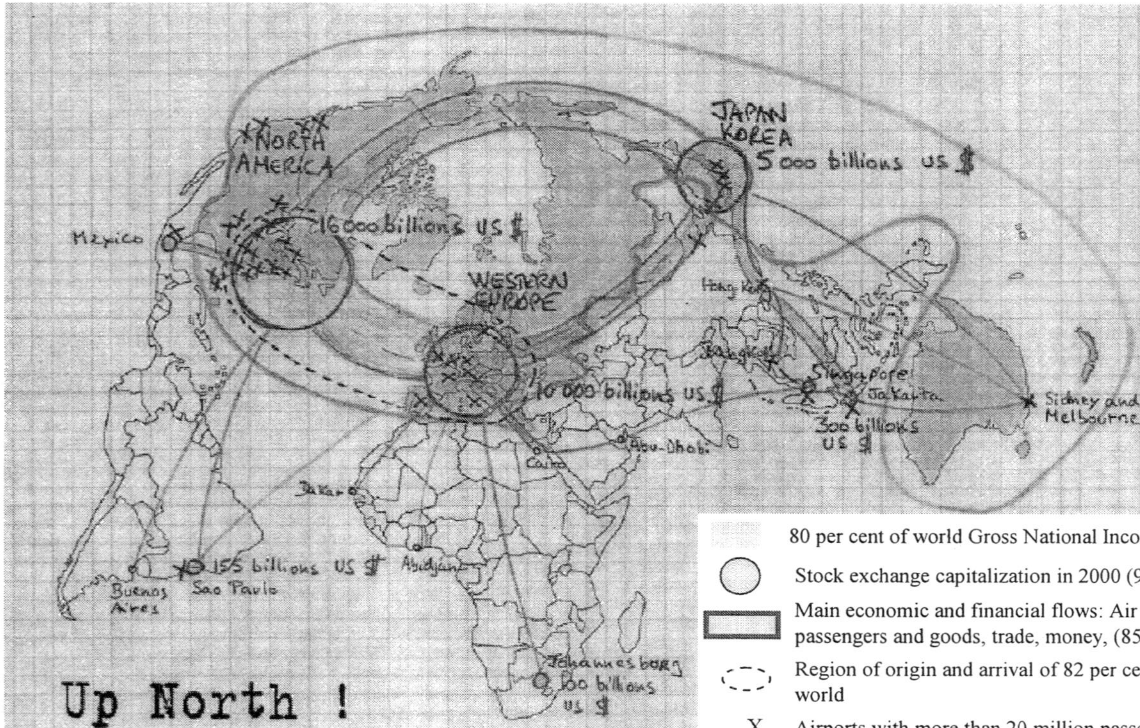

- 80 per cent of world Gross National Income
- Stock exchange capitalization in 2000 (98 per cent in the North)
- Main economic and financial flows: Air traffic traffic – passengers and goods, trade, money, (85 per cent in the North)
- Region of origin and arrival of 82 per cent of tourists in the world
- X Airports with more than 20 million passengers per year

Source: Philippe Rekacewicz, le Monde diplomatique, Paris, 2004

Figure 3.1

The concentration of economic and political power in a handful of countries is a commonly acknowledged fact. Eight countries (United States, Canada, Germany, France, United Kingdom, Italy, Russia, and Japan) account for about 66 per cent of global GDP, 44 per cent of global trade, 80 per cent of permanent UN Security Council seats with vetoing power and 96 per cent of active nuclear warheads. Most of the global Gross National Income, stock exchange capitalization, economic and financial flows and tourism are concentrated in the former colonial powers.

The concentration of economic activity mapped in Figure 3.1 is reflected in the biophysical transformation of the Earth. For example, those same eight countries account for 47 per cent of current and 58 per cent of cumulative greenhouse gas emissions in the period 1950–2000. In the case of climate change, the use of the global atmosphere for the industrial development of OECD countries is now well established.

In the case of ecosystem change, the relation is similar to climate change but needs a more nuanced explanation. In the period 1950–1990, large-scale transformation of ecosystems moved from OECD countries to their former colonies (Figure 3.2). By 1950, 70 per cent of those biomes largely located in the industrialized regions had been converted. Some of this conversion was to urban systems, but most has been to cultivated systems (Figure 3.3). Croplands expanded rapidly in Europe after 1700, and in North America and the former Soviet Union particularly after 1850 (Millennium Ecosystem Assessment, 2005: 32 and 26). Little change occurred in those biomes after that, and little is expected to occur in the future. The ecosystem change measured since 1950, as well as change expected in the coming decades, is largely concentrated in non-OECD regions.

> It is not possible to estimate accurately the extent of different biomes prior to significant human impact, but it is possible to determine the 'potential' area of biomes based on soil and climatic conditions. This figure [Figure 3.2] shows how much of that potential area is estimated to have been converted by 1950 (*medium certainty*), how much was converted between 1950 and 1990 (*medium certainty*), and how much would be converted under the four MA scenarios (*low certainty*) between 1990 and 2050. (Millennium Ecosystem Assessment, 2005:32).

Past conversions of biomes outside OECD regions need to be understood in the context of colonization between the eighteenth century and 1950. Unlike what happened with the benefits of the ecosystem transformation that took place in Europe, Japan and North America which primarily stayed in those nations, the benefits of ecosystem transformation in the rest of the world during that period flowed both to local populations and to distant colonial powers. The current flows of ecosystem benefits (and very likely future ones), in turn, should be analyzed in the context of globalization. The scale of global ecosystem transformation corresponds to the concentration of political and economic capacity in the hands of global powers, and the transformations in the 'peripheral' regions are to a large extent driven by consumption in OECD countries.

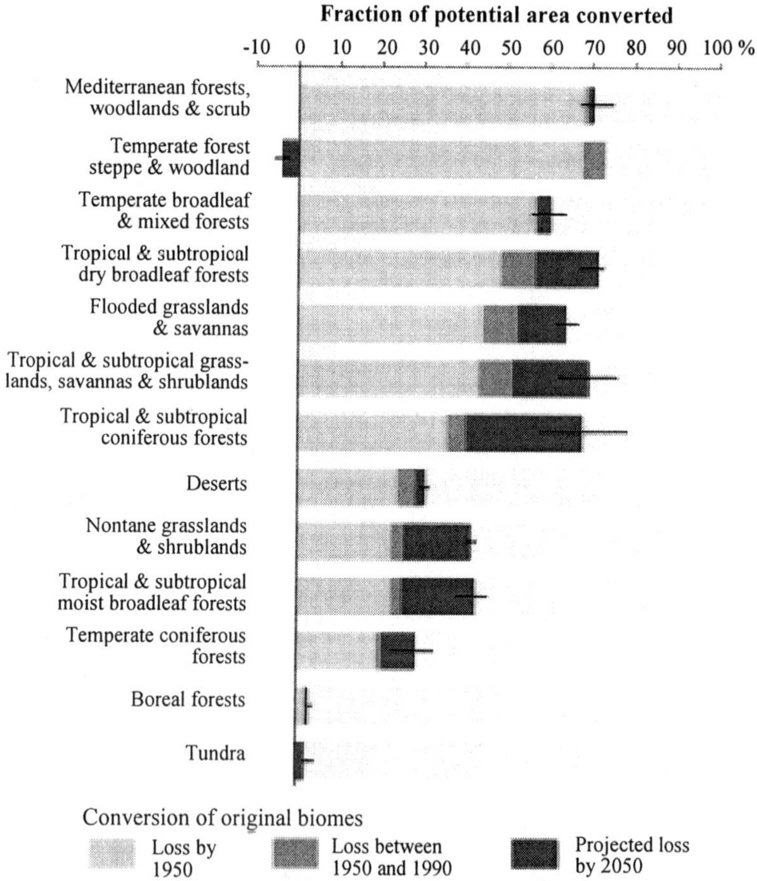

Note: According to four MA scenarios for 2050 projections, the average value of the projections under the four scenarios is plotted and the error bars (black lines) represent the range of values from the different scenarios

Figure 3.2 Conversion of terrestrial biomes

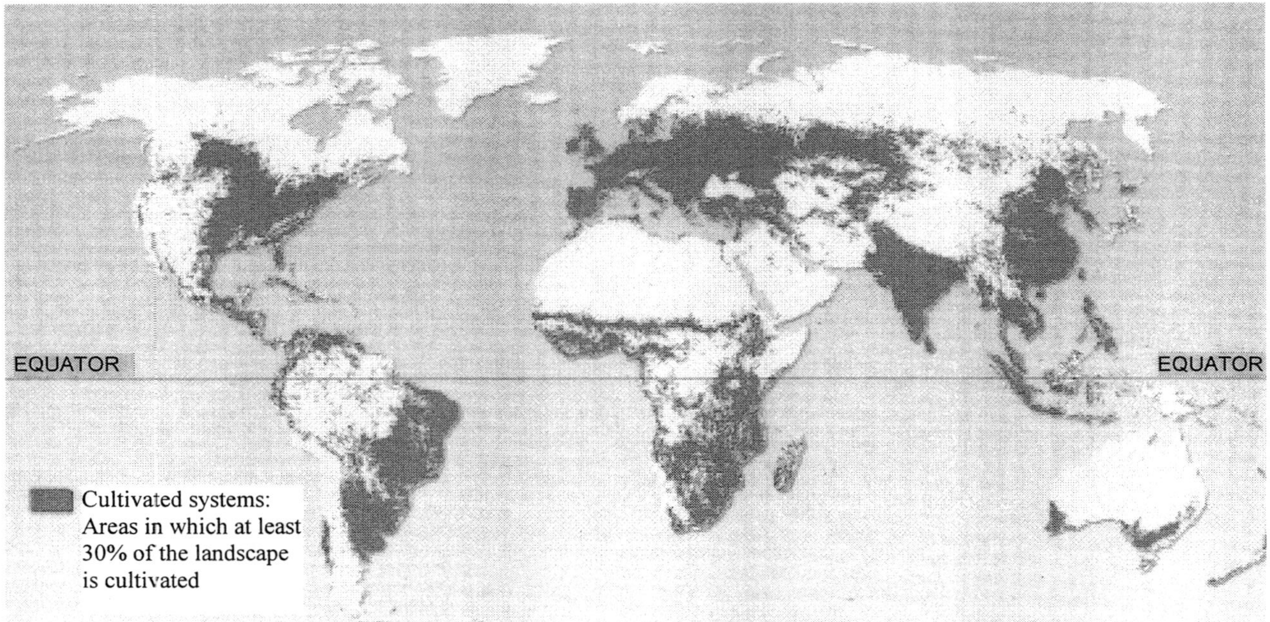

Figure 3.3 Extent of cultivated systems in 2000

These interpretations apply to other Millennium Assessment findings such as disruptions in nutrient cycling and the impoundment of freshwater behind dams; the transformation of natural systems corresponds with the shape of the socioeconomic processes that cause them. Generally, the benefits of ecosystem services flow in the same direction as the social, economic and political benefits. This flow is determined by globalization and leads to a concentration of ecosystem service benefits, and ultimately power, in particular sectors and regions.

Consider the Assessment finding that 'the amount of water impounded behind dams quadrupled since 1960, and three to six times as much water is held in reservoirs as in natural rivers' (Millennium Ecosystem Assessment, 2005: 2). These dams are primarily located in OECD countries and China (see Table 3.1). Outside these regions, dams were often built with loans from foreign multinationals.

> Foreign assistance has stimulated large investments in dams in developing countries, by providing financing – more than $4 billion per year during the peak of lending in 1975–1984 – and leading financing arrangements. Large developing countries with many large dams (including China, India and Brazil) have established internal capacity to build large dams, although in recent decades they have often used external finance and equipment to build larger projects (World Commission on Dams, 2000: 191–2).

Whether it is electricity, the food resulting from irrigation, water supply or flood control, the ecosystem services being harnessed by this altering of freshwater systems came to benefit urban sectors, thereby reinforcing existing patterns of wealth concentration.

Since the dawn of the industrial age, the societies located in regions that massively transformed their ecosystems without collapsing are the societies that subsequently became the centers of today's economic power. This power was then used to capture the benefits of ecosystem transformation in other countries. Today such processes of benefit accumulation in the regions where transformations are now mainly taking place (largely in the tropics) are hindered by the constitution of the globalizing process that configures a particular flow of benefits, a particular location of their accumulation, a concentrated distribution of opportunities and a new biophysical map.

Table 3.1 Top 20 countries by number of large dams

Country	ICOLD world register of dams 1998	Other sources	Percent of total dams	Cumulative percent
China	1855	22000	46.2	46.2
United States	6375	6575	13.8	60.0
India	4011	4291	9.0	69.0
Japan	1077	2675	5.6	74.6
Spain	1187	1196	2.5	77.1
Canada	793	793	1.7	78.8
South Korea	765	765	1.6	80.4
Turkey	625	625	1.3	81.7
Brazil	594	594	1.2	82.9
France	569	569	1.2	84.1
South Africa	539	539	1.1	85.2
Mexico	537	537	1.1	86.3
Italy	524	524	1.1	87.4
UK	517	517	1.1	88.5
Australia	486	486	1.0	89.5
Norway	335	335	0.7	90.2
Germany	311	311	0.7	90.9
Albania	306	306	0.6	91.5
Romania	246	246	0.5	92.0
Zimbabwe	213	213	0.4	92.4
Others	3558	3558	7.0	100.0
Total	25423	47655	100.0	

Source: World Commission on Dams (2000)

b. The distribution of vulnerabilities created by global ecosystem change corresponds with the distribution of socioeconomic vulnerabilities generated by globalization.

The above analysis leads to another political interpretation based on the Assessment findings: those societies who failed to accumulate the benefits of ecosystem transformation over the last 150 years will find it harder to do so over the next 50 years. In the context of global change, this means that developing societies face increasing levels of vulnerability. Vulnerability represents the interface between exposure to the physical threats to human well-being and the capacity of people and communities to cope with those threats (UNEP, 2002: 302). Consider the following findings:

- 'Most of the direct drivers of ecosystem change currently remain constant or are growing in intensity .... In all four Assessment scenarios, the pressures on ecosystems are projected to continue to grow during the first half of this century' and 'rapid conversion of ecosystems is projected to continue under all Assessment scenarios' (Millennium Ecosystem Assessment, 2005: 14 and 79). 'The biomes projected to lose habitat and local species at the fastest rate in the next 50 years are warm mixed forests, savannas, scrub, tropical forests, and tropical woodlands' (Millennium Ecosystem Assessment, 2005: 79);
- 'Actions to increase [the provision of] one ecosystem service often cause the degradation of other services' (Millennium Ecosystem Assessment, 2005: 6), and 'there is *established but incomplete* evidence that changes being made in ecosystems are increasing the likelihood of nonlinear changes in ecosystems (including accelerating, abrupt, and potentially irreversible changes), with important consequences for human well-being' (Millennium Ecosystem Assessment, 2005: 11).

Global ecosystem change as driven by globalization has effectively placed biophysical constraints on the development prospects of non-OECD societies. These societies must now design their development strategies in a biosphere that is more interconnected and fragile and with a larger population than in 1950, who not only have basic material needs, but also globalized consumption expectations that governments are under pressure to satisfy.

Table 3.2 (World Bank, 2003: 61) indicates the percentage of the population living on fragile lands[1] in the developing world. The data for Latin America is especially interesting because it hides the fact that Latin America is the most urbanized region in the developing world (UNEP, 2002) with 75 per cent of the population living in cities, and 38 per cent of poor (CEPAL, 2004) who can also be considered to live on lands with a different sort of 'fragility'.

Vulnerability to ecosystem change is particularly high in drylands. As the Millennium Assessment points out, these are the areas with the lowest natural productivity, the lowest per capita income and the largest demographic growth rates. The combination of ecological fragility and socioeconomic circumstances configure a situation of high vulnerability. Just as these are areas particularly marginalized from the benefits of globalization, they are the areas where the benefits of ecosystem change have not resulted in proportional benefits to the communities where the change has occurred.

Table 3.2 Regional distribution of people living on fragile land

Region	Population in 2000 (millions)	Population on fragile lands by region	
		Number (millions)	Share of total (percent)
Latin America and the Caribbean	515.3	68	13.1
Eastern Europe and Central Asia	474.7	58	12.1
Middle East and North Africa	293.0	110	37.6
Sub-Saharan Africa	658.4	258	39.3
South Asia	1354.5	330	24.4
East Asia and Pacific	1856.5	469	25.3
OECD group*	850.4	94	11.1
Other	27.3	2	6.9
Total	6030.1	1389	24.7
Total less OECD	5179.7	1295	28.9

* OECD: Australia, Austria, Belgium, Canada, Denmark, Finland, France, Germany, Greece, Iceland, Ireland, Italy, Japan, Luxembourg, Netherlands, New Zealand, Norway, Portugal, Spain, Sweden, Switzerland, UK, USA (23 original members).

Source: Average of CrESn and LandScan measurement methods

The distribution of vulnerabilities is also visible in the case of natural disasters that result from ecosystem change. It is too often the case that the capacity to deal with catastrophic events that result from poor ecosystem management is not built in the very places where the transformations are made, just as the benefits of such transformations do not accrue proportionately to those communities. The Parana basin in South America, for instance, has been transformed to enhance the provisioning of electricity (through dams), wood (causing deforestation), food (transforming wetlands), housing (in flood plains) and transportation (through dredging). The costs of the resulting disruption in hydrological flow are borne by the poorer among the riverine communities, as was seen in the 1997/98 floods in Argentina.

c. The increased social interconnectivity generated by globalization corresponds to the increase in interconnectivity among ecosystems.

Economic, cultural, technological and institutional developments that affect ecosystems translate into direct impacts through processes such as deforestation, soil erosion, species extinction and climate change – what the Assessment calls 'direct' drivers of change, and what the general public recognizes as 'environmental problems'. As societies have come into closer contact with each other through mass communication and the homogenization of consumption patterns, cultures and societal norms, a symmetrical process has occurred biophysically; increasingly changes in one ecosystem affect other ecosystems. Today, both direct and indirect drivers of change are globalized, the latter because of the current socioeconomic and political configuration of the world and the former because the scale of ecosystem change is such that the changes in one location increasingly ripple across ecosystems and political boundaries.

Climate change is perhaps the most dramatic illustration of this phenomenon, but higher levels of interconnection between ecosystems are also visible in the case of pollution, habitat change, invasive species and resource over-exploitation (the main 'direct' drivers of ecosystem change identified by the Assessment). In the case of the expansion of cultivated systems mentioned above, activities to obtain provisioning services involve habitat transformation, particularly conversion to agriculture. The Millennium Assessment scenarios project that habitat change will be concentrated in low income countries over the next 50 years, thus leading to higher pollution. The global flux of nitrogen, a common by-product of agricultural production, to coastal systems will increase by a further 10–30 per cent by 2030, with most of this increase again occurring in developing countries. This increase will continue to contribute to eutrophication of freshwater and coastal marine ecosystems. Also, because of habitat conversion to agriculture, genetic diversity has declined globally, particularly among cultivated species.

Trade-offs among ecosystem services and synergies between the direct and indirect driving forces of ecosystem change are becoming commonplace. As a result, the effective administration of territories is increasingly determined by social, economic, political and ecological interconnections. To take the case of the Parana river again, the electricity needs of the cities of Sao Paulo (Brazil) and Buenos Aires (Argentina) will have to be considered jointly with food production activities that encroach forests and wetlands in the basin (where export markets play a significant role) and assessed together with the natural hydrology of the system. Increasingly, a public authority wishing to administer its jurisdiction for the benefit of its people will have to deal with such complex webs of interconnections.

d. The scale and rate of ecosystem transformations correspond to the scale and rate of globalization.

The results of accumulation and inequitable distribution of benefits from global development and, more recently, globalization are not new. However, they are more visible today because of the increased understanding of the processes that constitute globalization and the larger levels of communication, interconnection and interdependence among societies. What is relatively new and more troubling is the scale and accelerated rate of these processes. Over the last 50 years, societies and economies have changed more rapidly than in any other comparable period of history. Wealth, for instance, measured as GDP, multiplied by six in just five decades. During this same time period, humans changed ecosystems more rapidly and extensively than in any comparable period of time in human history (Millennium Ecosystem Assessment, 2005).

2. Global Change Creates Local Environmental Conflicts

These local environmental conflicts are a result of tensions among different social, political and cultural systems. These tensions arise (a) across geographic and temporal scales, (b) among different social actors, (c) with different interests as well as (d) different power resources.

As globalization advances, environmental conflicts increase, with more people seeing their livelihoods and relationship with the environment threatened by a new global political and economic order.

Although environmental conflicts are often viewed as local phenomena, most involve concrete global dimensions that determine their structure. These include the driving forces that define the 'arena' of the conflict, the interests and actors at play and the strategies they pursue. Therefore, environmental conflicts must be interpreted in the context of globalization if they are to be successfully resolved.

As the Millennium Assessment shows, attention to 'multiple scales' is central to understanding global ecosystem change. Multi-scale integrated assessments as carried out by the Assessment are a right step in this direction. However, the dynamics of multiple scales demand a more assertive political approach to conflicts than offered by the Millennium Assessment. This must include acknowledging the diversity of actors, the complex issues involved and the respective power relations among stakeholders.

a. Environmental conflicts are essentially multi-scalar:
Environmental change over the last 50 years has been driven by global development as well as economic and institutional trends. Table 3.3 below provides an overview of environmental problems in Latin America (Rodríguez and Espinoza, 2002) and indicates some of these relations.

Table 3.3 An overview of environmental problems in Latin America

The nature of the problems	The problems
Problems related to the state of the environment	• Degradation and destruction of biodiversity, forests and land • Deterioration of the urban environment • Pollution of water and decrease in the availability of water sources • Increased vulnerability to natural disasters • Global environmental threats
Problems related to the structural contexts	• Development model based on fossil fuels and intensive use of natural resources, urban concentrations, inequity and crisis in the relationship with nature • Processes of re-organization of the region due to globalization (mega projects, mining and forest enclaves)
Problems related to processes of environmental management	• Reshaping of the public role of the State • Privatization, fragmentation and dismantling of national environmental systems • Short-term and improvised policies • Financial crisis

At a local scale, these problems are most commonly expressed in the following types of environmental conflicts (Rodríguez and Correa, 2005):

- confrontations over the conservation, use and management of protected areas;
- competition over the use of specific natural resources, such as land, water, forests, fisheries, and fauna, among others;

- struggles between local and regional, national and/or international interests over large-scale development projects such as mining, gas, roads, dams and oil;
- disputes over territorial management in the urban–rural interface; and
- struggles over the possession, titling and use of indigenous and black people's territories.

Almost invariably, globalization plays a central role in these conflicts. Very often, the economic, political, social and cultural drivers of globalization have concrete socioeconomic expressions in extractive activities carried out by globalized actors for the world market such as mining, fishing, tourism and energy generation, while local communities struggle with exclusion, exploitation, social fragmentation, cultural destruction and displacement.

Table 3.4 shows this pattern of accumulation and vulnerability in six case studies in Latin America. In addition to the environmental impact caused by the extractive or development activities, such activities can significantly modify the relationship of local and indigenous peoples with the environment, triggering unsustainable use of ecosystem functions.

Some conflicts do find negotiated solutions which contribute both to the sustainable use of natural resources and an equitable distribution of their benefits, but the vast majority do not. The aim of the following section is to understand why this is the case.

b. The diversity of stakeholders

Environmental conflicts involve a diversity of actors that must participate effectively in any successful resolution. Depending on the scale of the conflict, the composition of actors will vary. In most cases, the composition of parties involved in conflict resolution must include local, regional, national and international actors, such as local farmers, fishermen, cattle-ranchers, hunters, indigenous peoples, local councils, resource management agencies, academics, state governments, ministries, national private enterprises, transnationals, international human rights and environmental non-government organizations and multilateral organizations. Each of these actors is strongly influenced by the specific context in which they operate. Increasingly, addressing environmental conflicts requires dialogue and negotiation within and among different scales. Unfortunately, this is not always acknowledged, and negotiations often take place at the national headquarters of government agencies far removed from where the conflicts occur. In addition, such negotiations often involve only formal leaders, ignoring the fact that different social groups have different cultural and organizational traits that determine how they address and resolve conflicts.

Table 3.4 Social and environmental impact of globalization in six Latin American case studies

Case	Source of conflict	Form of global market economy	Indicator of environmental change	Indicator of social impact
Salinas de Manaure, Colombia	Large-scale salt extraction and comercia-lization	Salt transnationals	Destruction of two coastal marshes, exceptional and unique ecosystems of the semi-arid Colombian Guajira	Exclusion and exploitation of the local economy Indigenous social fragmentation
Galapagos, Ecuador	Fishing	Japanese fishing enterprises	Depletion of marine resources: sea cucumber	Social unrest Impossible to maintain concerted agreements
Yacuiaba, Rio Grande, Bolivia	Gas exploitation	Oil and gas transnational	Deforestation	Social stratification Changes in life style of indigenous people
Machupichu, Peru	Tourism	Tourists Tourism companies and operators	Waste accumulation Anarchic tourism management	Inequity in benefits from tourism Exclusion and exploitation of local economy
Los Pelambres, Chile	Copper mining	Mining transnational	Long-term water and land pollution	Erosion of local farming economies Displacement Social fragmentation
Cochabamba, Bolivia	Privatization of water	Transnational water use and management	Change in the use and management of water resources	Lack of respect for local definitions of water use. Erosion of local traditions and costumes.

Source: Rodríguez and Correa (2005)

Conversely, there are cases where important local, national or international actors have been omitted from conflict resolutions resulting in tenuous agreements that contribute little to the sustainable use of ecosystems.

It is also important to understand that within any social group in conflict there is rarely cohesion and consensus on visions and interests. Hence, communication and dialogue must adjust to the dynamics of conflict among the parties involved and avoid the creation of linear processes.

c. The diversity of views

Because the global economy and the dominant model of development have such a strong presence in almost every corner of the world and because at first sight many environmental conflicts spring from a competition over access to natural resources, there has been a tendency to propose material solutions to conflicts. This is especially true for business and governments that typically resort to the use of financial compensation, economic incentives, community development projects and funds, jobs and even new natural resource practices when attempting to resolve conflicts.

Such an approach ignores the fact that most conflicts over ecosystem services involve struggles over intangibles, such as identity, authority, different cultural constructions of nature and its use and different definitions of development. Effectively addressing these conflicts typically requires non-material solutions. For instance, conflicts that emerge around large-scale developments such as dams, gas and oil pipelines and road building are often rooted in complex ideological and political tensions. These tensions are often the product of a clash between the homogenizing nature of the nation-state as a model of political organization and the struggles of indigenous peoples to legitimize their cultural differences. Such tensions tend to be expressed through clashes of views over, among other things, models of political organization, definitions of citizenship, visions of development and meanings of land ownership. Instead of confronting these issues, governments and mining and development companies typically prefer to base their negotiations around solutions that legitimize their own views of development. Typical solutions include compensations and new community infrastructure such as schools, health units, or community development funds. In the medium- and long term, this gives rise to the environmental and social impacts described in Table 3.4.

Another case where conflicts have been inappropriately addressed is in protected areas. Many of these conflicts arise because of very different visions of nature and its use. The refusal of local populations in protected areas to eliminate certain traditional practices considered highly damaging to the environment, such as slash and burn agriculture, can be the result of the failure by conservationists to inquire about the environmental logic behind local natural resource use practices. Such failure reflects a generalized trend within protected area management to underrate local forms of knowledge and beliefs about nature and its use. What prevails instead is an overriding dominance of scientific justifications for protecting nature and explanations of environmental change and an assumed incompatibility between local people's livelihoods and conservation objectives. Such misunderstandings create major communication problems between conservation practitioners and local park inhabitants over natural resource management. (See Box 3.1 for an example of conflicts over different types of knowledge of fire in Canaima National Park, Venezuela.)

BOX 3.1 LOCAL VS. SCIENTIFIC KNOWLEDGE OF FIRE IN CANAIMA NATIONAL PARK, VENEZUELA

Since the 1970s environmental managers in Canaima National Park, Venezuela have attempted to eliminate local savanna burning practices due to a widely held belief that it causes a gradual reduction in park forest cover. Apart from attempting to teach the native Pemon new fire burning techniques, for decades resource managers have tried to introduce new farming alternatives in order to reduce the reliance on traditional burning practices, but managers have had little success. A recent study, however, has shown that far from being a destructive activity, Pemon savanna burning is in fact an important tool to control large destructive fires. The failure of resource managers to understand the purpose of Pemon fire burning practices and to value local environmental knowledge has been a key source of conflict. These failures are responsible for contributing to important alterations in local fire burning regimes which increase rather than reduce the occurrence of large-scale destructive fires in the park (Rodriguez, 2004). Similar conflicts have been observed in Africa and in Australia (Laris, 2002; Lewis, 1989 and Mbow et al., 2000).

Environmental conflicts in urban areas and in the rural–urban interface can also reveal struggles over visions. In this case the conflict arises over different definitions of local development linked to changes in the patterns of land use and regulation. One such example is the long-lived conflict in Valle de Lurin, in the outskirts of Lima, Peru. Here, the business sector, local authorities, religious institutions, farming communities and their organizations and environmental NGOs have been engaged in a protracted conflict over the expansion of extractive activities (mainly cement) and industrial plants on agricultural lands (Zucchetti and Lariviere, 2005).

Conflicts over the use of specific ecosystem services such as water, forests and fisheries frequently stem from clashes between different cultural and normative systems that seek to regulate their use. In these cases, the struggle is over authority. The mutual disregard by local, traditional authorities and public authorities of each other's ways of regulating the use of ecosystem services contributes to the unsustainable use of the resources and undermines agreements to control degradation. A case in point is the internationally known conflict in Cochabamba, Bolivia known as 'The Water War' where local farming and indigenous people fought and prevented a government attempt to privatize water use and the sewer systems, due to the Government's disregard of indigenous water use and management customs.

With the exception of the 'Water War', most of the conflicts mentioned above have not successfully resolved the core issues that give rise to the conflicts. The more powerful parties involved in such conflicts often strongly resist confronting the core issues in dispute since this may threaten their own position. However, unless these core issues are addressed, conflicts will continue to emerge and reproduce themselves in a cyclical manner, contributing to global ecosystem change and exacerbating existing social injustices.

d. Unbalanced power relations across scales:

A key, but often ignored, factor that reduces the likelihood of core issues being addressed in disputes is the relative balance of power among different parties in the conflict. Environmental conflicts typically involve different types of power relations that must be modified in order to overcome confrontation.

One type of power relation is the one we see in day-to-day interactions among actors during project and policy implementation. In this case power is exercised by the development of social networks that help win over to particular social or environmental projects others who share the same values and world views. This type of power relation is particularly prominent in conflicts that involve confrontations and day-to-day negotiations over new protected area management plans, joint management committees or urban zoning regulations.

Another way in which power is exercised is through the use or control of environmental knowledge – what is used, how it is generated and who has access to it. Such control can permit particular viewpoints and knowledge to prevail over others, as in the case of the use of fire in protected areas mentioned above. Leveling the power relations among parties requires investing the necessary resources to help those excluded from the dominant narratives, usually local people, to articulate their views of nature. This provides them with a stronger platform from which to engage in dialogue with other more powerful parties.

Finally, structural frameworks can also influence the balance of power among parties in a conflict. These include the way cultural, political and institutional structures shape the nature of relationships among actors. Structurally derived power is prominent in conflicts over large-scale development projects where the political and institutional frameworks of the nation-state on issues such as land rights and participation often limit negotiations. Changing these structural frameworks to level the power among parties, particularly in the dialogue and negotiation phases, is a major challenge in large-scale development conflicts.

III. CONCLUSIONS AND RECOMMENDATIONS

Neither global ecosystem change nor globalization are expected to diminish in the foreseeable future. The magnitude of these processes is such that some scientists now talk about the dawn of a new geological era.[2] The inertia inherent in the sheer magnitude of global change is such that 'the array of both indirect and direct drivers of ecosystem change is projected to remain largely the same' over the next 50 years (Millennium Ecosystem Assessment, 2005: 74). As a result, conflicts over natural resources will remain a central political problem in the near future.

But if the determinant influence of human beings on the biosphere is a given, the direction of that influence is the great uncertainty. The pessimism of reason, however, can be opposed with the optimism of the will. The Millennium Assessment is not particularly optimistic about future trends in ecosystem services, but does reflect a belief that a drastic change in human movement can have a significant positive impact in changing perceived future trends.[3]

The main conclusion of this chapter is that because of the contradiction between global change and individual and community needs, expectations and identities, political articulation will be required at multiple scales of global processes and community life projects. This articulation needs to happen in the context of a rights framework that directly strengthens the capacity of citizens and communities to pursue their life projects in freedom.

The challenge of achieving this will be to facilitate processes of political reorganization recognizing (a) the singular importance of each scale in terms of biophysical processes and the communities and identities at play; (b) the essential interconnection of human and natural systems with human systems driving ecosystem change at a global scale, but not altering the fundamental biophysical processes in nature; and (c) the importance of improving understanding and communication among global and local actors in contested issues about ecosystem change.

These processes of political reorganization require a large amount of social mobilization, a more proactive and more constructive involvement of civil society in public affairs, including the reaffirmation of local interests linked to ecosystem integrity. Nonetheless, the role of formal institutions and decision making processes cannot be underestimated, and the global community must develop institutions that:

- Enable citizens and communities at 'lower' scales to effectively respond to the challenge of globalization, diminishing the strategic value of violent tactics and negative identities;
- Facilitate the emergence of political organizations where human and natural dynamics are made compatible in the long term; and
- Promote transformative dialogues that reduce present patterns of accumulation and vulnerability

a. Enable Citizens and Communities at 'Lower' Scales to Respond Effectively to the Challenge of Globalization and Ecosystem Change

International institutions should empower citizens and communities with actionable opportunities that enable them effectively to control the factors that influence their livelihoods and quality of life. To achieve this:

i. international processes need to facilitate the further articulation and consolidation of universal values and norms, such as justice and sustainability; and
ii. international institutions need to be redesigned to become effective instruments for individuals and communities at multiple scales. 'Citizens' or specific social groups, such as indigenous peoples, and not 'governments' or 'states' must be the fundamental political

category of the international institutional architecture. In other words, the United Nations Charter preamble of 'We the peoples' needs to be re-signified.

The 'Rio institutions' are a case in point. The large number of conferences, declarations and, in particular, treaties that were catalyzed by the 1992 UN Conference on Environment and Development (such as the Convention on Biological Diversity, the Convention to Combat Desertification, the Convention on Climate Change, the Forest Principles, Agenda 21 and the Rio Declaration) are largely designed as bureaucratic instruments with little or no relevance to citizens and local communities. These instruments place the nation state as the exclusive political agent and have thus become largely ineffective at achieving their goals (such as stopping biodiversity loss, distributing the benefits of biodiversity, preventing desertification or mitigating climate change). International environmental institutions need to be designed with 'citizens' as the central subject to empower, not 'governments'.

b. Facilitate the Emergence of Political Organizations where Human and Natural Dynamics are made Compatible in the Long Term

One of the most important findings of the Millennium Assessment in terms of its implications for institutional design is that

> it is difficult to assess the implications of ecosystem changes and to manage ecosystems effectively because many of the effects are slow to become apparent, because they may be expressed primarily at some distance from where the ecosystem was changed, and because the costs and benefits of changes often accrue to different sets of stakeholders (Millennium Ecosystem Assessment, 2005: 11).

This finding, and the many others that describe time scales, inertia, resilience and non-linear changes makes a very fundamental point: political and natural systems operate on radically different time scales. The lack of recognition of this basic fact undermines the effectiveness of responses to ecosystem change. Bridging these two temporalities requires ingenuity at various scales. The Assessment focuses on the need for 'improved management'. This chapter emphasizes the need for new political organizations that are better designed to acknowledge a biophysical temporality that cannot be changed by societies.

In particular, international norms need to broaden the definition of the category 'nation-state' to enable and facilitate (bio)regional articulations.

The national political context needs to become more permeable to trans-border communication between communities thereby allowing the development of collective identities with a sustainable rationality.

c. Develop Transformative Dialogues that Reduce Current Global Patterns of Resource Use that lead to Unequal Distributions of Benefits and Vulnerability

Global ecosystem change requires processes of deliberation that help improve communication and understanding among actors about the causes of environmental degradation. This requires ensuring that the perspectives of people positioned differently in political power structures are taken into account, particularly the views of those most directly impacted by ecosystem change. NGOs and the academic community can play an important role in this process of facilitating this dialogue by helping local groups articulate their perspectives. These groups also have roles as strategic agents that can access and provide relevant information about the environment, public interests and threats to culture and biodiversity that can be of use in deliberation and decision-making processes. However, the extent to which this relationship between local communities and NGOs or the academic community actually develops will depend on the level of trust that exists among them. Without the necessary level of trust, such processes of external support could be seen as a way of facilitating the imposition of external agendas, thus closing, rather than opening, the opportunities for dialogue over global change.

NOTES

1 'Areas that present significant constraints for intensive agriculture and where the people's links to the land are critical for the sustainability of communities, pastures, forests and other natural resources' (World Bank, 2003: 59)

2 'There is no evidence that the Earth System has previously experienced these types, scales, and rates of change; the Earth System is now in a no-analogue situation, best referred to as a new era in the geological history of Earth, the Anthropocene' (Steffen et al., 2004).

3 For example: 'It is a major challenge to reverse the degradation of ecosystems while meeting increasing demands for their services. ... this challenge can be met ... although the changes required are large and not currently under way' (Millennium Ecosystem Assessment, 2005: 92).

REFERENCES

CEPAL (2004), *Panorama social de América Latina 2004*, Santiago, Chile: CEPAL.

Giddens, Anthony (1999), *Runaway World. How Globalisation is Reshaping our Lives*, London: Profile Books Ltd.

Laris, P. (2002), 'Burning the seasonal mosaic: preventative burning strategies in the wooded savanna of southern Mali', *Human Ecology*, **30** (2), 155–86.

Lewis, H. (1989), 'Ecological and technical knowledge of fire: aborigines versus park rangers in northern Australia', *American Anthropologist*, **91**, 940–61.

Mbow, C., T. Nielsen and K. Rasmussen (2000), 'Savanna fires in east-central Senegal: distribution patterns, resource management and perceptions', *Human Ecology*, **28**(4), 561–83.

Millennium Ecosystem Assessment (MA) (2005), *Ecosystems and Human Well-being: Synthesis*, Washington, DC: Island Press.

Rodríguez, I. (2004), 'Conocimiento indígena vs. científico: el conflicto por el uso del fuego en el Parque Nacional Canaima, Venezuela', *Interciencia*, **29** (3), 121–9.

Rodríguez, I. and H. Correa (2005), 'Lecciones, enfoques y retos en la transformación de conflictos socio-ambientales. "La experiencia del Programa CyC"', in H. Correa and I. Rodríguez (eds), *Encrucijadas Ambientales en America Latina. Entre el Manejo y la Transformación de Conflictos por Recursos Maturales*, San José: Universidad Para la Paz.

Rodríguez, M. and G. Espinoza (2002), *Gestión ambiental en América Latina y el Caribe. Evolución, Tendencias y Principales prácticas*, Washington, DC: Banco Interamericano de Desarrollo, Departamento de Desarrollo Sostenible, División de Medio Ambiente.

Sachs, Wolfgang (1993), 'The political anatomy of sustainable development', in W. Sachs (ed.), *Global Ecology. A New Arena of Political Conflict*, London: Zed Books.

Steffen, W., A. Sanderson, P.D. Tyson, J. Jäger, P.A. Matson, B. Moore, F. Oldfield, K. Richardson, H.-J. Schellnhuber, B.L. Turner and R.J. Wasson (2004), *Global Change and the Earth System: A Planet Under Pressure*, New York: Springer.

UNEP (2002), *Global Environment Outlook 3: Past, Present and Future Perspectives*, London: Earthscan.

World Bank (2003), *World Development Report 2003. Sustainable Development in a Dynamic World: Transforming Institutions, Growth and Quality of Life*, Washington, DC: World Bank.

World Commission on Dams (2000), *Dams and Development: A New Framework for Decision-making. The Report of the World Commission on Dams*, London: Earthscan.

Zucchetti, A. and R. Lariviere (2005), 'Conflicto y colaboración en un valle peri-urbano. El caso del Programa Valle Verde en la cuenca del río Lurín, Lima, Perú', in H. Correa and I. Rodríguez (eds), *Encrucijadas Ambientales en América Latina. Entre el Manejo y la Transformación de Conflictos por Recursos Naturales*, San José: Universidad Para la Paz.

4. Putting the assessment into action: Six steps for governments

Lailai Li

Ecosystems, ecosystem services, human well-being and the interactions among them form a complex open and dynamic system, with subsystems functioning independently and interdependently, contributing to the system equilibrium. There has been little theoretical breakthrough in classical structural functionalism,[1] the main theory attempting to define and explain system dynamics. This chapter is set within this framework and supplemented by empirical information. After summarizing the challenges presented by the Millennium Ecosystem Assessment, the chapter makes six recommendations for policy approaches based on the Assessment:

1. provide justice by making protection of livelihoods a premise of all decisions about economic development and ecological degradation;
2. institutionalize ecosystem impact assessments to legitimize development programmes;
3. provide incentives to encourage all stakeholders to pursue sustainable ecosystem management through existing institutions;
4. remove subsidies, tax pollution and waste, and shift to environmentally and ecologically responsible investment;
5. communicate knowledge, including use of scenarios, to help the public change its production and consumption behavior; and
6. work for political reform that leads toward sustainable development.

The chapter also emphasizes the important role played by inter-governmental agencies in influencing the national government to incorporate ecosystem management into development strategics and urges increased investment in education to communicate and disseminate the knowledge generated by the Millennium Assessment. Education, [2] including investment in piloting and replicating the best practices of the Assessment's conceptual framework (see the Norgaard and Janetos chapters in this volume for more information), is an effective means to fundamentally change the attitudes of those who alter ecosystems and who are affected by ecosystem change.

SUMMARY OF CHALLENGES PRESENTED BY THE MILLENNIUM ECOSYSTEM ASSESSMENT

The Millennium Assessment is intended to facilitate decision making and future research and assessment of issues and actions related to ecosystems and human well-being. However, the question of who will take what action to ensure the proper usage of the Assessment's findings is left by the authors for their audience to determine.

'Ecosystems and their Services around the Year 2000' not only summarizes the conditions and trends of the ecosystems that support life on Earth, but also provides tools and benchmarks for future assessments. If successfully communicated, the messages of the Assessment's evidence and analysis would make a great impact on the structure and behaviour of decision making as well as the production and consumption behaviour of business and the public. This raises the question, however, of who needs to hear this information and how these messages can be communicated effectively. There are many research projects that were done on the environment prior to the Millennium Assessment that have remained on the shelf with limited influence, although most were not as comprehensive as the Millennium Ecosystem Assessment. Among them are the series of Global Environment Outlooks (1, 2 & 3) published by the United Nations Environment Programmeme (UNEP) in 1997, 2000 and 2002 as reports of global environmental assessments. They include reviews of environmental conditions and policy along with information on human vulnerability to environmental change. In addition, almost every inter-governmental agency, for example the World Bank and the United Nations Development Programme (UNDP), publishes human or world development reports containing messages similar to the knowledge and information included in the Millennium Assessment. However, none of these reports have been used to the extent that the amount of investment – time, money and human resources – made in creating the knowledge should warrant. For example, in China environmental damage is not calculated into gross domestic product (GDP), especially at the provincial or other local levels. This is a lesson to which the Millennium Assessment organizers should pay attention.

The results of the four scenarios presented by the Assessment's Scenarios Working Group illustrate the future alternatives and trade-offs of different ecosystem management strategies associated with policies, institutions and governance of a society. The report states that for the results of three out of the four scenarios to come about, substantial changes are required in laws and governments managing ecosystems and human well-being. The Working Group concludes that these required changes are not happening. This raises the question – what are the necessary conditions for creating the changes to establish these three scenarios?

Societies differ by the prioritization they give to a set of factors, for example economic, political, technological, social and cultural, and the patterns of interactions between these factors. Based on these differences, the conditions to bring about the change discussed above can be determined. Therefore, across societies the changing conditions are different. For example, there are two sharply different attitudes toward the building of large dams in the policies adopted by different countries: some nations are stopping large dam projects due to their unexpected ecological impacts, while others are speeding up dam construction for energy purposes. These policies, arising from alternative national priorities, will certainly result in different impacts on ecosystems and disparities in access to ecosystem services across countries and across stakeholders within a country. The discussion that has grown out of the Assessment's scenarios would be more relevant to the general public, professionals and practitioners, especially decision makers, if the analysis also reflected the ability of economic, political, social and cultural factors to shape the pathways toward various scenarios. Another question to consider in the scenario analysis is how values or cultures play a latent, but critical, role in shaping a society in both the short and long term. Do they contribute to the scenarios, and if so, in what way? The question is not answered in the Millennium Assessment, but perhaps that report was not trying to answer this.

The Millennium Ecosystem Assessment sub-global assessments' findings are the most relevant for ecosystem stakeholders, those who affect or are affected by ecosystems, and to the general public, particularly those who work at all levels in the fields of environment and development. This relevance is because the assessments provide not only information, but also tools and instruments of analysis related to stakeholders' and decision makers' practices. More importantly, the Assessment process is itself a capacity-building process for participants from the targeted areas (see Norgaard's chapter in this volume for more information on the Millennium Assessment's conceptual framework). The sub-global assessments cover diverse areas, many of which are ecologically vulnerable with the drivers of ecosystem change arising out of the political processes.

In many countries today, the political systems operate according to the interests of the major contributors to national GDP and state revenues such as large 'developers', for example, large dam builders, conventional energy producers, large mining companies, major automobile companies and road builders, and those who finance these projects. It is, therefore, not surprising that development strategies in these countries are biased toward developers, whose projects often have great ecological impacts, and against groups who lose their livelihoods from such large-scale 'conventional development' and are thus driven to 'participate in unsustainable resource management regimes'. Unless a set of systemic legal responses coupled with education is adopted at the country level regulating 'development' behaviour and enhancing awareness on these issues, the trend of ecological deterioration will not slow down. The Millennium Assessment does not address this reality – the mainstreaming of development lacking justice.

Solutions to local problems may negatively affect global ecosystems; similarly, solutions to one country's problems may cause problems for other countries. For example, at the community level, development of biomass can alleviate an energy shortage problem, but it is simultaneously criticized at a global level for generating greenhouse gas (GHG) emissions, which worsen climate change. As an illustration of the latter point above, development projects on the upper stream of the Mekong River in China caused many cross-border debates. Although any trade-off made should be considered in the context of global sustainability, in reality what rules or whose rules are applied in making trade-offs is the result of a political process. Conflicts, including violent ones, are not uncommon, and the consequence of them is often the deterioration of ecosystems; two possible solutions need to be developed.

One is the adoption of a participatory approach to decision making, reaching consensus through dialogue amongst stakeholders at all relevant levels in order to maximize benefits and minimize costs. The other is the development of an effective methodology, tools or instruments with which decision makers 'calculate' costs and benefits to optimize trade-offs and at the same time educate the public. The participatory approach has been criticized as an overly simplistic tool. It is a simple method, but the people who use it are the ones who can make a radical difference. A large cadre of social developers is needed with a vision of, commitment to and capacity to implement sustainable development. It is not only the use of these tools, but also how they are used, that makes society better or worse.

It should be mentioned that in the present political context, decisions relating to the definitions of national, regional and global security are still largely a result of power struggles rather than of a participatory process or scientific analysis. This is the biggest obstacle to sustainability. The value of justice, which is missing in most contemporary political discussions and from the Millennium Assessment as well, can be used to challenge the unsustainable strategies which are legitimized through a combination of political power and economic interests.

In addition, one of the great values of the Assessment's sub-global assessments lies in the capacity-building exercises that were built into the whole process. How to capitalize on this value in order to encourage greater participation and effective application of the Millennium Ecosystem Assessment methodology is a no less challenging task toward global sustainability than the results of the Assessment themselves. As a long-term solution, education will take a long time to implement and take even longer to show its successes.

ANALYSIS AND RECOMMENDATIONS

Provide Justice

Justice focused on the protection of the livelihoods of those vulnerable to both economic development and ecological degradation is the basic premise on which a political decision should be made. Guaranteeing basic access to resources should be given the same level of attention in policy making as technical data collected by experts using 'scientific methodology'.

The issue of justice raised here is the need to mainstream social development concerns into the decision-making process, thereby ensuring that the marginalized and disadvantaged move from the periphery of the development process to its centre. It has been shown that ecosystem degradation harms the poorest people and is a principal factor causing poverty. At the same time, the politically 'legitimized' development programmes mentioned earlier also deprive many of the poor of their livelihoods, which marginalizes them even further.

Justice here means that plans and programmes should move toward the ultimate goal of responsive human development, basing success of outcomes on their ability to guarantee rights. In presenting the four scenarios, the Millennium Assessment focused on analysing the trade-offs between services of different ecosystem management strategies rather than simply saying which form of management would be better and which one less so. Given the complexities and dynamics of ecosystems, an action or strategy taken to fix one problem may cause a set of other problems, many of which are not apparent at the early stages of decision making and thus are difficult to identify. Because of this inability to understand all of the facets of a decision before making it, decision making to an extent is also value-led. This basic value of policy making should be justice.

The value premise of justice, the issue of protecting the livelihoods of local people at the community level, should be accounted for in cost–benefit analysis. Today, arguments made by experts, along with their supporting data, are the main voice at the decision-making table. This is particularly problematic because when contracted by different stakeholders, different arguments are made by experts with different sets of data, all of them equally 'convincing', but proving entirely different conclusions. As a result, decisions are made by those with more political power, those who rarely pick up the voices of the vulnerable groups. This reality will continue until the value of justice is mainstreamed into the decision-making process.

Today, the pursuit of economic interests by large industries and corporations is the strongest driver of ecosystem deterioration. These industries are also the main contributors to national economic factors, such as GDP, employment rate and income per capita. When national governments use these financial indicators as the base of the country's objectives, related economic activities assume political legitimacy. As a result, not only are conservation efforts victimized, but also the livelihoods of peripheral groups are put at risk, particularly those who depend directly on the ecosystems which are being degraded for their living. At the regional level, this pattern of events can result in local disputes and conflicts over resources.

Justice is the value that should underlie decision making, with particular attention being paid to securing the poor's livelihoods. The current structure of decision making produces results that are biased and risky. Unless this valuation is institutionalized and built into governance of society, the traditional system will remain. Understanding, acknowledging and adhering to this value will fundamentally benefit all human beings.

Institutionalize Ecosystem Impact Assessments

Improving governance and institutional arrangements by empowering the demand side: institutionalizing ecosystem impact assessments to legitimize development programmes – a main driver of ecosystem change creates procedural justice for fair results; adapting participatory approaches to decision making to mitigate the negative impacts, and encouraging NGOs to bridge the gaps.

Increased attention to levels of participation, access to information, equality, non-discrimination and equity is necessary to add the systematic consideration of justice to the decision-making process. What is needed is governance that is characterized by participation, transparency and accountability. In most countries, government, which commands economic, political, institutional and social resources, plays a critical role in making changes through its policies, institutional functions and dominant position in society. Under such circumstances, one way to facilitate the change toward sustainability is to nurture or empower the demand side of the equation. The 'demand side' refers to the general public along with government and the large actors in the market. Empowering the demanding actors can be achieved by undertaking the steps outlined in the next few paragraphs.

First, ecosystem impact assessments should be institutionalized by making such assessments a legal requirement at the national level for all 'development' projects and programmes. Procedural justice, to a great extent, leads to just results. Since the 1970s, many countries have adopted EIAs (environmental impact assessments) as a procedure to legitimize 'development' projects. Although an EIA is not a 'panacea' for pollution, it has created a process toward finding a 'panacea' by involving multiple stakeholders in dialogue and debate over the issues and concerns that might otherwise have been in the control of a few powerful stakeholders. Since an EIA law was adopted in China, for example, both the public hearings and question forums being held on major development projects and the legal assistance being given to pollution victims have started to challenge large development projects. The Millennium Assessment findings establish a rationale and a solid methodology for conducting ecosystem impact assessments. It is the right time for such assessments to be made into legal procedures attached to specified legal responsibilities. Inter-governmental agencies have a certain level of authority to propose and promote the enforcement of ecosystem impact assessments at the national level and should take advantage of their ability to do so.

There is a significant distinction between an environmental impact assessment as, for example, defined by the Chinese EIA law, and an ecosystem impact assessment based on the definition of 'ecosystem' given in the Millennium Ecosystem Assessment report. The former is focused more on direct, overt and visible phenomena which can be easily assessed in a discrete manner through an EIA. An ecosystem refers to a more complex system; thus its impact assessment requires a systemic or holistic approach. For example, a large dam with the potential to have great impact on all four ecosystem services could easily pass an EIA if its engineering design and implementation process were environmentally sensitive. In contrast, with an ecosystem impact assessment, after determining how the project will alter the ecosystem locally, the impact assessment would also consider the values generated by the proposed project that would alter the values provided by the local, and also more widespread, ecosystems.

The four types of ecosystem services discussed in the Millennium Assessment at scales from local to global serve as a methodological framework for assessing the ecosystem impacts of a large development programme. A main challenge of the ecosystem assessment is the valuation of the ecosystem services, but this process should not be delayed until there is definitive knowledge on valuation. Initial assessments can be done now using the Assessment's work as a foundation. Application of this framework by experts in many countries to evaluate ecosystems would partially overcome the challenge.

One major problem of EIAs is that they are primarily conducted by experts with little participation from other stakeholders. Participation is a key component and a capacity-building process for both the demand and supply side of decision making on 'development'. Broad participation helps to ensure that short- and long-term impacts are identified, feasible objectives are established, negative impacts are mitigated and benefits maximized and that the plan will be implemented as decided. Participation at the local level ensures local benefits, but can cause damage to neighbouring or larger ecosystems; enhanced participation at multiple scales can help to avoid damage of this kind. The purpose of participation should be to create synergies, bridge gaps, and facilitate equitable dialogue and negotiation, not to create a process that benefits local interests alone. Practice of this ideal participatory approach at the grass-roots level has happened with effectiveness and success. Examples are the development of biomass to resolve energy shortages and enhance livelihoods in rural Guangxi, Shaanxi and Tianjin in China. In these examples, the modern knowledge and indigenous knowledge were successfully combined and effectively used resulting in sustainable management of the local ecosystems and improvement of livelihoods.

Another step that needs to be taken involves the role of non-governmental organizations (NGOs). NGOs have emerged as a new actor in global politics, not only to challenge the conventional institutional arrangements at local, national, regional and international levels, but also in response to a functional requirement of society to meet a new unprecedented challenge to human society–ecosystem degradation. Well-educated, professionally connected and with strong commitments to values, most NGOs and their members work closely and directly with their target populations at the site of the problems. Their value is not that they perform better than other stakeholders, but that they have the ability to link different stakeholders, especially vulnerable groups, in dialogue and negotiations. As a result, NGOs are more effective in using and mobilizing available resources to fill gaps left by governments and markets. NGO participation, just like EIAs, is not a 'panacea' to solve all problems, but it has created a unique approach to addressing problems.

A challenge here, however, is that to most of the public and government, NGOs are still marginal. For political and institutional reasons, the national government, inter-governmental organizations and major international aid agencies tend to exclude NGOs from their decision making. Multilateral agreements are institutionally defined as transactions between governments only, and attached to them are often a great amount of resources. Policies 'supporting' NGOs are vague and often are quite discouraging in substance. 'In principle we are supporting NGOs. However, NGO participation is not a procedural component under the framework we signed jointly with the national government, nor is any input designated for this purpose', said an UNDP officer from a field office in response to the request of a local NGO to participate in a Global Environmental Facility (GEF)-supported country framework. In an attempt to build institutional arrangements that encourage participation of all stakeholders in decision-making, that are sensitive to the voices of vulnerable groups, and that are committed to carrying out decisions made, the role of NGOs should be mainstreamed at all levels, and in particular at the national level. In the Chinese context, a main reason for this to be done is that NGOs are able to raise different voices from different perspectives since a large number of NGOs work with the poor at the grass-roots level dealing directly with ecosystem degradation and poverty. One testimony of success is the participation of a local NGO in the social impact assessment of China's West–East Gas Pipeline, as required by the donor (see Box 4.1).

Provide Incentives

Incentives should be provided to encourage all stakeholders to pursue sustainable ecosystem management. There is great potential for improvement of existing institutions. Therefore, the focus should be on making the best use of existing resources before investing in new ones. A systematic study of the existing institutions should be carried out to identify problems and gaps.

BOX 4.1 A CHINESE CASE FOR NGO INVOLVEMENT WITH LARGE DEVELOPMENT PROJECTS

The West-East Gas Pipeline is one of the major projects in China's 10th 5–year National Plan. Starting in Tarim Basin in Xinjiang, it runs 4000 kilometres through eight provinces, arriving in Shanghai, as shown in the map below. Petro China, a large Chinese oil company, owns the pipeline. Shell funded a social impact assessment (SIA) when deciding on whether or not to make the investment. Because of the results of the SIA, which urged establishment of a transparent compensation mechanism, the compensation programme was extremely successful. As found later in an audit, the compensation process was transparent, responsive and prompt. In addition, a development fund was established to finance development programmes in pipeline affected areas. All these would not have happened without the SIA.

The SIA had an incredible impact in part due to NGO involvement. The Institute for Environment and Development (IED), an NGO in social development, was one of the six institutions engaged in the SIA of the pipeline. This was the first time that the Chinese government and a large corporation had engaged an NGO in a major development project.

The SIA was also the first of its scale to be conducted through a partnership between business, government and an NGO. Due to its success, a promising beginning toward greater justice, social impact assessments of other development projects in east and southwest China have followed.

Successfully managing ecosystems sustainably depends on having a coherent conceptualization of the issues of concern, that is, economy, social equity and environment, but largely these issues are still treated separately.

For example, in many countries, having a strong economy is regarded as a precondition for environment protection and is perceived as creating the resources needed for resolving all other problems. The existing institutional arrangement, which is mainly sector-based and economy-driven, was formed from such a perception and has reinforced this idea and its related behaviour.

It is difficult to design one mechanism that can facilitate the needed changes in institutions and governance, policies and incentives, social factors and behaviour toward global sustainability due to the complexities of many recognized challenges. However, it is very clear that all of these listed players would adopt or initiate proactive changes if they were to see the benefits of the change. Encouragingly, such changes have begun to happen. Large industrial players, such as Shell and BP, have invested a great deal of resources and successfully developed 'protocols' for sustainable energy which pave the way for new energy use that has little pollution or negative ecosystem impacts. These efforts thus far have primarily remained as protocols, however, because not only are they not yet generating 'profits', but they are further fed by the 'profits' made from the more conventional commercial business. At the local level, farmers and NGOs have developed many protocols of sustainable communities by enhancing livelihoods through improving and conserving ecosystems. Unfortunately, these practices often do not receive sufficient attention or inputs from decision makers because their achievements are not measurable in GDP terms. A policy that has been discussed, but so far remains missing, is one that gives incentives encouraging these practices.

The Clean Development Mechanism (CDM) of the Kyoto Protocol provides entrepreneurs with incentives, in this case the business opportunity, for making greenhouse gas (GHG) emission reductions. There are obstacles, however, to effective implementation. There is insufficient understanding and low awareness of CDMs among decision makers in all sectors and at all levels of society. The other bottleneck for implementation is that the transaction costs are too high and the procedures for use are too complex for thousands of SMEs, who are a major GHG emitter, as well as for the larger number of land users who could play an important role in carbon capture and storage through proper land use. There may already be education programmes that could enhance awareness and understanding, but obviously they have not reached all target populations effectively. Also needed to make the CDM process successful is a financial mechanism to help fund the emission reduction initiatives of SMEs and land users at the community levels.

There are already many policies and programmes and a tremendous investment made to improve ecosystem management, particularly at the international level. Many of these inputs have yielded benefits, but it should not be forgotten that these international investments have typically resulted in the Matthew Effect[3] – the disparity between rich and poor is intensified, resulting in a worsening of ecosystem degradation. Inter-governmental agencies, such as the World Bank, the United Nations Development Programme (UNDP) and the Global Environmental Facility (GEF) in command of resources should shift their focus from creating and financing new initiatives, to studying the institutions already established to identify the gaps between the original designed goals and the actual performance and impacts of the institutions. This can be done by studying best practices and developing success indicators to recommend improvements. By doing this exercise, organizations will have a strategy on how to fill institutional gaps.

Powerful international organizations, like the World Trade Organization (WTO) and the World Health Organization (WHO), have great potential to influence global production and consumption toward sustainability due to their power of regulation in the world. The question remains, however, as to whether these organizations are committed to the cause of global sustainability. Nowadays, for example, the focus of the WTO is on trade disassociated from social and environmental concerns.

The frameworks and protocols signed at the international level serve as guidelines which pressurize national governments. The Millennium Ecosystem Assessment findings provide another solid base on which international agreements can be developed so that national governments will incorporate sustainable ecosystem management into their development strategies. At the national level, however, because performance is measured by GDP growth, many government officials are devoted to attracting business investment, which drives the deterioration of ecosystems. Political reforms within countries can correct the behaviour of governments, but discussions on big issues at international levels also influence national governments. If the inter-governmental agencies were to assume responsibility, such needed influence could come about. These processes may be protracted, like the Kyoto Protocol, but the efforts are still worthwhile, since a protracted negotiation process at least shows that a serious attitude is being taken by national governments.

Finally, the current system of monitoring and evaluating (M&E) the performance of the existing inter-government agencies is inadequate. In the past, greater attention has been given to the development of and investment in new programmes than to examining whether the programmes work as they were designed to and how satisfactory the investment of physical and human resources has been. Even if M&E is improved, the amount to which the results are actually used for aid programmes remains questionable.

Again, the main point here is to make good use of existing institutions and to develop a strategy for improvement by studying the gaps between actual outcomes and the value and goals shared commonly. There is an urgent need to understand what already exists and what is missing before further investment in new actions is made. This is common sense. If the international institutions led by outstanding individuals with expertise and vision and command of resources cannot do it, no one else will be able to do so.

Remove Subsidies, Tax Pollution and Waste

Remove subsidies, tax pollution and waste, and shift to environmentally and ecologically responsible investment which will not only help the development of an economy featuring natural capital, but also alter production and consumption behaviour toward sustainability.

The use of taxes as economic incentives has been shown to be very effective as an instrument to re-adjust wealth distribution and increase equity. Without this levelling of wealth, social stability can be problematic. The Millennium Assessment, by using impressive evidence, emphasized the value of ecosystem services. The public would easily understand the economic implications of declining ecosystem services if a price tag were to be put on the value of these services, but 'most of the ecosystem values the scientists identified ha[ve] never been economically measured' (Hawken et al., 2000). Fortunately, there has been a lot of recent research in this area based on the concept of 'natural capital'. All of this work has supported a move toward taxing the consumption of ecosystem services.

The Millennium Ecosystem Assessment emphasized the removal of subsidies to respond effectively to increasing demands for ecosystem services; taxing consumption is the other side of the same 'coin'. A first and practical step in tax reform is to shift away from taxing labour and income to taxing pollution, waste, carbon fuels and resource exploitation, as suggested by the authors quoted above. However, in reality most of these damaging behaviours are currently heavily subsidized by most countries under the premise of increasing GDP, the country objective as set by government. If taxes were reformed as outlined above, though, the government would provide incentives and nurture markets for development of businesses that reduce pollution and save resources. When such a tax shift is implemented, there has been a tendency toward increasing demand for labour which works in resource conservation, reducing waste and decreasing the demand for resources. 'Every dollar of taxation that is added to the cost of resources or waste, one dollar is removed from taxes on labour and capital formation' (Hawken et al., 2000). In other words, a tax of this nature encourages the expansion of employment to enhance resource productivity, thereby reducing environmental and social problems for which the government is ultimately responsible.

A tax shift also regulates the consumption of ecosystem services while improving the economy of society, leading to a more equitable distribution of the benefits and costs associated with using ecosystem services. Of course, an adequate education programme and transparent government are required in order for the tax shift discussed here to be successful and for its benefits to be fully realized.

Along with taxing pollution and waste, a shift in investment from financing conventional manufacture of goods to financing activities that conserve ecosystem services should be encouraged. For example, instead of financing a fossil fuel power plant, an investment could be made to finance activities that save the same amount of energy as produced by the power plant. Investment programmes like this are already underway, but more resources and effort should be put into their expansion and improvement.

Willingness to pay heavy taxes for non-financial benefits has always been a problem in European countries, where social welfare is part of the taxation system and is challenged by the public. It would not be surprising if taxation for ecological benefits met similar challenges. Although based on a new concept to encourage sustainable development, lessons from the European experience are worth learning.

Investment will be needed while piloting this new tax scheme to educate and convince the public. There are existing examples of how this could work. In the early 1990s in China's Gongcheng County, a few households tested investments in biomass through government subsidies in order to stop logging and improve livelihoods. The pilot was successful and was replicated throughout the county in the following years, affecting a population of 276 000 with more than 20 ethnic groups. After the pilot phase, local farmers were willing to pay to the level that government subsidies were largely replaced by individual funding. Inter-governmental agencies, which have both institutional and financial resources, can make the resources available so that countries and organizations committed to piloting good concepts through an open and transparent bidding process can in fact pursue these policies.

Communicate Knowledge

Scenarios would help the public change its production and consumption behaviour if the analysis also explains the necessary conditions of change toward these scenarios. Rich knowledge and information has been developed to explain the present and future conditions of ecosystems linked to human well-being; what is needed urgently is communication and dissemination of this knowledge to the public in order to change behaviour toward sustainability.

The current strategy of prioritizing GDP growth while only reacting to existing environmental problems will lead to a doomed future; this is outlined in the Millennium Assessment scenarios. Three of the Assessment's scenarios, the three with the most positive outcomes, require significant changes in policies, institutions and practices to mitigate the negative consequences of the growing pressures on ecosystems. These necessary changes are not happening. This inaction is not due to lack of knowledge, but lack of understanding of this knowledge by decision makers and the public. To promote change, three categories of knowledge need to be understood:

- scientific knowledge that builds a new vision of development and establishes new development objectives;
- applied knowledge that helps to 'calculate costs and benefits' of ecosystem management strategies and to understand 'how to get there' (see Irwin's chapter in this volume for more information); and
- best practices that help to make effective use of the invested resources in the sustainable ecosystem management and livelihood enhancement.

Scientific knowledge, such as that generated through and well illustrated in the Millennium Ecosystem Assessment, provides a new vision and sets up the goal of development. The use of a larger framework with scales across time and space, coupled with the new provision of data, will change the behaviour of the general public and decision makers. The scientific knowledge provided in the Assessment enables people to choose from alternatives to protect their long-term interest which they would not otherwise know.

So far throughout the Assessment process, greater attention has been put into generating knowledge than disseminating and communicating it. A survey of elementary and secondary school curricula in China in 2003 by the Institute for Environment and Development (IED) found that knowledge of ecosystems, and their connection to human well-being, is not integrated into school education. In universities, courses on the environment or ecosystems are offered primarily in professionally related departments or taught as an optional or extra curriculum. Although many universities are playing an active role in research on these issues, the results are not adequately integrated in their teaching process. In China, the education system still operates within a conventional framework in order to meet the requirement of the exam system which gives little incentive to schools to incorporate lessons on ecosystem services into formal education. Efforts on the ground to mainstream environmental education come up against the rigid school curriculum at all levels. Is it a responsibility of the United Nationals Educational, Scientific and Cultural Organization (UNESCO) to help break the teaching 'regime' through its significant power of influence?

The use of applied knowledge that helps to facilitate decision making, for example, methodologies and instruments to calculate the benefits and costs of trade-offs, is urgently important. For example, in China this type of knowledge has the power to help millions of small entrepreneurs identify new business opportunities and calculate how much profit could be made through investments in conservation. Government decision makers will also be able to calculate the economic losses and/or gains from conserving ecosystems on GDP. Applied knowledge can also help farmers, especially young farmers, to understand the resources contained in local ecosystems, the value of such resources, and how to use these resources as a key for a promising future.

There are training programmes that attempt to teach people this applied knowledge, but existing programmes are not adequate because they often do not reach the necessary audience. Further, even if the training is accessible, participation is low because the content of the training is disassociated from the reality of problems trainees are facing every day. In most cases, training is focused toward telling people 'what should be done and where the destination is' rather than on 'how to do it and how to get there'. One possible explanation for this is that the experts teaching these training programmes are typically from academia with little on-the-ground problem-solving experience. 'How much more money can CSR (corporate social responsibility) help my company to earn or save?' 'What exactly should my company do to achieve this goal?' These types of questions are always asked by training participants from business, but the training experts seemingly have few 'resources' to answer these realistic questions.

The third category of knowledge comprises the best practices of ecosystem management and livelihood improvement, though all have shortcomings, happening at all levels – local, regional and international.

Over 70 responses assessed by the Millennium Assessment and summarized in its reports reflect both the values and challenges of these practices. Communicating and disseminating best practices requires great effort and often is quite challenging because solutions to ecosystem degradation are site-specific. Therefore, the communication, dissemination and investment to pilot the best practices in new locations must be a process of capacity building.

Rich knowledge in all of these three categories is continuously developed and updated, but unfortunately it is not well communicated to the public and decision makers. It has been more than three decades since the Stockholm conference, but to many people and governments GDP is still the key indicator of development, and increasing GDP is regarded as the solution to society's problems. GDP grows quickly in developing countries with large populations, stimulated by economic restructuring programmes and rather low economic status. This only serves to intensify the demand for ecosystem services, causing increasing pressure on the ecosystems.

Some issues are communicated well to people around the world, unlike information about the environment. Probably the best examples are the varieties of new theories of economics that have appeared over the last decade, for example, those of development, trade, currency, and so on. These theories are taught in schools and universities and are fascinating to policy makers.

The institutions and individuals who helped to disseminate the knowledge of economics effectively can play the same role in communicating the information produced by the Millennium Ecosystem Assessment. For example, over last three years, IED has started to promote environmental education in Chinese schools for fifth to seventh graders. The textbooks have been pilot tested with teachers and students, and have shown themselves to be quite successful and useful. The main obstacle for mainstreaming ecosystem services into Chinese education is that the entrance exams for higher education do not require environmental knowledge, and as a result, the schools are very reluctant to change the existing curriculum.

Obviously, national governments have a key role in instituting change, and many local NGOs are promoting change in their own capacity. The question remains, however, of what role international institutions, like UNESCO, have, if any, to play in national education.

Successful communication and education will fundamentally change the human behaviour that threatens the environment. In the Millennium Assessment, education and communication are assessed and found to be effective response options for improving ecosystem management, but the lack of financial and human resources remain as large obstacles. The situation could be improved, however, by using existing resources and inputs more effectively. In practice, it has been seen that many of the successful practices in education happen not because of increased financial investment, but because of the strong commitment and enhanced capacity of the institutions and individuals engaged in the educational programmes.

Finally, the impact of educational training lies not only in the knowledge being delivered, but also in how the messages are being delivered. For example, experts sent out by inter-governmental agencies often spread the value of 'justice', 'equity', and 'participation' in a way that is often seen as authoritative and non-participatory for the locals. Such behaviour not only creates resistance to implementation, but also sets an example for the younger generation of experts who most likely will then take the same attitude to 'educate' others, offsetting the inputs and efforts made for the value-led goals. In addition, the media have great influence on the drives for demands and pressures on ecosystems and ecosystem services. For youth, the media may be the most influential actor in this area; this represents an opportunity for the media to influence youth by playing a role in reversing current degradation trends. Similarly, if the media do not use the value of a free press and its unique capacity of reaching the public effectively for socially responsible causes, any efforts made by education will be greatly offset.

Work for Political Reform

Political reform is needed for change toward sustainable ecosystem management. Issues of common concern heard in heated debates form a platform which defines the discussion of political reform. This in turn affects and is being affected by changes in other aspects of society. Recent application of scientific knowledge to decision making is changing this political process toward a consultative and learning process.

In discussing political reform, the main concerns of government, still the most influential decision maker at the country level and in terms of affecting international policies, need to be examined. In most countries, the government's primary concerns are those associated with issues like national sovereignty and security, GDP growth, increasing energy production, expanding trade, and protecting domestic markets. Achievement of these objectives is believed to eventually lead to greater national assets, stronger national defence, higher employment, enhanced income and greater social stability.

This belief is based on the indicators that are used to measure both national development and government performance, primarily GDP. One problem with using GDP for this purpose, however, is that it has marginalized factors of livelihood improvement and poverty reduction from national accounting in many countries because livelihood improvement often involves little consumption in the market. For example, the practice of 'household ecological habitat', employed on grassland suffering severe desertification, has not only improved the livelihoods of local poor farmers as a productive source of food, housing and cash, but has also helped local ecological conditions recover, so that local farmers are no longer displaced by the fast desertification process. Unless the protection of livelihoods and ecosystems is mainstreamed in national accounting practices, there are few incentives for large investments into conservation behaviour, such as the 'household ecological habitat', at the policy level.

While 'political reform' as a proposal could sound 'offensive' to some governments, many are willing to change if action would help resolve issues of concern to them. Communication and education can help decision makers to re-conceptualize the issues, see the linkages between ecosystems and the issues of concern, and visualize the future with a fresh perspective. Underlying this new vision is the value of justice.

To cope with the challenges of ecosystem deterioration and human well-being, political reform will have to happen in all countries sooner or later. The reform is a comprehensive process stimulated by the challenges of human adaptation of the physical environment, the 'demand'. Reform also affects and, at the same time, is being affected by social and cultural factors. This change will have to happen hand in hand with the reform of other aspects of society – shifting of value premises, trends of legitimizing ecosystem impact assessment, making decision making more participatory, expanding the role of NGOs in bridging the 'demand side' with the 'suppliers' of development change, and so on.

As summarized in the Millennium Ecosystem Assessment, political reform, featuring the adoption of new laws, policies and institutions for ecosystem stewardship, has taken place or is underway in many countries. When dealt with as focuses of policy, the issues of ecosystems, ecosystem conservation and human well-being have formed a very open platform for accommodating political reform in many countries, including those with strong traditions of centralized planning. For example, in China holding public hearings on the environmental outcomes of development projects over which the public are concerned represents a breakthrough in conventional decision making in Chinese government.

Finally, the role of education and communication should not be neglected as facilitators toward the necessary political reform. Scientific knowledge not only guides decision making, but also changes the decision-making process from a top-down approach to a process of consulting with experts and learning from other stakeholders. Applied knowledge can also help decision makers at the national level, for example:

- to calculate which is more cost-effective, investing in water-saving agriculture or diverting water over 1000 kilometres away from the south of China to the north; and
- to decide which is more cost-effective, investing in energy saving, developing renewable energy or trying to control global energy resources by all means, including wars.

Education and communication should also, of course, be geared to educate the public, who form the demand side of institutional change in the push for change regarding decision makers.

NOTES

1 Structural-functionalism drew its inspiration primarily from the classical sociological theories of Emile Durkheim and Bronislaw Malinowski which were inherited by Parsons in the twentieth century. It looks on social systems as having certain needs and society as a system of social structures (that is, economic, legal, educational, gender). If needs are met, it is by the social structures. The structures are thus functional in the sense that they help society to operate. Interconnections exist within and among these structures, and individuals and groups are constrained by these structures.

2 Education, as mentioned here, does not only mean the delivery of messages or knowledge, but also, and more importantly, the process by which the messages or knowledge are being delivered. The values advocated by education such as justice, equity and the participatory approach are also built and spread throughout the process.

3 The Matthew Effect, though applied broadly to different sectors of academia, generally claims that 'the rich get richer and the poor get poorer'.

REFERENCE

Hawken, P., A. Lovins and L.H. Lovins (2000), *Natural Capitalism: Creating the Next Industrial Revolution*, Boston: Back Bay Books.

5. From assessment to action: Operationalizing environmental sustainability at the national level

Guido Schmidt-Traub and Albert Cho[1]

The Millennium Development Goals (MDGs) are the world's shared framework for sustainable development (Box 5.1). Extracted from the Millennium Declaration that was signed by 189 countries in 2000, the Goals lay out a comprehensive set of objectives for economic, social and environmentally sustainable development to be achieved by 2015. Critically, the Goals include time-bound, quantitative targets and map out the responsibilities of rich and poor countries alike. They have been reaffirmed repeatedly in international forums, including the 2002 Johannesburg World Summit on Sustainable Development, the Monterey Conference on Financing for Development, the 2005 Gleneagles G8 Summit, and the World Summit that took place in New York in September 2005. For the first time, the world has a common set of objectives for achieving sustainable development that is supported by developing and donor country governments, civil society, international organizations and private sector companies alike.

While some countries have made tremendous progress towards reducing poverty in its many forms, many are still off-track in terms of meeting the Millennium Development Goals. Most of sub-Saharan Africa, but also countries in other regions, will not achieve the Goals unless progress is accelerated dramatically over the next ten years. As shown by the United Nations (UN) Millennium Project,[2] the good news is that the tools and knowledge exist to achieve the Goals within existing commitments. The MDGs are stretch goals, but achievable if governments, international organizations, donors, and civil society work together intensively over the coming decade.

Virtually no part of the world appears to be on track for meeting MDG 7, which calls for achieving environmental sustainability and reversing the loss of environmental resources. The recent publication of the Millennium Ecosystem Assessment – a unique effort to assemble the best scientific expertise on environmental issues – is therefore extremely timely. The Assessment underscores the importance of the environment for income generation, gender equality, health, nutrition, and many other dimensions of human well-being. Moreover, the Assessment shows that most ecosystem services are threatened with substantial and potentially irreversible declines. The case for action to improve the management of the environment has never been stronger, nor has the link between development and the environment ever been clearer.

BOX 5.1 THE MILLENNIUM DEVELOPMENT GOALS AND THE ENVIRONMENT

The eight Millennium Development Goals (MDGs) lay out a comprehensive set of outcome targets for sustainable development derived from the recognition that the reduction of poverty in all its forms requires not only a genuine partnership between rich and poor countries, but also integrated responses that address the interrelated challenges of poverty, hunger, disease, gender inequality, lack of education, and environmental degradation. The MDGs are national goals that need to be concretized, operationalized, and achieved in each country.

Goal 1 Eradicate extreme poverty and hunger
Goal 2 Achieve universal primary education
Goal 3 Promote gender equality and empower women
Goal 4 Reduce child mortality
Goal 5 Improve maternal health
Goal 6 Combat HIV/AIDS, malaria, and other diseases
Goal 7 Ensure environmental sustainability
Goal 8 Develop a global partnership for development

Goal 7 on environmental sustainability, and its Target 9, which calls for 'integrating the principles of sustainable development into country policies and programs and reversing the loss of environmental resources', place the environment at the center of development. Some critics make the charge that a separate environment goal isolates environmental sustainability as a 'sectoral' issue, and hampers the mainstreaming of the environment into development policies. Quite the opposite is the case. The explicit identification of environmental sustainability as a development objective provides a critical litmus test for all development strategies. Unless development strategies provide measures to achieve environmental sustainability, they are not sufficiently ambitious to achieve the Millennium Development Goals.

World leaders at last year's World Summit rightly shifted the focus towards implementation. Specifically, the summit resolved that 'by 2006, [all countries should] implement comprehensive national development strategies to achieve the internationally agreed development goals and objectives, including the Millennium Development Goals' (UN Millennium Project, 2005a: paragraph 22a). What does the shift to implementation mean for the environment? What can countries do to achieve MDG 7 on environmental sustainability, and to what extent does the work of the Millennium Assessment help in responding to this challenge?

This chapter attempts to answer these questions by drawing on the findings of the Millennium Ecosystem Assessment and the UN Millennium Project. Focusing on the needs of developing countries, where the pace of environmental degradation has been greatest, the international community should support national development strategies based on the MDGs that include practical and ambitious strategies for reversing environmental degradation. The chapter proceeds in five parts. First, the key lessons from the Assessment are distilled on the main drivers of and responses to environmental degradation. Next, a brief assessment is offered on why progress towards environmental sustainability has been insufficient in most developing countries. The third part of this chapter lays out an operational framework for achieving environmental sustainability as part of national MDG-based development strategies. The fourth section describes how the international community must organize its technical work to support developing countries in achieving the MDGs. A final section outlines how countries can get started in operationalizing environmental sustainability.

I. KEY LESSONS FROM THE MILLENNIUM ECOSYSTEM ASSESSMENT

The more than 1000 scientists brought together by the Millennium Ecosystem Assessment have gathered the best available scientific knowledge to assess the state of the environment. Their mandate was to 'assess the consequences of ecosystem change for human well-being and to establish the scientific basis for actions needed to enhance the conservation and sustainable use of ecosystems and their contributions to human well-being' (Millennium Ecosystem Assessment, 2005a). To this end, they were asked to consolidate and review existing data rather than to generate new scientific results or put forward recommendations for action. Several notable achievements of the Millennium Assessment stand out.

First, the Assessment has developed a compelling analytical framework focusing on ecosystem services that links the environment with human well-being. Most importantly, it puts an end to the perception that development objectives are inherently at odds with environmental objectives. Moreover, the Assessment's framework of direct and indirect drivers is easily comprehensible to policymakers and makes it possible to identify and assess trade-offs and synergies between development strategies and the environment. By making these linkages explicit, the Millennium Assessment helps us understand how changes to the environment affect ecosystem services that in turn influence human well-being.

A second achievement is the Assessment's rigorous evaluation of the lack of sustainable ecosystem management and implications for human well-being. Building on previous assessments of the world environment (for example UNEP, 2002), the Millennium Assessment finds that 'humans have changed ecosystems more rapidly and extensively than in any comparable period of time in human history … and this has resulted in a substantial and largely irreversible loss in the diversity of life on Earth' (Millennium Ecosystem Assessment, 2005a). Marshalling evidence from local, regional and global analyses, the Assessment shows that man-made environmental change is a growing threat to human well-being. The Assessment's scenarios illustrate how environmental degradation may worsen over coming decades unless corrective action is taken urgently.

Third, the Millennium Ecosystem Assessment provides a review of response options to environmental degradation, concluding that environmental management must be integrated into development policy. It identifies key drivers of ecosystem change and appraises response options on the basis of their effectiveness in stemming the decline of ecosystem services. For example, to conserve biodiversity, the reports suggest that the designation of protected areas and international cooperation through multilateral environmental agreements are 'promising' responses to the key drivers of adverse environmental change. Policy changes that are considered necessary to improve the state of the environment include the implementation of market-based instruments, participatory and adaptive processes in environmental management, and the need for coherence between environmental strategies as well as trade and macroeconomic policies.

In spite of these achievements, the Millennium Ecosystem Assessment has not put forward a plan of action to achieve environmental sustainability. The Assessment's limited mandate has prevented it from laying out the sorely needed operational framework for implementation. While the Assessment proposes many response options, these do not add up to outlining the operational strategy that the world needs now.

II. WHAT IS HAMPERING PROGRESS TOWARDS ENVIRONMENTAL SUSTAINABILITY?

Operational strategies for achieving environmental sustainability need to be based on a sober assessment of why progress has been so unsatisfactory thus far. Why does environmental degradation continue unabated despite the unprecedented attention paid to the environment by rich and poor country governments, civil society, science, and the private sector since the 1992 Rio Summit? Why do so few development strategies in low-income countries, such as Poverty Reduction Strategy Papers (PRSPs), address environmental issues in satisfactory detail?

The most frequent explanation – lack of political will – is often incorrect and falls short of explaining the lack of progress. Contrary to the commonly held perception, many policymakers in developing countries do appreciate the importance of the environment for their country's development. For example, in its country-level advisory work, the UN Millennium Project has collaborated closely with Heads of State, Ministers of Finance and Planning, and other senior decision makers who often join their environment colleagues in describing environmental degradation as a central developmental challenge. These leaders are aware of the problem, but each of them struggle with the difficult question of how to halt and reverse environmental degradation.

There is therefore a need to dig deeper to understand why so little has been achieved since Rio. Building on chapter 3 of *Environment and Human Well-being*, the report of the Millennium Project Task Force on Environmental Sustainability (UN Millennium Project, 2005a), this chapter identifies six factors that have contributed to this lack of progress.

1. Insufficient Focus on National Implementation Mechanisms and Coordination Across Sectors

National implementation mechanisms are critical for achieving Goal 7. Many ecosystem services respond to national and local strategies and interventions. The degradation of freshwater ecosystems and aquifers, deforestation as a result of agricultural extensification, or destruction of coastal ecosystems due to destructive fishing techniques require local solutions that need to be coordinated at the national level. Second, responses to environmental challenges need to be integrated across sectors and actors. Only national and local governments are able to ensure the needed degree of coordination and consultation. Finally, the cost and complexity of international cooperation can make it difficult to implement the 'first-best' coordinated responses to managing biomes and ecosystems that extend beyond national borders. Second-best solutions relying on national implementation mechanisms can therefore be more appropriate and immediately implementable responses to transboundary environmental challenges. Even where international coordination is necessary, experience shows that it can only work if underpinned by strong national institutions or mechanisms.

Meanwhile, environmental policymaking suffers from insufficient coordination and leadership at the national level. In comparison to health, education, infrastructure development, and other development challenges, environmental policymaking is not integrated into national development strategies, such as the Poverty Reduction Strategy Papers. Instead, much of the debate focuses on environmental initiatives at the local and international level. Both are important, but Goal 7 and the other MDGs cannot be met without operational *national* strategies for achieving environmental sustainability.

Environmental policy is subject to a particularly large number of international agreements, conventions, and implementation mechanisms, such as the multilateral environmental agreements (MEAs). One rationale for focusing on international conventions is that environmental challenges and their drivers tend to transcend national borders and therefore require internationally coordinated responses. Good examples for such challenges include climate change, threats to some marine fisheries, acid deposition, or the degradation of transboundary watersheds.

Yet while the intention of international treaties is to support national implementation, they often have limited impact on national policy processes. Environmental ministries create the national plans of action required by international treaties, but many of these plans gather dust on shelves and do not significantly influence national policy, particularly in sectors that affect the environment. All too often they are prepared by outside consultants who have no stake in their implementation so that the documents become formalities to be satisfied under the terms of the international agreements instead of operational strategies to improve environmental outcomes. Sadly, this is the opposite of what is needed. As underscored powerfully by the Millennium Ecosystem Assessment, countries need to integrate environmental policies into their development strategies instead of preparing stand-alone strategies addressing specific challenges.

At the other end of a spectrum, a large number of projects focus on improving environmental management at the community level. Many have been successful and have taught us important lessons. However, very few have been designed to answer the basic question of how to scale up interventions to the national level. As a result, many lessons learnt from local projects do not apply to the national, let alone regional level. While pilot projects have been instructive in developing new approaches to environmental management, renewed focus must be placed on the problem of moving to scale.

Due largely to this insufficient focus on the national level, national development strategies tend not to address environmental challenges in sufficient depth or detail. For example, a recent World Bank assessment of Poverty Reduction Strategy Papers (PRSPs) found that environmental issues were rarely integrated ('mainstreamed') into national development policies (Bojo et al., 2004). Achieving environmental sustainability requires a strengthened focus on national implementation mechanisms. The Millennium Assessment recognizes this problem, but falls short of proposing specific national mechanisms to achieve environmental objectives.

2. Lack of Operational Objectives and Prioritization

Few country strategies contain environmental objectives or outcome targets. They aim to achieve 'environmental sustainability' without specifying what this means in practice and how progress can be tracked. For example, Bojo et al. (2004) show that most PRSPs lacked concrete environmental outcome objectives, and Bojo and Reddy (2003) find that few PRSPs present information on baselines and targets associated with MDG 7. Similarly, a recent review of national MDG reports found that only 23 percent of countries had developed national and sub-national targets for reversing the loss of environmental resources (Ghanime and Lee, 2005).

Instead, environmental policy frequently focuses on process targets, like the adoption of national reporting systems or Integrated Water Resource and Management (IWRM) strategies – if they include targets at all. Marginal changes to institutions and policies dominate strategies with little regard to the long-term outcome targets that are to be achieved. This hampers the effectiveness of environmental strategies since without clear, time-bound, and quantifiable targets and objectives, it is impossible to prepare long-term strategies for improving environmental management, identify the needed level of human and financial resources, and track progress towards environmental sustainability. This lack of operational targets for the environment stands in sharp contrast to areas like public health, where countries aim at specific and measurable objectives, such as cutting maternal mortality rates by three quarters over ten years, or reducing the incidence of acute respiratory infections by a certain percentage.

Operational outcome targets for the environment are urgently needed and should specify clear benchmarks, such as what thresholds need to be met to maintain critical ecosystem services, or which levels of resource use cannot be surpassed without threatening the viability of ecosystem services. For example, countries can specify the maximum share of internally renewable freshwater resources to be used for human consumption or identify maximum levels of biological oxygen demand in a particular freshwater ecosystem. Once such objectives are in place, they can form the basis of a coordinated national strategy to achieve MDG 7 – provided these targets are quantified, time-bound and measurable.

In many areas, such as water and air quality, international norms already exist and can guide national strategies. Elsewhere, country-specific objectives will need to be identified. Establishing such targets requires careful scientific analysis, but as the Millennium Assessment demonstrates, the lack of adequate data, our limited understanding of the full complexity of the environment, extreme site specificity, and the essentially political nature of trade-offs between environmental protection and other social objectives make it impossible to derive outcome targets on the basis of science alone.[3] Hence, an inclusive and concerted political process is needed to reach agreement on targets among a broad number of stakeholders by drawing on expert assessments and public participation, as described by the Task Force on Environmental Sustainability (UN Millennium Project, 2005a).

3. Limited Understanding of what Works

A related obstacle is that interventions for environmental management remain poorly understood. There are no comprehensive catalogues of best practices for improving environmental management that are backed up by rigorous evidence on the effectiveness of interventions. For example, a country wishing to identify the most effective responses to combat desertification will not have an obvious place to turn to for operational guidance and advice.

Again, a comparison with public health is instructive. The Lancet Child Survival Series is a landmark in public health (*The Lancet*, 2003). It provides a comprehensive analysis of the key determinants of child mortality and puts forward strategies for addressing each one of them. The authors link inputs – such as immunization, treatment for major childhood illnesses, long-lasting insecticide-treated bednets to prevent malaria transmission, investments in water supply, sanitation and clean energy services – with quantitative reductions in child mortality. Moreover, the effectiveness of each intervention is quantified on the basis of extensive epidemiological studies. In short, this series presents a practical guide for addressing a complex public policy challenge in developing countries. Governments can use it to devise their own strategies for meeting the Goal of cutting under-five mortality rates by two thirds.

The contrast with environmental policymaking is stark, as underscored by the Millennium Ecosystem Assessment. Links between inputs – policy reforms or investments – and environmental outcomes remain poorly understood and tend not to be quantified. The Assessment only offers qualitative assessments of responses without providing any quantitative appraisal of their effectiveness (c.f. Millennium Ecosystem Assessment, 2005b). Moreover, in contrast to the Child Survival Series, very little, if any, empirical evidence is available to support the qualitative assessments. Even if fully implemented, it is unclear if these response options would substantially improve ecosystem management. Existing knowledge provides only a weak analytical foundation for scaling up investments in environmental sustainability and justifying the increased financial resources required to strengthen ecosystem management.

4. Neglect of Direct Investments in Environmental Management

Environmental strategies tend to focus on the need for major policy changes to improve environmental management. Both the Millennium Ecosystem Assessment (2005b) and the UN Millennium Project's Task Force on Environmental Sustainability (UN Millennium Project, 2005a) identify a set of key policy reforms that are needed to address the drivers of environmental degradation. They include the resolution of various types of market failures, changes in land tenure systems, changes in subsidy programs, and expanded environmental education. These policy reforms are critical, but they are not enough to improve environmental management.

Policy reforms must be complemented by direct *investments* in environmental management, such as the scaling up of environmentally sound agricultural extension services, afforestation, the construction and operation of water treatment facilities, and technologies for the abatement of air pollution (c.f. DFID, DGFD, EC, UNDP, World Bank, 2002). Perhaps most importantly, developing countries require more environmental scientists and experts to contribute to public policy. These experts need access to world-class training and better remuneration, which in turn requires higher wages and more funding. Unfortunately, the Millennium Assessment, like most country strategies, focuses almost exclusively on policy reforms and institutional changes, and neglects direct investments in environmental management.

5. Lack of Operational Budgets or Environmental Management

Countless national environment strategies do not even contain a budget, and therefore cannot be implemented. The absence of explicitly costed budgetary line items is a major reason why environmental issues do not figure prominently in national development strategies prepared under the leadership of ministries of finance or planning. Only if the budgetary implications of environment strategies are developed in detail will it be possible to include them in national budgets, which are after all the ultimate expression of national priorities.

Once again, the comparison with public health shows the way forward. In its 2001 report, the World Health Organization's Commission on Macroeconomics and Health (WHO, 2001) put forward the first comprehensive assessment of the cost of reducing the incidence of major diseases and maternal mortality rates. By linking this needs assessment with public budgets, the Commission made the investment shortfall explicit and identified what needed to be done to meet internationally agreed health objectives. This work has had a major impact on the international debate around public health and was instrumental in leveraging greater financial resources.

A partial explanation for the dearth of environment budgets lies in the lack of data and public finance expertise among environment experts that would allow them to link their recommendations to real budgets. Where financial analyses of the cost and benefit of investments in environmental management exist, they typically lack rigor, tend to be highly aggregate, and are not expressed in budgetary terms. Due to its limited mandate, the

Millennium Assessment was unable to close this gap. Likewise, the UN Millennium Project did not succeed in developing rigorous estimates of country-level investment needs for environmental management due to the dearth of available data.

6. Capacity Constraints and Lack of Financial Resources

A further obstacle to achieving MDG 7 is the shortage of human and financial resources. Many developing countries have limited technical expertise and capacity to implement environmental programs. The resulting capacity constraints include weak institutional capacity for monitoring environmental problems, lack of nationwide inventory of ecosystems, insufficient environmental specialists, and data limitations (Ghanime and Lee, 2005). In most low-income countries these constraints derive in large measure from strikingly low levels of financing available to environmental institutions, particularly to support adequate salaries and other recurrent costs.

The UN Millennium Project (2005c) and the Commission for Africa (2005) have shown that public investments need to increase sharply to meet the Millennium Development Goals and that poor countries cannot afford these investments on their own. Of the $110 per capita that a typical low-income country will need to invest each year to reach the MDGs by 2015, some $70 will need to be provided as official development assistance. The G8's decision to increase development assistance by $50 billion by 2010 and the commitment of the EU15 countries to increase their development assistance to 0.7 per cent of their GNI by 2015 are therefore very encouraging.

III. NATIONAL MDG-BASED DEVELOPMENT STRATEGIES

To overcome these constraints and achieve the Millennium Development Goals, countries need to develop effective national frameworks of implementation, as resolved by world leaders at the 2005 World Summit (UN, 2005b: paragraph 22a), following the recommendation of the UN Millennium Project (2005b 2005c). We recommend that these MDG-based strategies (i) be outcome based; (ii) derive from consultative processes; (iii) ensure full integration of environmental strategies with the social and productive sectors; (iv) map out needed investments through to 2015; and (v) link to real budgets.

An MDG-based approach takes the MDG targets as the starting point and asks how they can be achieved by 2015. To this end, it maps out the requisite investments and policy changes over the next ten years (Figure 5.1).

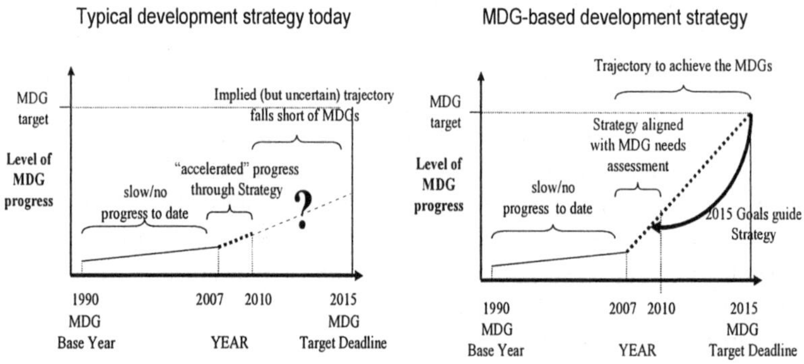

Figure 5.1 The principle of an MDG-based national development strategy (UN Millennium Project 2005a)

Based on this medium-term needs assessment, a short-term (3–5 year) MDG strategy can be derived. While the principle of grounding short-term action in a medium-term assessment may seem obvious, this is not how development strategies in most countries are prepared today. Current practice focuses on short-term strategies that are not derived from an analysis of how the MDGs can be achieved by 2015. The resulting strategies are insufficiently ambitious and cannot map out the long-term investments needed to overcome capacity constraints in areas such as human resources, infrastructure and management systems.

In using the MDGs as development targets, it is important to note that many investment areas are not explicitly mentioned in the Millennium Development Goals, even though they are critical for achieving each Goal. For example, the Goals do not refer to energy even though modern energy sources are necessary to fight poverty, improve healthcare, and reduce pressure on the environment. Sexual and reproductive health and transport are two further examples of such 'missing MDGs' that must be part and parcel of MDG strategies. They must all be addressed in MDG-based strategies if the Goals are to be achieved.

An effective environment component of MDG-based development strategies will consolidate previously disconnected environmental policies and strategies into a single integrated strategy with clearly specified objectives and responsibilities. It will also look broadly across sectors to ensure that the environment is mainstreamed throughout the development strategy. The process of creating such an MDG-based development strategy can be thought of as having five steps (UN Millennium Project, 2006). It should be emphasized that no blueprints or cookie-cutter approaches exist for integrating the environment into national development strategies. Therefore, the process described below must be tailored to the needs and priorities of each country.

Step 1: Launch an Effective and Inclusive Process

National governments can begin by establishing a national MDG Working Group to launch and coordinate the work. In practice, this will entail convening key ministries, including planning and finance ministries, and committing them to the process. It will also involve recruiting and training planning experts to oversee the needs assessment process. To integrate the environment into the national strategy, an environment group can be set up to handle the environmental needs assessment process. Throughout, it will be important to involve all critical stakeholders in the process of preparing the MDG-based development strategy (Box 5.2).

BOX 5.2 THE IMPORTANCE OF CONSULTATIVE PROCESSES

An inclusive and participatory process must underpin the crafting of MDG-based development strategies. To avoid a top-down process or a one-size-fits-all approach, broad participation is important at every stage in the design of environmental strategies (UN Millennium Project, 2005b):

Setting objectives. Not all stakeholders may agree initially on the key priorities for the national strategy, and an inclusive and participatory process can help develop consensus around the critical challenges to be tackled. Consultations and stakeholder dialogue can help countries identify shared environmental objectives.

Selecting interventions. Broad participation can facilitate the identification of environmental interventions. Different stakeholders bring diverse knowledge and perspectives to the table to reflect the full array of available options. Furthermore, participation is important because interventions to address environmental issues invariably affect the interests of various stakeholders.

Implementing the strategy. Environmental strategies will affect a wide range of stakeholders who should therefore be actively consulted and involved in the implementation process. Participation can inform discussions about how to sequence interventions to reach various regions and social groups, how to manage any adverse consequences, and other key dimensions of implementation. Stakeholders, including local communities, can also be active participants and agents in implementation and should be included in the roll-out of interventions.

Step 2: Compile and Review Existing Strategies

A careful analysis of lessons from the past can be a good starting point to avoid repeating work that has already been carried out. In most countries, existing strategy documents, such as National Environmental Action Plans or the national strategies developed under MEAs, contain much of the information needed to prepare an MDG-based development strategy. Countries will be likely to draw heavily on past work in three areas: environmental targets, interventions and supporting data. Existing environmental targets, including those derived from international commitments, may help inform the choice of national outcome objectives. Likewise, past work on the effectiveness of various interventions can guide the design of environmental strategies. Finally, existing data and analysis collected on the national environment will be a critical input into strategy formulation. World Bank Country Environmental Analysis papers, already completed for a number of countries, are just one of many potentially useful resources (World Bank, 2004).

Step 3: Carry out an MDG Needs Assessment

Needs assessments provide an analytical framework for preparing *outcome-oriented* environmental strategies by addressing the question, 'What will it take to achieve the MDGs?' Because MDG 7 does not provide a comprehensive set of targets, achieving this Goal will require countries to operationalize environmental sustainability by establishing concrete, quantifiable objectives for reversing the loss of environmental resources. To do so, the environment working group can start by identifying environmental priorities, such as halting desertification and land degradation, protecting a specific freshwater ecosystem, or reducing particulate emissions in a major city.

For each priority area, countries will then develop quantified outcome targets to be achieved by 2015 with clear interim milestones set before. To guide operational strategies effectively, targets must be quantified, time bound, and refer to measurable *outcome* objectives, such as the percentage of land covered by forest, specific reduction in the rate of biodiversity loss, ambient air quality levels, or reductions in the pollution of freshwater resources. For example, Bolivia's Poverty Reduction Strategy Paper includes timebound and quantified targets for increased income based on sustainable wildlife management (Republic of Bolivia, 2001), while Vietnam specifies targets for household and industrial wastewater treatment and forest coverage (Socialist Republic of Vietnam, 2002). The process of setting environmental targets is described in detail in UN Millennium Project (2005a).

A common concern is that the complexity of environmental sustainability cannot be captured in a limited set of outcome targets – no matter how well they are chosen – and that our knowledge about the physical environment is too incomplete. This argument is a classic case of making better the enemy of the good. The 1000+ scientists contributing to the Millennium Ecosystem Assessment have shown that there are limits to our knowledge of the environment, but that enough is known to start improving the management of core ecosystems today. As many others argue (for example Juma, 2004), no country should be told to wait for perfect information before taking action to stem environmental degradation. Enough is known to start implementing practical solutions now. In addition, many knowledge gaps can only be filled through 'learning by doing' and the use of adaptive environmental strategies.

The needs assessment helps countries identify what policy reforms and investments in environmental management are needed. Examples include direct investments in environmental management, but also investments in capacity, such as human resources, systems for monitoring and evaluation, as well as research capabilities to develop local solutions to environmental problems. Identifying these policies and investments requires drawing upon the best available knowledge to determine which actions need to be implemented and how they should be sequenced.

Step 4: Develop a Medium-term Framework for Action

We recommend that needs assessments be translated into a medium-term (for example 10-year) investment framework that charts out the country's path toward achieving the Goals. This may take the form of a vision or long-term strategy document, already prepared by many countries. The medium-term framework needs to accomplish at least three tasks:

- prioritize and sequence interventions across all sectors;
- identify the required policies and institutional reforms to support the

investments; and

- assign responsibilities among key actors.

The medium-term framework can then serve as the basis for elaborating a short-term (for example 3–5-year) MDG-based development strategy (Figure 5.1) (c.f. UN Millennium Project, 2006).

Step 5: Craft a Short-term MDG-based Development Strategy

We emphasize the importance of deriving a short-term (for example 3–5 year) strategy from an assessment of medium-term needs, since most environmental strategies and the capacities to deliver them require medium- to long-term strategies for scaling up. To be operational, such a short-term strategy needs to be anchored in a financing strategy and expenditure framework that links planned investments with budgets. For example, an MDG-consistent medium-term expenditure framework (MTEF) can lay out how resources will be generated and disbursed.

While MTEFs are becoming more common in key social and infrastructure sectors, few environmentalists know of them. We are not aware of a single country that has a medium-term expenditure framework for the environment. Changing this will require robust environment strategies that are properly costed. In addition, institutions charged with environmental management must develop the public finance expertise required to review and participate in the design of MTEFs. Many low-income country governments will probably require support to develop these technical skills.

IV. ALIGNING INTERNATIONAL SUPPORT WITH MDG-BASED DEVELOPMENT STRATEGIES

Many countries will require enhanced international support to develop and implement the environment component of MDG-based development strategies. As requested by the Secretary-General (UN, 2005a), the UN system, multilateral environmental agreements, and other international organizations should align their technical work and financial assistance to support the preparation and implementation of MDG-based national development strategies. Several types of support stand out.

Technical Advice

A particular challenge is the identification of effective outcome targets to guide the development of environmental strategies. Providing targeted advice on how to set such targets and identify the interventions needed to achieve them could be an important focus of the UN system's work at the country level and the programmatic work supported under MEAs. Similarly, UN organizations and the World Bank need to strengthen their country-level advice on planning and programming the investments and policy changes required to achieve time-bound environment targets. Both local and international environmental NGOs can also play an important role in this area.

Financing

Most low-income countries will require substantial increases in official development assistance if they are to meet the MDGs (Commission for Africa, 2005; UN Millennium Project, 2005b; World Bank, 2005). Multilateral and bilateral donors will therefore need to increase aid levels over the coming years to reach the internationally agreed target of 0.7 per cent of rich countries' gross national income in development assistance by 2015. Aid needs to be delivered largely through direct budget support or other high-quality delivery mechanisms and must be accompanied by comprehensive debt cancellation to many poor countries (UN Millennium Project, 2005b). Transparent and rigorously costed environment strategies offer opportunities to explicitly target and ring-fence funding towards environmental sustainability.

Capacity Building

A third challenge to the international system that has not been discussed in detail in this chapter lies in training decision makers, technical staff and local communities in developing countries (c.f. UN Millennium Project, 2006). A comprehensive MDG needs assessment can help identify key capacity constraints and ways in which they can be relieved over the next ten years to 2015.

Documenting and Sharing Best Practice in Ecosystem Management

National governments and the international community should launch a major effort to collect and consolidate best practice in environmental management, including a rigorous assessment of the links between inputs and outcomes in environmental policy. The aim must be to establish a robust knowledge base akin to *the Lancet's Child Survival Series*. Perhaps the United Nations, World Bank and the MEAs can be tasked with supporting efforts to develop such catalogues of best practice for their respective areas. These catalogues could take the form of databases similar to those in which the World Health Organization collects and evaluates best practice for fighting disease.

Further scientific research and technology development is needed to improve environmental management techniques. An interesting example is the Millennium Village initiative led by Columbia University's Earth Institute, Millennium Promise and the UN Development Programme (UNDP). It supports and monitors integrated environmental management at the community level for different agro-ecological zones in Africa.[4] More research of this kind needs to be carried out, perhaps even as a follow-up to the Millennium Ecosystem Assessment. International organizations and conventions need to participate actively in this work to ensure that lessons learned inform environmental policymaking on the ground.

Likewise, more rigorous research on effective methods of implementation is needed. Lessons suggest that local ownership, which requires participation, consultation and training, is a critical success factor in implementing key interventions. We now need to understand better how these different elements can be brought together to ensure the success of scaled-up national programs.

V. GETTING STARTED

Thanks to the Millennium Ecosystem Assessment, the world has clear evidence on the importance of ecosystem services for human well-being and the extent to which ecosystem services have declined in recent years. The case for action to improve environmental management has never been clearer, and it is matched by a growing recognition among policymakers that environmental sustainability lies at the heart of achieving the Millennium Development Goals. As this chapter argues, countries need to adopt outcome-oriented environment strategies to achieve environmental sustainability. These should in turn be integrated into national MDG-based development strategies.

The experience of the UN Millennium Project shows that many developing countries want to develop and implement environment strategies as part of their MDG-based development strategies, but they lack the technical and financial support needed from the international community. This provides an opportunity for MEAs, international organizations and civil society organizations to join forces to support developing countries in preparing MDG-based environment strategies. Such an initiative requires ongoing coordination – perhaps assumed jointly by the UN Environment Programme (UNEP) and the UN Development Programme (UNDP) – and should focus on countries where the political leadership is committed to implementing a strategy for achieving environmental sustainability. MEAs could organize themselves to provide targeted technical assistance in their areas of expertise and start the compilation of databases of best practice.

If launched successfully, the first MDG-based environment strategies can be prepared during the course of a year. Not only will this work support progress towards environmental sustainability in the countries under consideration, but it will also provide invaluable lessons in how countries can manage their environment as a central part of their development strategies.

NOTES

1 Guido Schmidt-Traub is Associate Director at the UN Millennium Project. Albert Cho is a consultant at McKinsey & Co. This chapter does not represent the views of the UN Millennium Project, the United Nations, or McKinsey & Co., but of the authors alone. The chapter was produced with funding from the World Resources Institute (WRI).
2 The UN Millennium Project was commissioned by UN Secretary-General to put forward the best strategies for achieving the MDGs. Its reports were submitted to the Secretary-General in January 2005, and are available at www.unmillenniumproject.org.
3 It is a pity that despite the caliber of the scientific expertise garnered by the Millennium Ecosystem Assessment it does not endorse specific outcome targets, put forward additional objectives, or outline approaches that countries could use to identify new objectives.
4 See www.millenniumvillages.org for more information.

REFERENCES

Bojo, J. and C. Reddy (2003), 'Poverty reduction strategies and the Millennium Development Goal on environmental sustainability: Opportunities for alignment', *Paper no. 92, World Bank Environmental Economics Series*.

Bojo, J., K. Green, S. Kishore, S. Pilapitiya and R.C. Reddy (2004), 'Environment in poverty reduction strategies and poverty reduction support credits', *World Bank Working Paper 30890*.

Commission for Africa (2005), *Action for a Strong and Prosperous Africa: Report of the Commission for Africa*, London, March.

Department for International Development (DFID), Directorate-General for Development (DGFD), European Commission (EC), United Nations Development Programme (UNDP) and World Bank (2002), *Linking Poverty Reduction and Environmental Management: Policy Challenges and Opportunities*, Washington, DC: World Bank.

Ghanime, L. and L. Lee (2005), 'Environmental sustainability in 100 Millennium Development Goal country reports', *UNDP Energy and Environment Group*, New York, http://www.undp.org/fssd/docs/mdg7countrepsumm.doc.

Juma, C. (2004), 'Technological innovation and sustainability: Implementing the Millennium Declaration and the Millennium Development Goals', Paper presented at the High-Level Roundtable on the Implementation of the Millennium Development Goals, New York.

The Lancet (2003), 'Child survival', collection of articles in *The Lancet*, vols. **361** and **362**.

Millennium Ecosystem Assessment (2005a), *Millennium Ecosystem Assessment Synthesis Report*, Washington, DC: World Resources Institute.

Millennium Ecosystem Assessment (2005b), *Responses Assessment: Ecosystems, Ecosystem Services and Human Well-being: Understanding Response Options and Strategies, Summary for Decision-makers*, Washington, DC: World Resources Institute.

Republic of Bolivia (2001), *Poverty Reduction Strategy Paper*, La Paz, Republic of Bolivia, online at http://www.imf.org/external/NP/prsp/2001/bol/01/033101.pdf.

Socialist Republic of Vietnam (2002), *Comprehensive Poverty Reduction and Growth Strategy*, Hanoi, Government of Vietnam, online at http://www.imf.org/External/NP/prsp/2002/vnm/01/053102.pdf.

United Nations (UN) (2005a), 'In larger freedom: toward development, security and human rights for all', *Report of the Secretary-General of the United Nations for Decision by Heads of State and Government in September 2005*, New York, United Nations, 21 March.

United Nations (UN) (2005b), '2005 World Summit outcome', *General Assembly Document A/Res/60/1*, New York: United Nations.

United Nations Environment Programme (UNEP) (2002), *Global Environmental Outlook 3*, London: Earthscan.

United Nations (UN) Millennium Project (2005a), 'Environment and human well-being: A practical strategy', *Report of the UN Millennium Project Task Force on Environmental Sustainability*, New York: Earthscan.

United Nations (UN) Millennium Project (2005b), 'Innovation: Applying knowledge in development', *Report of the Task Force on Science, Technology, and Innovation*, New York: Earthscan.

United Nations (UN) Millennium Project (2005c), *Investing in Development: A Practical Plan to Achieve the Millennium Development Goals*, New York: Earthscan.

United Nations (UN) Millennium Project (2005d), 'Needs assessment models', online at http://www. unmillenniumproject.org/policy/needs03.htm.

United Nations (UN) Millennium Project (2006), *Preparing National Strategies to Achieve the Millennium Development Goals: A Handbook*, New York: United Nations.

World Bank (2004), 'Dominican Republic – Environmental priorities and strategic options: Country environmental analysis', Caribbean Country Management Unit, Environmentally and Socially Sustainable Development, World Bank, June.

World Bank (2005), *Global Monitoring Report 2005: Millennium Development Goals: From Consensus to Momentum*, Washington, DC: World Bank.

World Health Organization (WHO) (2001), 'Macroeconomics and health: Investing in health for economic development', *Report of the Commission on Macroeconomics and Health*, Geneva: WHO.

6. Environmental policy priorities for the World Bank

Robert Goodland

INTRODUCTION

There should be little debate on the Millennium Assessment's diagnosis that ecosystems worldwide are being used unsustainably, and that ecosystem services are being increasingly impaired. Previous international diagnoses are too consistent and frequent. Any questions about the Millennium Assessment's diagnosis are more likely to be a delaying tactic to protect business as usual. Environmental deterioration has been documented for decades, although not as comprehensively and recently as by the Millennium Assessment. It is valuable in that it confirms and corroborates previous documentation of ecosystem degradation. The Millennium Assessment has come at a crucial time: every significant indicator of global environmental health is heading in the wrong direction.

Lack of information on ecosystem degradation is not the constraint. The Millennium Assessment does not identify those responsible for damage. 'Shame-and-blame' is needed, and over the long haul. Some may claim that this is impossible in a UN report, but the World Bank has started to blacklist corrupt corporations. Shame-and-blame in the form of league tables is provided by Transparency International, and by pollutant-release inventories in several countries. Moody's and Standard and Poor's stock market ratings are akin to shame-and-blame, although not on the environment. Action plans on what needs to be done to prevent the dire predictions of the Millennium Assessment are likely to be hotly debated.

TARGET DECISION-MAKERS

Few decision-makers will read the Millennium Assessment. The approach of focusing on the ecosystems is fine for natural scientists, but less so for decision-makers, most of whom focus on sectors of the economy, rather than on individual ecosystems. Ministers of Energy are prime targets for persuasion, for example. Energy is possibly the most important sector to transform if we are to achieve sustainability because energy touches all sectors and all people. Phasing out of coal, leaving it to the private sector, is the first goal. The Bank needs to work with governments to ensure that external costs, especially greenhouse gas emissions, are fully internalized. Second, adopt the World Commission on Dams recommendations before moving ahead with hydroprojects. Third, accelerate natural gas as a bridging fuel. Fourth, focus on vastly accelerating renewable energies: wind, solar, biomass (for example, cellulose alcohol). Fifth, stringently keep up with energy conservation and efficiency (for example, building codes).

By design, no recommendations were included in the Millennium Assessment, which is one of the voids this review helps to rectify. Decision-makers need actionable recommendations, applied findings, and best judgment. Despite the fact that the Millennium Assessment deliberately avoided recommending policies, the ecosystem assessors are often the most knowledgeable and experienced group to suggest how to reduce risks and shift to a more sustainable course. Although the Millennium Assessment team has disbanded, it remains a valuable asset in the formulations of policy. It should be possible to ask each ecosystem-assessing group to recommend policies to prevent the deteriorating trends they document so well. The ecosystem assessors probably feel strongly about 'their' ecosystem and would be delighted to help with recommendations.

Eventually the recommendations, actions, or policy options concerning each ecosystem need to be codified into a single 'Policy Priorities Report'. That is all most decision-makers will look at. Even the 'Policy Priorities Report' will need a 25-page summary booklet. This chapter focuses on the necessary policy priorities.

Because decision-makers are mainly sector-oriented (for example, Ministry of Transportation), the recommendations also should be sector-based, such as on agriculture, transport, mining, energy, education, health, and so on. The recommendations need to be ranked or prioritized in several ways. Short, medium and long term is an obvious cut; another useful cut would be to identify who needs to act, the constituencies for change: governments, Civil Society Organizations, the UN, labor, individuals, professionals, corporations, academia, the World Bank, and so on. As such a report would be lengthy, it could focus on energy, with sections on energy priorities in other sectors such as agriculture and transport, while leaving the other sectors to annexes.

Most decision-makers will start with the recommendations within their sphere of influence, and if their interest is piqued, may go back to find the underlying rationale in the ecosystem assessment which led to the recommendations of interest. Few will read the whole just in order to identify the recommendations or actions of concern to each decision-maker. The policy specifics that Chapter 1 extracts from this and other reviews contained in this book need to be targeted to specific decision-makers.

The Millennium Assessment says that radical changes are necessary, but does not specify what these changes are. The Assessment's 'four scenarios' (for example, technological innovation in the Techno Garden scenario) come closest to prescribing changes, but are far from direct prescriptions. As a diagnosis the Millennium Assessment is reliable; its weakness lies in its lack of prescription. Comparisons between the Millennium Assessment and the International Panel on Climate Change (IPCC) are inevitable. If the government-led IPCC and the UN Framework Convention on Climate Change are moribund, why is the Millennium Assessment more likely to succeed? Why is the UN's Environment Program the only member of the UN family dependent on voluntary funding? How many more reports does it take before we take our predicament seriously? Another report, more analyses, annual score cards, the next 'collapse' prediction like the excellent one of Jared Diamond (2005) all help, but lack of diagnosis is not at all the problem.

THE ANTI-ENVIRONMENT MOVEMENT

Today's struggle over the United States' energy policy shows how polarized society has become. The lobby now gutting environmental and social prudence is formidable but ever more short-sighted with every new environmental score card.

The lobby for environmental recklessness is extremely powerful. One strident anti-environmentalist, supported by the World Bank and presumably by the US administration, summarizes neo-conservative foolhardiness. Fixing the Bank:

> will require facing down the activists who have forced the bank to adopt excessive rules on the environment, corruption, and the protection of indigenous peoples. It will involve standing up to these lobbies when they gang up against particular bank projects .... If the institution's new president is to succeed, there must be no more fiascos of this kind. The bank must not

> be forced to drop projects because of baseless activist campaigns, and it must be given political cover in Congress and in European parliaments so that it can streamline its safeguards (S. Mallaby, 2005).

In this example, the anti-environment lobby is against policies caring for vulnerable ethnic minorities and against policies reducing corruption. In addition, 'streamlining safeguard' policies means gutting them still further. Streamlining means devolving responsibility for social and environmental quality to individual governments and absolving the Bank from major responsibility. The opposite prescription is what is needed. The Bank's safeguard policies need strengthening in order to protect the weak and the environment, thereby improving the effectiveness of economic development. Without these safeguard policies, the Bank reverts to its 1960s goal of amount of money lent, regardless of the social and environmental damage it causes. Insertion of the weasel word 'baseless' to qualify activist campaigns does not protect Mallaby as he does not distinguish between baseless and non-baseless campaigns, not even by examples.

Our predicament is so grave that both grass-roots and top-down approaches will be necessary. If we choose to save the environment, or at least to slow the rate of degradation, national governments need to lead the way. However, the share of Official Development Assistance allocated to the environment has fallen since the mid-1990s from 20 per cent to 9 per cent, according to the OECD. The US has abdicated its environmental leadership.[1] Although unfortunately the World Bank also appears to have chosen to abandon its environmental leadership, to its credit, it has not yet totally abandoned its priority of direct poverty reduction. National governments are influenced, at least in developing countries which have the most to lose and who are bearing the brunt of degradation, by the United Nations and specifically by the World Bank.

This chapter focuses on the implication of the Assessment for the President of the World Bank Group, because Bank practices are enormously powerful in guiding others, especially developing country governments. The Bank is well positioned to shift from being part of the problem to becoming part of the solution. Other reasons for focusing on the Bank are noted below.

IMPLICATIONS OF THE MILLENNIUM ECOSYSTEM ASSESSMENT FOR THE WORLD BANK

This section briefly notes some of the Bank's contribution to environmental degradation and introduces what it could do to reduce ecosystem over-use. Is the environmental degradation documented by the Millennium Assessment worsened by the Bank, or is the Bank financing more conservation than damage? On balance, subtracting the relatively modest environmental grants (averaging about $600M p.a.) of the Global Environment Facility, it seems clear that the mainstream Bank finances more ecosystem damage than it cures. This section focuses on how the World Bank could promote environmental prudence.

Basically, the World Bank seems unwilling or unable to change its core 'business as usual' strategy. If the Bank can reject – then actively undermine – advice from its biggest shareholder,[2] as well as the advice that it pays for, such as the Morse Commission on India's Narmada scheme (1992), the World Commission on Dams (2000), and the Extractive Industry Review (2004), why is the Millennium Assessment likely to be more readily accepted? Essentially, the Millennium Assessment diagnoses much the same trends dealt with by both the World Dams Commission and the Extractive Industry Review: all three are consonant. Despite the efforts of one of the Millennium Assessment leaders, Climate Panel chair and Bank staffer, Bob Watson, the Bank resumed financing coal mining and coal-fired electricity generation in 2003 after a de facto moratorium of more than a decade. The Bank tolerated Bob Watson working for the Millennium Assessment for four years, but will that be another epicycle? Or will the Bank internalize the Assessment's findings? The key message to the World Bank President from the Millennium Ecosystem Assessment Policy output should be: 'don't be distracted by all the bells and whistles that have accreted to the Bank's structure and portfolio over decades; instead, focus on integrating an ecosystems-based approach into mainstream Bank policy and operations.'

The mainstream Bank undermines the Kyoto Protocol by financing coal and subsidizing oil, but is willing to house the epicycle of the Global Environmental Facility (GEF) which finances renewable energy and carbon sequestration. Major Bank financing of coal burning is folly even in the face of relatively minor GEF finance to undo such damage. The ratio between loans exacerbating climate change (about $28bn since the Rio Earth Summit, 1992) to grants that seek to ameliorate it may exceed 17:1.[3] The mainstream Bank preserves the flawed model of development by adding more and more epicycles such as the Global Environment Facility, Clean Development Mechanism, several carbon trading mechanisms, forest alliance, and global water partnership while preserving its core business approach.

- Millennium Ecosystem Assessment Finding: 'Tropical forest support over half of all known biodiversity which underpin goods and services to humanity'. The Bank's ten-year-old policy of not financing logging of tropical forests was rescinded in 2002.[4] Deforestation continues at 10 million Ha/year. This huge loss of tropical forests is the second leading cause of global warming after fossil fuel combustion.
- The ten-year decline in financing destructive dams that degrade rivers and impoverish millions of people was reversed in April 2005.[5]
- The Bank has backed off Green Accounting (environmental resource accounting; revision of the system of national accounts).
- The Bank is actively financing conversion of Amazon forest for industrialized soy monoculture, despite Compliance Advisor Ombudsman's (CAO) negative report (2005) on the damage caused.
- Bank-financed private sector forces in the Guatemalan Marlin goldmine (owned by Canada's Glamis Corp.) shoot and kill indigenous protestors with impunity. Violence and intimidation against impacted peoples are becoming pervasive worldwide.
- The Bank is hostile to the will of impacted people, and refuses to adopt 'free prior informed consent'. Are there any limits to how 'industry-friendly' the Bank has become? The Bank's suite of environmental and social safeguard policies are themselves being systematically gutted.[6]
- The Bank's funding for direct poverty reduction is giving way to indirect trickle-down growth. The pro-poor approach would be to balance (a) direct poverty reduction with (b) trickle-down growth at least annually, as amplified below.
- The flurry of Bank interest in sustainability following Brundtland and the UN's Millennium Development Goals has stalled and risks being reversed.

These facts point to 'how' to implement improvements. Taking the above bullet points in order, these 'how' measures are amplified below: extend the moratorium on financing logging in old growth forests; embrace the findings of the World Dams Commission; resume work on Green Accounting and internalize environmental externalities; cultivate soy on savanna lands, not by converting Amazon forest to soy; adopt a set of Human Rights policies to prevent violence against the poor impacted by development projects; adopt the best practice policy of free prior and informed consent before going ahead with any development project; annually recalibrate the balance between direct poverty reduction and trickle-down growth; resume work on environmental sustainability, make it into time-bound targets, offer incentives for reaching sustainability.

GENERAL POLICY FUNDAMENTALS

The broad outlines of the required policies have been repeated consistently.

- Refocus on sustainability; think through what sustainability is in every sector. Sustainability and poverty reduction should be made the paramount goals of development, and all other priorities should flow from these two overarching goals.
- Directly reduce Southern poverty: health, nutrition, education, water supply, training, job creation, domestic processing, value added.
- Reduce Northern overconsumption: promote sufficiency or 'enough is best': make frugality a societally laudable goal. The Bank should join and encourage the movements in this direction, even though its main role is Southern development. Reducing Northern overconsumption is an effective means of supporting the South.
- Accelerate the transition to renewable energy. Agree on long-term dollar amounts to be invested annually in renewables. Reverse the ratio of fossil to renewable lending from its current 90:10 within ten years. Shadow price the release of each ton of greenhouse gas at any non-zero price, such as $5 or $10/ton. This will accelerate the use of gas and renewable energies and the phase out of coal. Leave coal and oil to the private sector, but work with governments to internalize the costs of GHG emissions in project selection.
- Accelerate local population stability. Governments need to control illegal immigration. Government services and ecosystem services are stressed by the dissipative nature of population, by sheer numbers consuming the irreducible minimum of food, clothing and shelter – especially the overconsumers. When two World Bank clients are neighbors and one cares for its citizens (for example, Botswana), while the other (for example, Zimbabwe) cares less about its citizens, the well-run neighbor risks being swamped by its autocratic failed-

state neighbor. The Bank has an important role in such cases, improving management in badly-run countries, while protecting well-run countries from forces outside their borders.

- Continually recalibrate being 'industry-friendly' with being labor- and environment-friendly. This is out of balance at the moment as the WBG has become overwhelmingly industry-friendly and almost hostile to labor (for example, against freedom of assembly and collective bargaining; unwillingness to adopt UN ILO's core labor standards) and against environmental prudence. Force transnational corporations to cease environmental damage.
- This chapter leaves to others the issue of unregulated trade between asymmetric partners. The International Monetary Fund's prescriptions are frequently damaging to society (for example, bread riots), and to the environment; so far the Fund has resisted any environmental advice or staff, although it did hire its first social specialist a couple of years ago. The World Trade Organization has yet to catch up with the International Monetary Fund.
- The Bank should invigorate improvements in governance by grants or by other means. Improved governance is essential, especially transparency and the prior informed consent of the people potentially impacted by a development project.

The Millennium Assessment proves that the current style of development is harming our life-support systems and cannot continue much longer. This means that economic development should focus more on moving up the value chain, value added, domestic processing, job creation, and local markets, rather than keeping less developed countries as suppliers of raw materials to be exported to industrial countries where the processing and manufacturing are more efficient. Production mainly for one's own people, while exporting to buy essentials one cannot produce remains a low impact fundamental. The World Bank is ideologically opposed to log export bans. Indonesia rejected its advice, banned log export and now has one of the world's most profitable and efficient plywood industries, with major forward and backward linkages. Indonesia now exports valuable knock-down furniture, doors, window frames, moldings, and dimension timber. Similarly, in Mozambique, the World Bank tried to stop the export of high value roasted cashew nuts. The Bank tried to force Mozambique to export the unbeneficiated cashew fruits to India, which is efficient at cashew nut beneficiation.

SPECIFIC WORLD BANK POLICIES

The means to implement these policy fundamentals have also been repeated over the years. They are listed here in abbreviated fashion.

1. Millennium Development Goals

- Focus on the Millennium Development Goals. Keep the Millennium Goals as the top priority. Prioritize direct poverty reduction above indirect trickle-down 'growth'. Prioritize development above growth, although both are needed in different measures (see Schmidt-Traub and Cho's review in this volume for more information).
- Stop intensifying poverty through the forcible displacement of people from big dams and other infrastructure. If people cannot be made at least modestly better off promptly after their displacement, the project should not go ahead.
- Rebalance direct poverty reduction by de-emphasizing indirect and leaky trickle-down economic development projects.[7]
- As the Assessment points out, about 90 per cent of the planet's disease burden falls on the poor in developing countries. Get serious about prevention of communicable diseases, especially AIDS (education, curing sexually transmitted diseases), malaria (treated bed-nets, destruction of vector breeding sites, new low-cost *Artemisia*-related chemotherapy)[8] and tuberculosis.
- Direct focus on malnutrition, micronutrients, supplements, immunization, maternal health, free primary education for all.
- Emphasize job creation and training.

2. Population Stability

- Boost girls' education: promote gender equality in primary education. Reducing teen pregnancies also reduces poverty.
- Facilitate mothers' easy choice of family size and spacing. Deferring the first pregnancy is demographically powerful in reducing population pressures. Parents choosing later pregnancies should get all the encouragement they can. Boost maternal and child health and pre-natal nutrition.
- Women must have equal rights in owning and inheriting land and other property, as well as equal access to credit.
- Work with opponents of population stability (for example, some religions) by focusing on doctrine-based stewardship and family

responsibility. It is unacceptable to oppose family planning without a social policy to provide care for children once they are born. Family planning is better than unwanted, deprived children.

3. Economic Priorities

- *Internalize externalities*: Systematically account for the costs of environmental and social impacts. Benefit/cost ratios systematically undermine environmental services because accountants use zero in the absence of a precise cost; benefit/cost analysis needs to use reasonable estimates where precision is unavailable. Resuscitate green or environmental resource accounting (UN System of National Accounts).
- *Repeal perverse subsidies*, especially for fossil fuels (estimatedUS$20–235bn annually), agriculture, old-growth forest logging, capture fisheries, and corporate welfare, which are taxes on citizens. Such perverse subsidies reduce resources available for everything else.
- *Green Taxation*: This must be revenue neutral by phasing down taxes on what we want to encourage (for example, income, employment), while raising taxes on what we want less of (for example, coal, pollution). Much environmental prudence can and should be gained by economic incentives rather than by command-and-control regulation. Tradable quotas and schemes for trading carbon and sulfur hold promise when prudently designed and enforced. Phase out value added, employment and payroll taxes. Impose tax only after subtracting savings. Balance taxes on consumption, pollution, waste and depletion, rather than on job creation and income (see Bateman's review in this volume for more information).
- *Mandate annual scrutiny of national budgets*: This will identify perverse subsidies, and environmentally harmful expenditures.
- *Pro-poor economic policies* will not reduce poverty unless accompanied by policies reducing social injustice and political inequity.
- *The precautionary principle*: Now that we have let the world ecosystems become so damaged, and now that the risks have increased of non-linear responses to further stresses on ecosystems, more prudence is essential than before. The Precautionary Principle needs to become the enforced norm, especially in the case of irreversible impacts. Since it was formulated by the OECD in the

1960s, reaffirmed in the 1992 UN Rio Declaration, and adopted by the EU in 1999, 'better safe than sorry' has prevented much damage, but is still too easy to circumvent. A recent example is the leaching of endocrine disruptors from common plastics which then in humans.

- *Payment for ecosystem services*: Introduce systematic mechanisms to ensure payment for ecosystem services through incentives for stewardship of natural resources. Such schemes are now well-developed, and have been shown to generate economic benefits that are much greater than costs even before environmental benefits are calculated.
- *UN harmony on poverty and sustainability*: Encourage the World Trade Organization to support United Nations and Millennium Development Goals of environmental sustainability and poverty reduction. At the least, the World Trade Organization and the International Monetary Fund should cease undermining UN, MDG and World Bank poverty and sustainability goals.

4. Transport Sector

- Emphasize mass transit, metro, rail, ride-sharing and non-motorized transit
- De-emphasize automobiles and highways
- Raise fuel efficiency (mpg), road user fees, congestion charges, annual permits commensurate with vehicle weight, parking fees, and internalize carbon emission costs. Lower maximum speeds and promote 'no sulfur' clean diesel. Raise incentives for ultra-light cars, hybrid and electric vehicles. Convert natural gas to liquids to blend with gasoline and diesel. Promote the development of hydrogen fuel-cells as a promising fuel for trucks and buses. Promote cellulose-to-ethanol biofuels. Amory Lovins's *Winning the Oil End Game* amplifies such priorities.[9]
- Mandate recyclability of vehicles when they are replaced.
- Rescind tax exemptions on aviation fuels

5. Energy Sector

- As the Assessment points out, global energy demand is projected to grow by 50 per cent in the next 25 years. Amory Lovins's prescription in *Winning the Oil Endgame* is the solution to the energy predicament and is highly recommended as the guide for the World

Bank, instead of its risky response to the G8's mandate of focusing on coal technology.

The only way to survive without unacceptable impacts is to have the following twin priorities:

(a) *Conservation of Energy*: Keep conservation of energy stringently up-to-date (that is, doing more with less energy; permit investment in new generation only when it is lower cost than the next conservation measure). Energy consumption in OECD countries has to stabilize and, according to the European Union policies, to decline.
(b) *Renewable Energy*: Diversify out of fossil fuels and into renewables. Some measure of energy independence and diversification out of fossil fuels has the big advantage of creating massive numbers of permanent jobs. In tandem with conservation, the overriding goal is to accelerate the transition to renewable energy systems, especially wind and solar,[10] including biofuels, tidal and wave energy. The European Union's goal of generating 22 per cent of its electricity from renewables by 2010 is commendable, but must go farther. Recycling wastes (for example, bagasse) and the best geothermals also have a role, as do mini, micro and 'run-of-river' hydroprojects that do not flood large areas, nor displace people. Natural gas is by far the 'least of all evils' for an interim bridging fuel, although there are more reserves (about 70 years) than the earth can afford to use. Even so, gas imports will increase, so gas pipelines have to be built with low impact.[11]

- Offer consumers the choice of purchasing renewably-generated electricity, over coal-fuelled and nuclear energy.
- Get the price right, and foster a market-based approach. Remove subsidies for coal, oil, nuclear energy and damaging hydro. Stop suspending normal economics in the case of pollution, especially of greenhouse gas emissions. As soon as the massive costs of damage to human health, the global climate, many other ecosystems, acid drainage and waste disposal and the forcible displacement and impoverishment of oustees are internalized by application of good economics, coal's phase out will accelerate. Coal energy appears cheap only because standard economics is suspended for coal cost/benefit calculations. China, the US and India need the strongest incentives to internalize coal pollution costs. Encourage the trends reducing the energy intensity per unit of GDP, especially in

developing countries as they strive towards manufacturing-led growth.

- Control greenhouse gas emissions, sulfur and nitrogen oxides, persistent organic pollutants and particulates by means of incentives. Prices for emissions must be set realistically. There is consensus that carbon dioxide emissions should not be valued at \$0/ton in the market. \$5/ton of CO_2 emitted is a very conservative estimate; more realistic estimates exceed \$20/ton. Effective pricing measures should start at \$5 and be raised gradually as more scientific evidence accrues. Better pricing of emissions would level the playing field with renewables. This single measure, the application of standard economics to greenhouse gas emissions, would be a powerful incentive to accelerate the transition to renewables, and to foster conservation. Severance taxes or a carbon tax at coal mine-mouth and oil well-head are said to be more efficient and less costly than gasoline taxes.
- Encourage co-generation wherever possible, as well as linking independent generation to energy grids by net reverse-metering, by requiring utilities to purchase privately-generated power.
- Finance solar electricity, electrolysis of water to produce hydrogen, and desalination of water. Encourage fuel cells and hydrogen from solar or wind. Renewable credits trading and production tax credits have a useful role.
- Large-scale hydro that kills river systems is not renewable; mandate the recommendations of the World Commission on Dams.
- Follow the recommendations of the independent Extractive Industry Review 2004 of the World Bank Group's oil, gas and mining portfolio (www.eir.org, 7 vols; Goodland 2004).
- Civil society should openly debate their opinion of coal and accept the much more benign alternatives, such as well-sited wind farms. By all means design bird- and bat-friendly turbine blades.

6. Agriculture Sector

- *Agriculture for nutrition*: Set as a national priority the refocusing of agriculture towards nutrition to prevent malnutrition (starvation and obesity).
- *Water conservation*: Follow the Stockholm International Water Institute's excellent prescriptions on water for food production.[12] Communities must own their water so that they can invest in it and conserve it. Strenuously conserve water especially in irrigation

(precision agriculture, computerized drip irrigation to replace sprays); tax irrigation and industrial water as a user cost to prevent waste; conserve watersheds and aquifers; harvest rain and store as much water as possible (for example, in watersheds and underground); promote water recycling. Wetlands are worth twice as much intact as they would be converted to industrial agriculture. Provide incentives for small-scale water providers. Reduce leakages in water pipes. Encourage the use of water-efficient appliances. Phase out new sewers using quality water. Desalination is becoming more necessary; there is scope for the use of solar-based desalination on desert coasts and hydrogen production.

- *Limit harvesting from the wild*: Phase out commercial harvesting of anything from the wild, particularly the bush-meat trade and the annual trade in 20000 wild-caught parrots and other exotic birds. If humanity wants to consume something, it should be brought into cultivation. Exceptions should be made for low-level collection of non-timber forest products for societies, especially indigenous peoples, depending on them. If, in the future, harvesting from the wild can be stringently controlled so it does not exceed regeneration rates, there may be scope to reconsider some wild harvests.
- *Tree plantations*: Accelerate tree plantations for timber, fuelwood, watershed protection, carbon sequestration, and aquifer recharging. Foster job-creation by increasing domestic processing and value added such as plywood, doors, windows, and knock-down furniture. Most institutions should have tree plantations around their property (for example, schools, barracks, mosques, sports facilities, town halls, community centers and hospitals).
- *Natural forests*: Ban export of undressed logs; promote value added and domestic processing in countries, even if it is inefficient at first. Promptly phase out logging of old-growth forests; foster conservation of those remaining. Never finance conversion of old-growth and tropical forests to agriculture. Foster the sustainable and certified timber movements.
- *Promote grain-based diets*: 'The human appetite for animal flesh is a driving force behind virtually every major category of environmental damage now threatening the human future – deforestation, erosion, fresh water scarcity, air and water pollution, climate change, biodiversity loss, social injustice, the destabilization of communities, and the spread of disease' (WorldWatch, 7/2004). Beef needs 15 cubic meters of water per kilogram produced, whereas grains and beans require about one cubic meter of water per kilogram. The

World Bank has a commendable livestock strategy[13] mandating a phase-out of large-scale grain-fed cattle ranch financing, but the International Finance Corporation claims it does not have to follow the World Bank's strategy, hence it is seeking to expand extensive cattle ranching in place of Amazon forest as of 2005.

- *Mixed agriculture vs. industrial monoculture*: Small-scale mixed agriculture creates more jobs at far less environmental and social cost than extensive industrial monocultures. Balance organic agriculture, recycling, mulch, green manures and legumes. Control excessive use of nitrogen fertilizer and fertilizer runoff. Prioritize investments in waiting, fallow, regeneration, and rehabilitation of degraded lands.

7. Oceans and Fisheries

- *Oceans*: Oceans, which are the largest ecosystem in the world covering 70 per cent of the earth's surface, are in deep trouble. The Millennium Assessment, however, under-emphasizes the gravity of the problem. The Assessment alludes to the oceans' problems in several places (for example, Climate & Air Board, CO_2 & Oceans, Coastal Board), but for decision-makers these are difficult to find. Overfishing is the main cause of the 50 per cent decline in fish species (for example, Japanese 'long-lines' which are 100 km long baited hooks reaching a 300 m depth). Cod, formerly the world's most productive fishery, has essentially vanished, putting thousands of families permanently out of work. Commercial fishing has depleted 90 per cent of the global abundance of big fish, and every single commercial fishery is declining precipitously worldwide. After overfishing, the second most damaging impact on oceans is pollution (for example, untreated sewage from the world's coastal cities, agricultural runoff from pig, poultry and cattle effluent, chemical fertilizer runoff), leading to widespread eutrophication. The areas of the oceans that are totally devoid of oxygen (hypoxic) due to pollution are rapidly expanding (for example, 15000 km^2 in the Gulf of Mexico alone). Dissolved CO_2 benefits photosynthesis of plankton up to an unknown point, yet primary productivity in oceans (+/- chlorophyll concentration) has fallen sharply in the last 20 years.
- *Ocean acidification*: Oceans absorb one third of atmospheric CO_2 which is one of the main reasons why climate change is not as acute as it would otherwise be. The world's oceans have already absorbed about half of the CO_2 produced by humans, mainly by the burning of fossil fuels, over the past 200 years. However, this massive benefit

of healthy oceans is in jeopardy because of ocean acidification. Acidic, warm water absorbs less CO_2 than regular ocean water. Carbonic acid (from dissolved CO_2), disassociates into bicarbonate $HCO^{3}-$, some carbonate ions, and some protons H^{+}, which is acidifying the oceans. 'Normal' ocean pH is 8.2; this is falling and is projected to drop to about pH 6.5 (clearly acidic). This is a massive change that is irreversible in our lifetimes because oceans are exceptionally buffered (for example, bicarbonate ions, carbonate rocks). Like the oceans, the atmosphere and ozone shield are also enormous, but they are not buffered at all. Ocean acidification is also expected to (a) slow calcification (important for organisms such as corals, shellfish, crustaceans, *Coccolithophorida* (a silicon-fixing *Chrysomonad* algae), foraminifera (the silication of diatoms), and even to decalcify calcareous structures, (b) reduce oxygen, (c) damage photosynthesis, hence decreasing food chains.[14]
- *Ocean fisheries*: Fish farming in ocean inlets, fjords and ocean pens causes more damage (for example, manufacture of fishmeal, eutrophication, antibiotics, infections such as by mites, accidental escapes harming wild populations) than benefits. Catching wild larval shrimp and conversion of mangroves for shrimp farms are also fraught with problems, although shrimp farms are starting to reduce their previous impacts. In the future, if ocean fisheries can be realistically controlled (though there is little evidence for it at present),[15] then wild fisheries management would become an option.
- *Freshwater fisheries*: Although not problem-free, aquaculture in ponds (without any mangrove destruction and with waste recycling) should be used, rather than harvesting wild fish from oceans. A food-chain based on photosynthesis in the pond could be sustainable.

8. Corporate Standards and Pollution

The World Bank should do business only with those corporations with adequate track records and policies on both Corporate Social Responsibility and on Sustainability. The Bank should fortify its system of black-listing corporations unwilling or unable to follow socially responsible and sustainable policies. These offending corporations should have their charters revoked and their listings removed from public stock exchanges. The World Bank could foster partnerships to promote legislation on the best available sustainable business practices.

Meaningful taxes should be placed on pollution, including greenhouse gas emissions, and exemptions for airfreight fuels should be ceased.

'Polluter pays' and fines must be enforced and include the costs of detection and prosecution. Fines should be set commensurate with damage caused and the costs of clean-up or remediation. Preventive fines, prison sentences for Chief Executive Officers, and seizure of corporate assets (for example, corporate Headquarters Buildings, trucks, equipment, factories) for repeat offenders must become standard. As with the less damaging narcotraffickers, environmental criminals' assets should be seized and auctioned, with the proceeds allocated to remedying the damage.

Escrowed performance bonds or insurance must also be made routine and set commensurate with remediation costs. Third-party monitoring (for example, Global Witness, Environmental Investigation Agency) should be routinely and systematically used for infractions of social responsibility, sustainability, pollution and international treaties (for example, CITES, Ramsar).

9. Urban Sector

- Plan urban development to preclude or sharply reduce the necessity for individual car ownership and to minimize consumption.
- Foster 'Green Architecture' by constructing energy-efficient buildings with solar roofs and smart windows and using low-impact materials.
- Foster urban planning that promotes community and comity by encouraging pedestrian traffic and bicycles, while simultaneously reducing the need for cars.
- Mandate tree-growing to improve air quality and reduce energy needs by city council rules and building codes.

10. Biodiversity

- Do not finance projects causing irreversible biodiversity or habitat loss. Fine-tune the Bank's Natural Habitats Policy commensurate with the increasing loss of biodiversity.
- Get serious about 'no-go' areas. If there are genuine tradeoffs, mandate much higher ratios of compensatory offset[16] financing in perpetuity with offshore escrow accounts, while avoiding falling into the greenwash trap.
- Make critical ecosystems a priority through direct conservation, bolstered by regulations (for example, performance bonds, environmental insurance) and financing.

- Where deep poverty and rich biodiversity coincide as in remote rural regions, foster poverty reduction through conservation. Out-migration is rarely a solution because there are too many people living in such areas to be adequately absorbed into mainstream economies.
- Codify and support traditional knowledge.

CONCLUSION

The Problem is Political, not Scientific

Environmental prudence and social justice are under siege in Washington. Political needs are currently more important than scientific ones. Political consensus-making is outside the scope of this chapter, but important steps would be taken if the non-US members of the OECD on the Bank's Board could lead a Bank reform process. Island nation states should also be supported to pressure the Bank to internalize greenhouse gas emission costs in an attempt to prevent sea-level rise. But such political actions, important as they undoubtedly are, fall outside the extent of this chapter. The chapter has set down the most important actions necessary to reduce poverty and protect our life-support systems. Some of these actions are admittedly unpopular, so any political follow-up should take seriously the consideration of palatability in phasing. Today's incremental approach to environmental prudence is not working effectively.

The policies needed to reverse ecosystem degradation are already well known.[17] Most of the remedies are contrary to current political correctitude. Either unpopular choices are faced now or we will face much less tractable choices in the future. Time has run out. Damage to our life support systems are mounting and risks intensifying. The anti-environmental lobby in power prefers to treat the earth as a business in liquidation. The crash, when it comes, will be that much worse, especially for those least able to cope.

The problem lies not in knowing which policies to adopt, but in the lack of political will to adopt and implement prudent policies. Much more effort should be allocated to raising political will and fostering adoption of already known policies rather than on more diagnoses. Anti-environmental conservatives control all three branches of the US Government, many governorships and most statehouses. US politics is dominated by the radical right much more now than it has been over the last three decades. 'Every man for himself' was understandable when human populations were low and natural resources seemingly inexhaustible, but this is no longer the case.

Many corporations no longer work primarily for the public good; they undermine labor and environmental safeguards, health and safety standards, fix prices, weaken national legislation, abuse the right of 'eminent domain', externalize as many costs as they can, and are so rich that fines are just a minor cost of doing business.[18] Transnational corporations outcompete many weaker governments. Maximizing profits for shareholders is becoming more pervasive than the public interest. Where shareholders are a subset of the impacted public, there are fewer problems. Where stakeholders are in OECD nations that are insulated from impacts and while corporate impacts are in developing countries, the asymmetry of costs and benefits has become unjust. The socially and ethically responsible investment movement is encouraging. Stock and bond market rating agencies (for example, Fitch, Moody's, Standard & Poor) need to accord some weight to corporate responsibility. Whatever policy is adopted to reverse environmental decline will have to ensure that transnationals start working for the public good once again.

Now that the world is overfull and has run out of absorptive (sink) capacity, we are all globally interconnected. Progressive liberals need to attract the disenfranchised, the disempowered, the dispossessed and the disengaged. The environmental movement's twin priorities of social justice – especially poverty reduction and protection of our life-support systems on which the poor depend more than the rich – mean changing the public's values so that war is a last resort and then only by broad coalitions of nations and the UN.

More than $20 million has already been allocated to create the Millennium Assessment's diagnosis by 1400 scientists in 95 countries; fostering political will to choose environmental prudence will cost much more and will certainly take much longer than the four years it took to compile the Millennium Assessment. Most environmental warnings have been top-down; the Assessment is a more grass-roots initiative, coming as it does with no governmental mandate. The problem is that governments have fewer obligations to listen. The Millennium Assessment's thousand scientists are a good start, but a popular mandate is needed to persuade political representatives.

Signs of Hope

Admittedly, the forces of anti-environmentalism are currently overwhelming. However, there is reason to be guardedly hopeful for the following reasons:

1. First, there is growing consensus among thoughtful people that anti-environmentalism is imprudent, short-sighted and risky. The United Nations, The Earth Charter, Parliamentarians, the Royal Society, the US National Academy of Science, the National Council of Churches, Archbishops, Government Chief Scientists[19], the Millennium Assessment, even the Pentagon,[20] and many other groups[21] now are all issuing much the same warnings. The most pragmatic proposal to combat climate change comes from the founder of WRI and the former head of UNDP, Gus Speth. His 'Ten-Point Plan'[22] is the most thoughtful and practical blueprint available. The earth needs prudence, care, social justice and risk reduction rather than recklessly seeking bigger profits even if they damage the world.
2. Second, hostility to prudent environmental management is centered in the US. Europe's social and environmental precautions are burgeoning (for example, Restriction of Hazardous Substances (heavy metals & persistent organic pollutants), Registration and Authorization of Chemicals, Waste from electrical equipment). Europe should be in the lead pressuring the World Bank.
3. Twenty-nine of the world's biggest commercial banks have signed on to the Equator Principles (a set of social and environmental precautions modeled closely on those of the World Bank Group) since they started in 2003. These were even featured in *Time Magazine* on 1 May, 2005.
4. Third, ethical investing and socially responsible corporate funds are burgeoning. Calvert Funds, The Sierra Club's Portfolio 21, Domini, Pax, Citizens, Women's Equity Fund, amplified at naturalcapital.org. If ethical or socially responsible funds eschew World Bank bonds, their prized AAA rating could slip, depriving them of their main advantage compared with Equator commercial banks, some of whose environmental policies are now less risky than those of the World Bank. The Morgan/Chase Bank plans to lobby the US Government to adopt a national policy on greenhouse gas emissions, is financing protection for vulnerable ethnic minorities, and is conserving extensive forest habitat. Home Depot, among other retailers, regularly carries certified forest products. Citigroup recently ordered Malaysian loggers to get its Papua New Guinea timber certified before Citigroup would buy it.
5. Fourth, the insurance industry is taking environmental risks extremely seriously, especially climate-related damage.

6. Fifth, consumer sovereignty is working. Toyota's Prius hybrid's success defied every single prediction. Environmental architecture is burgeoning. Wind energy is surging. Organic food has started to displace industrial toxic-laced food.
7. Civil Society Organizations (CSOs) seem almost unanimous in recognizing the gravity of the environmental crisis and that more care and risk reduction is essential. CSOs are burgeoning, energetic and increasingly effective, and seem to be the main springboard to persuade governments to follow a more prudent course.

Admittedly, the best technical solutions are not necessarily the best politically, but even the most unpopular actions listed here are essential. The impractical must be made practical, the unpalatable palatable. Political follow-up should articulate a vision of the future commensurate with the magnitude of the crisis.

The most important challenge is a clarion call to the President of the World Bank. If you want to leave a grand legacy for posterity to thank you for, it is fully achievable during your tenure. You can ensure your place in history if you internalize the environmental and social costs of development. The challenge is to act prudently before catastrophe forces us to. You will galvanize a veritable army of supporters who are just waiting for a signal from you. You personally, Mr President, can make all the difference. If you are determined to make the world a better place, your moral leadership on the priorities outlined above would achieve just that. Your personal actions could achieve lasting poverty reduction. Literally millions of people will be eternally grateful to you personally. You can earn your place in history on an epic scale. It is up to you. God Speed.

NOTES

1 Philippe Sands 2005, *Lawless World: America and the Making and Breaking of Global Rules*, London: Allen Lane. Najam, A. 2002, 'Financing sustainable development: Crises of legitimacy', *Progress in Development Studies*, **2**(2): 153–160.

2 The 2000 Allan Meltzer Commission of the US Congress as part of Congressional due diligence with respect to the IMF and World Bank.

3 Vallette, J., D. Wysham and N. Martinez 2004, *A Wrong Turn from Rio: The World Bank's Road to Climate Catastrophe*, Washington, DC: The Sustainable Energy and Economy Network, International Policy Institute and the Transnational Institute.

4 'Broken promises: How the World Bank Group policies and practice fail to protect forests and forest peoples-rights', Washington, DC. The Rainforest Foundation, Global Witness, CDM Watch, Down-to-Earth Stewardship, Forest Peoples Program,

Environmental Defense, World Rainforest Movement 2005. Also at: www.rainforestfoundationuk.org/PDFs/Broken per cent20Promises.pdf.

5 172 of the world's 292 biggest rivers are already dammed for hydro, more if irrigation dams are included. (C. Nilsson, 2005, 'Fragmentation and flow regulation of the world's largest river systems', *Science*,) **308**: 405–408.

6 'Legal note on free, prior and informed consultation', Sr VP and Gen. Counsel, W. Bank and Gen. Counsel IFC, VP and Gen. Counsel MIGA, 2 August, 2004, (Does not distinguish between 'consultation' and 'consent'.) 'Safeguard policies: A framework for improved development effectiveness', The World Bank, 2002, ESSD & OPCS.

7 *The Dictionary of Economics* (D. Rutherford, 2000, New York) 'The Trickle-Down Theory is defined as a theory of economic development which asserts that development should follow traditional growth patterns with prosperity gradually coming to the least (sic) deprived members of society. This theory has also been applied to advanced free-market economies in which the richer members of society through their spending raise the real incomes of the poor, rather than by removal of all income inequality, which would need a "flood"' (not a trickle).

8 *Saving Lives, Buying Time: Economics of Malaria Drugs in an Age of Resistance*, Kenneth Arrow and Hellen Gelband, 2004, Washington, DC: National Academies Press.

9 Amory Lovins et al. 2005, *Winning the Oil Endgame*, Rocky Mountain Institute and www.theoilendgame.com.

10 United Nations Environment Programme, Project SWERA (Solar and Wind Energy Resource Assessment) 2005. http://swera.unep.net.

11 *Extractive Industries Review*, 2003, Washington, DC, The World Bank Group 7 vols; also eir.org. Goodland, R. 2005, 'Oil and gas pipelines: social and environmental impacts: Best practice', Washington, DC: IAIA.

12 Stockholm International Water Institute (SIWI) 2005, 'Let it reign: the new water paradigm for global food security', Final report to the UN Commission on Sustainable Development #13. Also at: siwi.org. Also see: 'Making water a part of economic development: the economic benefits of improved water management and services', available at http://www.siwi.org/.

13 World Bank, 2001, 'Livestock development strategy: Implications for rural poverty, the environment, and global food security', Washington, DC: The World Bank.

14 Carl Gustaf Lundin 2005, 'Ocean & climate change', Gland, IUCN; 'Oceanography: Anthropogenic carbon & ocean pH'; *Nature*, vol. 425 of 25 Sept. 2003. The Royal Society: Royal Society Working Group on Ocean Acidification. 'Ocean acidification due to increasing atmospheric carbon dioxide', 30 June 2005 Ref: 12/05; The Royal Society: 'Rising ocean acidity provides watertight argument for cutting carbon emissions', 8 July 2005; Boris Worm & Ransom Myers, 'Impacts of Biodiversity Loss on Ocean Ecosystem Services', *Science* 2005, 29 July; cf: Carl Safina et al. 'Ocean fish recovery: staying the course', *Science*, 2005, **309** (5735): 707–708.

15 Ellis, R. 2003, *The Empty Ocean*, Island Press and especially: Clover, C. 2005. *The End of the Line: How Overfishing is Changing the World and what we Eat*, London: Ebury Press of Random House.

16 Ten Kate, K., J. Bishop and R. Bayon 2004, *Biodiversity Offsets: Views, Experience and the Business Case*, Gland, Switzerland, World Conservation Union IUCN [and] London, Insight Investment Management. Also at: biodiversityeconomics.org/pdf.

17 Daly, H. and J.B. Cobb 1994, *For the Common Good: Redirecting the Economy Toward Community, the Environment, and a Sustainable Future*, Boston: Beacon Press; and Daly, H. and J. Farley 2004, *Ecological Economics Principles and Applications*, Washington, DC: Island Press. Jozef Ritzen 2005, *A Chance for the World Bank*, Stylus Publ; and Jeff Sachs 2005, *The End of Poverty: Economic Possibilities for our Time*, NY: Penguin Books.

18 Of the world's biggest 100 national governments + transnational corporations, a huge 76 were corporations (in 2002, the latest year for which figures are available).

19 Sir David King, KG, FRS, Chief Scientist to the HM's UK Government: 'Climate change is the most severe problem that we are facing today – more serious even than the threat of terrorism'. The academies of Britain, France, Russia, Germany, America, Japan, Italy and Canada, along with those of Brazil, China and India, combined to demand that the G8 summit: 'Acknowledge that the threat of climate change is clear and increasing.' *Fide*: Bruce Albert, President of the US National Academies, demurred (6 July 2005).
20 Andrew Marshall, Director (since 1973) of the Pentagon's Office of Net Assessment, supported by Royal Dutch/Shell's former Director of Planning, Peter Schwartz, dated 2003.
21 For example, The Bretton Woods Project, the '50 Years is Enough Campaign', the Development GAP, Oxfam, FOE, Environmental Defense, NRDC, BIC and many others. See also: 'Kept in the dark: a report on parliamentary scrutiny of the IMF and the World Bank', Action Aid International et al. April 2005, and: 'Reform of International Institutions' (World Bank, IMF, WTO) at reformwatch.net; ubunto.upc.es.
22 www.redskyatmorning.com/afterword.html. Gus Speth 2004, *Red Sky at Morning: America and the Crisis of the Global Environment*, New Haven: Yale University Press.

REFERENCES

Diamond, Jared (2005), *Collapse: How Societies Choose to Fail or Succeed*, London: Allen Lane.

Mallaby, Sebastian (2005), 'Essays–Saving the World Bank', *Foreign Affairs*, 84 (3):75.

Goodland, R.J.A. (2004), *Sustainable Development Sourcebook for the World Bank Group's Extractive Industries Review: Examining the Social and Environmental Impacts of Oil, Gas, and Mining*, Washington DC: World Bank Group, 189 p.

PART II

Using new tools and skills

7. A scenarios approach to developing ecosystem management strategies

Mark E. Bateman

The Millennium Ecosystem Assessment presents a balanced view of the interaction between modern man and ecosystems. By presenting four categories of ecosystem services (provisioning, regulating, supporting, and cultural), the Assessment captures most conceivable human interaction with ecosystems. The Assessment also does a reasonable job of presenting the interactions and relationships between the four services.

Decision-makers need not accept the Assessment's premise that all four ecosystem services are equally deserving of protection and management resources, but they should take seriously the challenge that current policy priorities do not adequately account for the interactions among ecosystem services. Decision-makers should review the implied priorities of their policies and develop appropriate ecosystem management strategies based on these reconsidered priorities.

The Millennium Assessment makes clear that provisioning services drive many of the ecosystem management issues that challenge decision-makers. Decision-makers should find ways to rein in continued expansion of provisioning services and make a priority of the maintenance and improvement of the regulating and supporting services provided by ecosystems.

The Assessment included in its evaluation of the current state of the world's ecosystems four plausible scenarios for their future state in 2050 based on divergent approaches to ecosystem management. This chapter imagines a world where an intentional blending of these divergent approaches seeks to meet the challenges of ecosystem management between now and 2050.

Presented in three parts, this chapter first evaluates the four scenarios to project a 'most likely' development path given current trends in governance as well as the Assessment's findings regarding trends in ecosystems. Second, it proposes a series of policy initiatives. These initiatives are drawn from findings of the Assessment in that many are based on policies that contributed to the successes of the four scenarios.

Ten policy initiatives are then offered, forming a catalyzing strategy for greater change in ecosystem management. The final section of the chapter looks at how the strategy might affect the state of ecosystems by 2050.

1. THE IMMEDIATE FUTURE OF ECOSYSTEM MANAGEMENT

The Millennium Ecosystem Assessment explicitly states that its four scenarios are not predictive. Instead, it calls them 'plausible' scenarios. It is not difficult to look at the world's current political and economic systems and assess the status quo based on key elements of these four scenarios. A little additional thought suggests a likely trajectory that ecosystem management may follow over the next 50 years – a 'status quo projection' (this could also be called 'business as usual'). This projection is the most plausible future given the trajectory of current public policies and approaches to governance. Since the Millennium Assessment bases the four scenarios on combinations of proactive/reactive ecosystem management and local/global systems and decision-making, an assessment of these issues identifies several elements that could define the Status Quo Projection.

While there are a few environmental issues that are tackled on a global scale, the current system is one based largely on local (national) decision-making. One can almost look at this as the environmental fallout of the Treaty of Westphalia (1648) which recognized the supremacy of national decision-making for national affairs. The primacy of a nation's right to make decisions regarding activity within its borders has a profound effect on the current state of ecosystems and on the options for new approaches to management.

Ecosystem management has historically been reactive – develop now, clean up later. Human nature suggests that decision-making will tend toward reactive. Even simple things like eating are done on a reactive basis – the feeling of hunger drives people to eat. Therefore, it should not be a surprise that policy makers pursue ecosystem management on a reactive basis – first identifying a problem, and then finding and implementing a solution. Local policy makers may adopt a policy in what they feel is a proactive manner, but what they are doing is following the example of policy makers elsewhere who adopted the same policy in reaction to a specific situation that needed a solution. For example, rules that bar development within certain distances of wetlands may seem proactive, but policy makers have adopted them because construction damaged or destroyed too many wetlands in other areas.

If the status quo is a combination of reactive management techniques and national level decision-making, it falls into the Millennium Assessment scenario 'Order from Strength.' Unfortunately, based on the Assessment's analysis, if this status quo is maintained for the next 50 years, the probable outcome is that the vast majority of the services currently provided by the world's ecosystems will be severely degraded. Further, the Assessment analysis shows that of the four scenarios developed, Order from Strength is likely to lead to the worst outcome in 2050 (see Box 7.1).

BOX 7.1 SCENARIOS EFFECT ON ECOSYSTEM SERVICES

By assigning scores to the sub-categories within each of the ecosystem services described in the MA, this author created a numerical score for each category of ecosystem services within each scenario (3 = increase in service, 2 = remains the same as in 2000, 1 = decrease in service). The aggregate scores were then converted to symbols. Scores of 1.0–1.5 were assigned a single minus sign. Scores of 1.51–1.99 were assigned a double-minus sign, 2.01–2.5 were assigned a single plus sign and 2.51–3.0 were assigned a double plus. Numerically representing the changes to ecosystem services allowed for ranking and better comparison of the aggregate outcomes of all four scenarios. Such aggregates are a blunt tool to facilitate discussion. The reader should remember that these aggregates hide detail and blur nuance. For industrialized countries, Adapting Mosaic scored higher than the other three scenarios in all but provisioning services, where it tied with TechnoGarden. For developing economies, Adapting Mosaic again scored higher than the other scenarios except for provisioning services where Global Orchestration and TechnoGarden tied for the highest score. Order from Strength had the worst average in all cases.

	Global Orchestration		Order from Strength		Adapting Mosaic		TechnoGarden	
	Reactive/Global		Reactive/Local		Proactive/Local		Proactive/Global	
	Industrialized	Developing	Industrialized	Developing	Industrialized	Developing	Industrialized	Developing
Provisioning Services (e.g. food)	+	++	–	– –	++	+	++	++
Regulating Services (for example climate)	+	– –	– –	– –	++	++	+	++

Today's status quo taken on a flat trajectory leaves the world in Order from Strength in 50 years, but systems constantly evolve, making the assumption of a flat trajectory inaccurate. What is the likely evolution of ecosystem management over the next 50 years, and how does it relate to the Millennium Assessment scenarios?

When possible, proactive decision-making is preferable to reactive (crisis management) decision-making. The barriers to making proactive choices are often the lack of ideas and aversion felt toward paying the cost of implementing protection against something that may not happen. These barriers will continue to influence the evolution of ecosystem management, but given the current concern regarding environmental issues and the extensive time that creative people are putting into thinking about problems and solutions associated with ecosystem management, it is likely that a number of proactive environmental policies will emerge. The global communications infrastructure will assist in the dissemination of these ideas, so it is probable that a number of countries will adopt proactive policies, particularly in Western Europe, some progressive developing economies (for example Costa Rica), and some US states.

While it is possible that such a shift will be made toward proactive policies, it is unlikely that there will be a significant shift toward global decision-making. While other nations can put pressure on a country, only the national decision-makers can determine the actions and policies of the country. It is national decision-makers who negotiate multilateral treaties and who can choose to leave them.[1] The emergence of regional blocs of countries does not disprove this point. The North American Free Trade Agreement (NAFTA) represents a single-issue association negotiated successfully by sovereign nations. The European Union (EU) is really the creation of a larger 'country' (almost a 'United States of Europe') for purposes of interacting with the rest of the world. Neither example represents a significant step toward a 'world government' capable of coordinating global ecosystem management.

Local decision-making with proactive ecosystem management describes the Millennium Assessment scenario 'Adapting Mosaic', the path that plausibly creates the best results in 2050. Therefore, it seems that there may be a natural tendency toward implementing this scenario. However, the Adapting Mosaic system will not fully unfold in the next 50 years. Adapting Mosaic assumes that all localities are making proactive ecosystem management decisions. Some decisions will work and some will fail, but all are aimed at managing ecosystems proactively. Over the next half century, it is more likely that some countries will work to adopt proactive management strategies while others will continue to be reactive.[2]

Therefore, the likely trajectory for ecosystem management over the next 50 years leads to a result somewhere between the scenarios of Order from Strength and Adapting Mosaic. This means that the probable outcome of the Status Quo Projection is somewhere between the Assessment's imagined worst possible outcome and its imagined best possible outcome (See Box 7.1).

2. CREATING A NEW STRATEGY

While we have predicted the likely trajectory based on the Millennium Ecosystem Assessment scenarios, there is obviously an opportunity to put ecosystem management on a different development path by creating a New Strategy. If policies can be identified to overlay on the likely trajectory, the opportunity arises to enhance strengths and correct weaknesses, improving the overall ability of the system to manage ecosystems productively. Policy makers must take care to analyze the effect on ecosystem services of policies, including those ideas presented in this chapter, before adopting them.

This chapter will construct a new strategy with ten policy initiatives in three parts. First, critical strategies will be borrowed from successful aspects of the Millennium Assessment's four scenarios, thereby incorporating the Assessment's massive research and analytical effort. Second, ways to incentivize reducing consumption, a clear driver of many of the negative findings within the Assessment, will be considered. Finally, policies will be identified that are associated with key actors, including companies and financial markets, seeking to address their fundamental role in the future of ecosystems. While no one of these policies is a 'silver bullet' that will remedy all that is wrong with the management of ecosystem services, each policy initiative improves the framework within which future decisions and policies can be pursued.

Borrowing Policies from the Scenarios

Several of the scenarios' storylines have something to recommend them and therefore something that may be appropriate to 'borrow' as a policy to create our New Strategy for managing ecosystems over the next 50 years.

Assessing economic value

As described above, the Status Quo Projection assumes an evolution of public policy over time, specifically an increase in proactive policies for ecosystem management. Because of this increase in proactive policies, it is assumed that the Status Quo Projection will exist somewhere between an Order from Strength world and an Adapting Mosaic world. However, important lessons can be learned from the TechnoGarden scenario (proactive ecosystem management combined with global decision-making) due to its heavy reliance on financial mechanisms to create its proactive policies.

These financial mechanisms include cap and trade schemes for pollutants along with other financial incentives meant to place a value on ecosystem services currently considered free. For example, the ecosystem service of climate regulation is currently thought of as 'free', and subsequently activities that degrade this service, such as burning fossil fuels that release CO_2, currently continue at no additional cost. Countries (particularly within the European Union) are now introducing systems to assign a cost to this degradation, and it can reasonably be assumed that more countries will follow suit on this topic and others.

The TechnoGarden scenario assumes that financial incentives managed on a global scale, combined with new technologies, will strike a balance that improves ecosystem management. The TechnoGarden Scenario has some specific drawbacks, however. The proactive manipulation of ecosystem management on a global scale can lead to wild swings in ecosystem performance when implementing new management policies without a full understanding of their impact. The Millennium Assessment concludes that ecosystem failures in such a highly managed (technology-based) world could be more dramatic and difficult to correct and harm more people than in other scenarios (MA, 2005a, p. 5). Policy makers should try to avoid the pitfalls of the TechnoGarden Scenario when developing market-based approaches to ecosystem management. Decision-makers need significant theoretical modeling of markets that reflect ecosystem service fluctuations to balance these risks with the benefits.

> **Policy Initiative One:** *Adopt mechanisms for determining and collecting the economic value of ecosystem services used or degraded in any economic transaction. Guard against using technology to 'over' manage ecosystems while seeking to maximize financial return based on economic incentives.*

Free trade

Pertinent lessons can also be learned from Global Orchestration (the Assessment Scenario based on a global decision-making model and reactive ecosystem management) due to its dependence on a system of global free trade and 'good governance' to improve human well-being,[3] particularly well-being supported by provisioning services. Global Orchestration generally provides a better level of provisioning services for developing economies than Adapting Mosaic (while tying with TechnoGarden for the highest score). It is obvious from the comparative information provided as part of the Millennium Assessment analysis that free trade provides a superior distribution mechanism for ecosystem services. The concept of free trade applies to goods, services, information and population migration.

Sustainability of the world's ecosystems is under great pressure from increases in population and increased per capita demand for ecosystem services. While all of the scenarios project significant population growth, Global Orchestration's is the lowest. The Millennium Assessment concludes that free trade contributes to slower population growth and higher income growth in both industrialized and developing economies. By using free trade as a critical global policy (and the free flow of information associated with it), decision-makers will take a significant step toward improving the distribution of provisioning services to all people and quite likely a step toward reducing population.

Policy makers will need additional analysis, however. While economists have studied free trade extensively, little if any of this analysis captures the effect on ecosystem services. Free trade also increases migration levels, but it is not clear how migration affects the overall distribution of people or how the distribution of people affects ecosystems themselves.[4] There is an argument that local ecosystems furnish provisioning services more efficiently. Proximity to services means the energy required to deliver those services is substantially less. Therefore, migrations that brings people and provisioning services closer together are net positive migrations (at least in terms of reduced energy demands). Migrations that increase the burden on the new ecosystem for provisioning services beyond its capacity are negative migrations, but what are the other effects of migration? What is the ability of a local ecosystem to perform its regulating and supporting functions if population patterns shift substantially?

> **Policy Initiative Two**: *Adopt free trade agreements as broadly as possible to include goods, services, information and population migration. Guard against overwhelming local ecosystems.*

Increasing data collection and communication

The Adapting Mosaic Scenario, as discussed earlier, assumes an increase in information flow, an important component to bring to our New Strategy. Pursuit of communication improvements should be a global priority. The Millennium Ecosystem Assessment is an important step in increasing the knowledge of the Earth's ecosystems. Unfortunately, the Assessment has no mechanism for keeping its initial data collection up to date. Better communication fosters continued data collection, improved trade (part of the free trade policy agenda), and improved learning. While both proactive and reactive decision-making can benefit from improved learning, proactive planning and management is difficult without improved knowledge premised on globally shared information – in other words, a world communication network.

Since much of the improvement in ecosystem management needs to take place at the micro level, communication strategies should assist in bringing more people into such a global communication/data network. While global management may have some advantages (as demonstrated in the Global Orchestration and TechnoGarden scenarios), the real understanding of what is happening to an ecosystem lies with the people who live in it. Improved communication can be utilized to transmit local understanding more broadly, not only to local populations, but also to local, regional, national, and global decision-makers.

Particularly in developing economies, one dramatic example of improvement in data collection has been associated with the adoption of cellular telephones as a primary communication device. Aid agencies, non-governmental organizations, and national decision-makers should take specific note of the powerful potential of cell phones (or any other inexpensive mobile communication technology) to improve the understanding and management of ecosystems.[5] People given this kind of tool for the first time have demonstrated an ability to innovate well beyond the ideas that national (or global) planners might have. The truly mobile nature of the technology allows for dispersion of the infrastructure to work within a local culture to meet its needs. It should also be remembered that cell phone functionality continues to expand; text messaging and Internet access have further enhanced the communication and information power available from cell phone technology.

Policy makers who have pursued communication expansion to date have not done so as part of an environmental agenda, but their policies have environmental implications. Just as cell phones and the Internet portend the expansion of democracy, they have the potential to improve environmental assessments and understanding. Widely distributed cell phones increase the opportunity to gather real time data worldwide on a variety of activities. Consider the convening power of the cell phone[6] applied to ecosystem management. Environmental non-governmental organizations (NGOs) might even consider distributing cell phones and paying for the service in exchange for regular reporting on environmental conditions. For example, recently in Argentina Greenpeace gave the local Wichi people on the Pizarro reserve mobile phones and lessons in how to text information to Greenpeace activists. The increased communication between activists and locals helped to halt the clearing and development of part of the 25000-hectare Pizarro nature reserve (Oberman, 2005).

When people are given phones by NGOs or other actors, they can not only communicate with the group, but also are free to use the phones in other ways beneficial to economic development. The possibilities for real-time ecosystem data collection and management are nearly limitless.

> **Policy Initiative Three**: *Finance infrastructure that supports mobile communication technology. Find creative ways to increase dispersion of the mobile technology across all populations. Create models for gathering, verifying, and using ecosystem data from local populations through this infrastructure.*

Energy research

TechnoGarden (the Assessment's scenario based on global decision-making and proactive ecosystem management) does well relative to the other scenarios when we compare ecosystem services. Significant goals within the TechnoGarden scenario include pursuit of energy efficiency and a 'preference for renewable energy resources and rapid technology change' (MA, 2005a: p. 8). If national decision-makers were to intentionally adopt a strategy of energy research to pursue these goals, TechnoGarden shows that the world's energy outlook can improve considerably. After the development of improved energy options, the free trade system and the free flow of information favored in our New Strategy (not to mention *current* mechanisms that promote the free flow of information and ideas) will promote the adoption of energy alternatives.

Research needs to define the impact of energy on standards of living and the Millennium Development Goals better (see Schmidt-Traub and Cho's review in this volume for more information on the Goals). The Millennium Assessment identifies energy demand and energy supply as important indirect drivers for the outcomes within the four scenarios. Energy in its most common current forms degrades regulating and supporting ecosystem services. Energy is also one of the large stressors on future provisioning services. Can energy's role in meeting development goals be reduced, or is it necessary to find alternative modes of producing energy that have less impact on ecosystem services? The exact nature of the relationship between energy availability and human well-being could be an important aspect of decision-makers' priorities.

Development of alternative energy sources offers the opportunity to fuel the push for free trade and increased standard of living while reducing the negative impact on ecosystem services (successful alternative energy sources will most likely improve regulating, supporting and provisioning services; air quality, water quality and climate regulation are all likely to improve with better energy options). The technical and economic challenges of developing these alternatives make critical a better understanding of the role of energy to human well-being.

> **Policy Initiative Four**: *Finance energy research and create and allow market mechanisms that adopt successfully developed technologies.*

Incentivize Reducing Consumption

As already mentioned, all four of the Assessment's scenarios envision increased population and increased per capita consumption. This pressure on ecosystems is tremendous, and ultimately we must find mechanisms to reduce the total need for the provisioning services of ecosystems.

Without resorting to unrealistic options that amount to drastic rationing (potentially implemented as a reduction in the standard of living among industrialized populations or as a prohibition on improving the standard of living within developing economies), decision-makers need to find policies that provide people with incentives to consume less. One possibility is a shift from income-based taxation to consumption-based taxation; another is rewarding local governments for reorganizing development plans in order to pay more attention to ecosystem issues.

Consumption taxes

There is a proposal in the United States that recasts the entire tax structure to focus on consumption-based taxes.[7] The current proponents of this policy are not seeking to improve ecosystem management, but as the Millennium Assessment points out, the greatest impact on ecosystems comes from decisions made with no environmental intention. Non-environmental laws with substantial environmental outcomes are quite common. The US Controlled Substance Act of 1972, for one, makes it illegal for anyone to be in possession of a controlled substance after it has been in the possession of the person to whom it was prescribed. As a direct outflow from this legislation, the US Drug Enforcement Agency requires that a patient's prescriptions be flushed away when the person dies; any other action is illegal. Environmentalists fighting to keep controlled substances out of the water supply have come up against this law with no resolution. However, with potential unintended environmental consequences arising from non-environmental legislation come not only challenges to environmentalist movements, but also the opportunity to improve the environment with policies conceived for other reasons. Without dwelling on the specifics of any one proposal, the concept is worth further consideration.

The Millennium Ecosystem Assessment argues that the market currently does not capture the real value of ecosystem services; taxing consumption offers an opportunity to place a price directly on the use of ecosystem provisioning services. Taxes assessed as a percentage of the purchase price of a product are proportional to the cost of delivering the service, in so much as the price charged to consumers reflects the cost to produce the good. By taxing only new consumer items, the tax provides an incentive for consumers to use items longer and, when possible, to purchase used goods. In his new book, 'FutureShop', Daniel Nissanoff discusses an emerging market, what he terms the 'auction economy', which will make this type of product use easier. Along with the technological development of Internet auction and classified sites, there has been a trend toward people buying higher end goods, ones that they formerly could not afford, with a specific eye toward eventually reselling the products online. Under a consumption-based tax system, people will be pushed away from frequent-use products and toward long service and used items, easily accessible and affordable under Nissanoff's new 'auction economy'. Subsequently, businesses will have an incentive to make and market goods that are of higher quality and greater durability – touting the consumers' ability to resell or their lower tax bill over the life cycle of a product (Nissanoff, 2006).

Proponents of the consumption tax rail against the inefficiency of taxing work and wealth. They argue that in the current tax system, the only way for someone to control the amount of tax they pay is to regulate their effort at adding value to the economy. By switching incentives within the tax code, individuals have control over the amount of tax they pay by making specific and informed decisions about their consumption. Environmentalists encourage informed decision-making about consumption and should therefore embrace this opportunity to remake one of the fundamental financial mechanisms in today's society.

There are certainly a number of complex issues that would have to be considered in such a radical change of tax structure – progressivity of such a system, the tax rate at which enough revenue is generated, and so on (these issues, and other challenging issues that will emerge in a full discussion, have solutions when not treated as 'line in the sand' political issues). Many people will express concern over the implications of such a plan for the poor. Various methods of protecting the buying power of the poor for essential goods and services are possible (for example exempting certain goods from the consumption tax, providing a tax rebate on income up to the poverty level, and so on). Additionally, environmentalists may argue that such a tax applied universally to consumption does not target specific problem areas such as fossil fuels. While there may be a strong argument for policies that address problem areas, no new policy that could practically be adopted into law will be perfect. The switch in basic focus from income to consumption is so fundamental that even traditional adversaries should not pass up this opportunity to work together. This policy area offers the opportunity to bring together disparate policy perspectives in pursuit of a fundamental shift in attitudes about taxation, consumption and income.

One additional element for policy makers to keep in mind is that a consumption tax has a multiplier effect on the costs embedded in a product, that is the higher the manufacturing cost structure of the product to begin with, the higher the tax on that product. There are three policy implications derived from this. First, government subsidies that reduce the production costs, and therefore the price for the consumer, will reduce the taxes paid on that product. If part of the goal in switching to a consumption tax is to reduce the pressure on provisioning services, then the combination of reduced price associated with the original subsidy and the reduced tax burden (the multiplier on the original reduced cost) is counter productive to this goal. Particularly for economic activities that draw on ecosystem services at a higher rate than is desired, serious consideration should be given to reducing or eliminating these subsidies – especially in a consumption tax environment. The Millennium Ecosystem Assessment gives the example of agricultural subsidies causing the overuse of ecosystem services for agricultural production (MA, 2005b: p. 95) – a non-environmental policy with environmental consequences. Elimination of the subsidy would be part of a solution; a switch to a consumption tax would help it even more.

Second, as a corollary to this, if there are activities beneficial to ecosystems that the government chooses to subsidize, the consumption tax again acts as a multiplier; by reducing the cost to the company with the subsidy, the government implicitly reduces the tax for the consumer, providing two discounts and thereby encouraging environmentally-friendly behavior. These subsidies should be encouraged.

Finally, if additional costs are added to production of products and services through market mechanisms such as cap and trade initiatives to address issues like climate change, a consumption tax again acts as a multiplier on these costs when they are passed on to the consumer, driving up the retail price and discouraging consumption of these products.

The need for provisioning services to meet the demands of more people and higher incomes per capita is a recurring theme throughout the Millennium Assessment and one that creates a drag on any attempt to improve the state of ecosystems to perform their other functions. Unfortunately, it is not clear what level of reduction in the demand for provisioning services is needed to maintain or enhance ecosystems. Decision-makers need some insight into where the initial target should be set. Is the global consumption at year 2000 levels sustainable (considering the other ecosystem services)? Ignoring for a second the different needs of industrialized and developing economies, would a goal of a global static level for provisioning services be a worthwhile goal? From an ecosystem perspective, should decision-makers be looking at an overall reduction in worldwide consumption using baselines from even earlier years?

Decision-makers should also require thorough modeling of the effects of a consumption tax on ecosystem services. How much does trading of used goods reduce the need for provisioning services? Will a consumption tax encourage companies to increase quality and durability? Will consumers take full advantage of such product longevity? Similarly, introduction of this concept into developing economies may slow the rate of growth in total consumption (though arguably not the increase in standard of living if product quality is improved), but it does not stop or reverse the overall increased use of provisioning services. Again, what problems for ecosystems does this absolute growth in consumption pose?

Ecosystem-based tax distribution

By assigning an economic value (within the tax distribution system) to the ecosystem services provided by protected forestlands, local planning authorities retain planning and economic development decision making. In 1989, Brazil adopted a new formula for the distribution of taxes to local governments that specifically takes into account the amount of land set aside for ecosystem and watershed maintenance. The May study concludes that Brazilian states have increased the number of protected areas based on this tax distribution scheme (May et al., 2002: pp. 2, 16).

Questions still need to be answered regarding the overall economic and ecological benefit of designating specific areas as protected. In addition, the economic value created by distributing taxes to local governments may not match the actual economic value of the ecosystem services preserved. Certainly 'unvalued' ecosystem services compare more favorably to other economic development than they do without such a system, but the playing field may not be level.

Policy Initiative Five: *Adopt a broad consumption tax as a replacement for income taxes.*

Policy Initiative Six: *Eliminate subsidies for environmentally harmful economic activities.*

Policy Initiative Seven: *Adopt subsidies for environmentally beneficial economic activities that meet other development and policy criteria.*

Corporations and Financial Markets

Corporations are central to the functioning of the world's economies. Businesses are the primary societal mechanism that does the work of distributing goods and services and production (in the terms of the Millennium Assessment, making use of the provisioning services of ecosystems).

Currently, the vast majority of corporations and financial markets do a disservice to long-term ecosystem management issues for three primary reasons. First, as discussed above, markets do not price the economic value of many ecosystem services appropriately, and companies take advantage of these 'missing' costs to make a substantial profit. In addition, investors and companies take a very short-term view of growth, profitability, investment return and sustainability. Finally, companies give investors an incomplete picture of their operations and financial position in relation to the world economy at the ecosystem level.

Reduce business and ecosystem uncertainty

Financial valuation of ecosystem services will increase in the coming years. In fact, the Status Quo Projection described earlier assumes as much, but the exact extent of the trend is difficult to predict. In this uncertainty, companies are hesitant to be proactive in implementing the investment and business models needed to thrive in a changed financial environment. This is an ongoing risk to many companies, but can also serve as a great opportunity for those companies that are aware that the structure of business will be changing.

Businesses should be able to see this trend toward greater financial valuation, and should demand that policy makers reduce business uncertainty by making decisions sooner rather than later. Current policy differences between the European Union and the United States on climate change create very different working conditions and pressures for business. Despite the current opposition to specific policies addressing climate change at the federal level in America, individual states are creating new policy environments. This strongly suggests that the policy environment will be changing in a dramatic fashion, but when and where is currently a significant business uncertainty. Opportunity and economic growth exist in technology shifts, and much of the coming change to ecosystem management will involve these shifts. This will be a challenge to some business models and industries more than others, but the economy as a whole should not suffer, since new companies and new business models will be engines for growth.

> **Policy Initiative Eight:** *Adopt business models that take into account new ecosystem management policy initiatives.*

Lengthening time horizons

Investors' fixation on short-term gain drives corporate hesitation to address long-term challenges. Companies are often hesitant to invest in new technologies that will lower their ecological footprint because the financial return will be too slow. Corporate valuations are theoretically calculations of the present and future value of assets and earnings discounted to today's currencies. Unfortunately, the speculative nature of the markets and the emphasis on capital gains (share appreciation), rather than profitability (demonstrated to investors through dividends), distorts efficient economic decision-making and long-term planning and investment.

The historic preferential tax status of capital gains over dividends has driven much of this kind of thinking. If tax policy expresses any preference between capital gains and dividends (which it would not with the adoption of a consumption tax as discussed above), it should be in favor of dividends. Such a change in emphasis offers an opportunity to lengthen time horizons in corporate decision-making, specifically promoting technology investments that will decrease resource use and pollution emissions.

A second factor driving short-term decision-making is the reliance of corporate boards on stock option plans as compensation for executives and the boards themselves. Many view stock options as a way of aligning the interests of management and boards with the investors for which they work and whom they represent. However, as recent financial scandals suggest, managers can manipulate the stock price through short-term thinking and questionable financial management. Companies should base incentive compensation on long-term profitability measures, not short-term share price appreciation. Corporate governance advocates are pursuing remedies for this financial disconnect, and those concerned with ecosystem management should seek to be part of these same conversations. Company management, boards of directors and investors must lengthen their time horizons by finding – and correcting – policies that promote short-term thinking. Managers looking at the long term will be more inclined to evaluate the position of the company relative to ecosystem performance and adjust accordingly. These changes will act as a catalyst for the change of perspectives among investors.

> **Policy Initiative Nine:** *Adopt long-term profitability policies such as tax preferences for dividends and compensation plans based on long-term profits.*

Increasing disclosure

The exact nature of the relationship between any given corporation and the ecosystems in which it operates is not well known.

While the theory of capitalism assumes perfect information, today's capital markets offer anything but. Investors currently labor in an era of incomplete information even though properly functioning capital markets should have an incentive to provide those with capital (investors) accurate information about those seeking capital (corporations). Since companies interact with ecosystems, both transforming ecosystems and being transformed by them, investors would benefit from a full account of these relationships.

Stock exchanges may be in the best position to represent shareholder interests on this matter. Exchanges manage listing requirements for companies to trade their shares and could require more complete disclosure of environmental information. The Johannesburg Securities Exchange (JSE) has considered a requirement that all companies publish a report based on the guidelines of the Global Reporting Initiative (GRI).[8] All stock exchanges should add rules that mandate the uniform disclosure of corporate interactions with ecosystems. GRI's guidelines may not be the right combination of factors on which to base ecosystem management (and many of the guidelines have little to do with ecosystems), but the GRI guidelines are widely accepted and understood as a starting point for consistent sustainability reporting.

Similar listing requirements adopted globally would fundamentally change the range of information available to investors and therefore have the potential of altering some of the basic assumptions currently driving capital markets. Even though exchanges only cover a small percentage of all business enterprises in the world, they are still an important starting place. In addition, disclosure by these companies of their entire operation, including their supply chains, would bring a larger percentage of business into the fold. This is not a short-term solution for any ecosystem management issue. An increase in information (even if instantaneous) will require time before it affects the knowledge base. Any eventual changes in assumptions driving investment decisions would not overthrow the capitalist system. Rather, they would change the information guiding its invisible hand.

Policy Initiative Ten: *Mandate comprehensive, comparable sustainability reporting by companies.*

3. THE WORLD IN 2050

Based on the Status Quo Projection and the ten Policy Initiatives presented here, what would the world look like in 2050?

New Strategy Policy Initiatives:

1. Adopt mechanisms for determining and collecting the economic value of ecosystem services used or degraded in any economic transaction. Guard against using technology to over-manage ecosystems while seeking to maximize financial return based on economic incentives.
2. Adopt free trade agreements as broadly as possible to include goods, services and population migration. Guard against overwhelming local ecosystems.
3. Finance infrastructure that supports mobile communication technology. Find creative ways to increase dispersion of the mobile technology across all populations.
4. Finance energy research and allow market mechanisms to adopt successfully developed technologies.
5. Adopt a broad consumption tax as a replacement for income taxes.
6. Eliminate subsidies for environmentally harmful economic activities.
7. Adopt subsidies for environmentally beneficial economic activities that meet other development and policy criteria.
8. Adopt business models that take into account new ecosystem management policy initiatives.
9. Adopt long-term profitability policies such as tax preferences for dividends and compensation plans based on long-term profits.
10. Mandate comprehensive, comparable sustainability reporting by companies.

Under the New Strategy, in 2050...

Nation states continue to be the basic unit of world organization, and national governments make primary decisions regarding the parameters of ecosystem management, both directly and indirectly. Some governments pursue proactive policies. Most of the rest are reactive, while a few ignore problems all together. Most of the world observes a large free trade zone for most products (Policy No. 2). This approach has improved distribution of goods, services and information, thereby raising the standard of living, particularly in developing economies over the last 50 years. This increase in standard of living and increased migration rates has slowed world population growth.

Corporations continue to be primary actors within the parameters established by national governments. Investors, regulators, national decision-makers and companies themselves have more information regarding the state of ecosystems and the demands placed on them largely because stock exchanges mandate comprehensive annual disclosure of sustainability information by companies (Policy No. 10). Global communication improvements (Policy No. 3), specifically cellular telephone technology and wireless infrastructure in developing economies, facilitate the free trade system and expand access to real-time data on ecosystem performance. Worldwide knowledge of disparate ecosystems, including upstream/downstream ecosystem relationships, improves as local communities begin to take more responsibility for their impact on ecosystems. These local communities widely communicate their observations, ideas, and challenges regarding their ecosystem. This massive increase in information improves understanding of ecosystems and allows development of targeted policy innovations.

Company perspectives are more long term. Corporate boards adopt longer-term, profit-focused incentive plans providing a framework within which a better understanding of the relationship between companies and ecosystems might flourish. Governments also modify tax structures for investors to favor dividends over capital gains (Policy No. 9) to encourage longer-term investments. This allows corporate managers to account for the environmental costs in decision-making. Both initiatives create virtuous feedback loops for companies, managers, investors and the environment.

Companies are also increasingly bold in adopting business models that take advantage of the new reality of ecosystem management (Policy No. 8), specifically including new market mechanisms for valuing ecosystem services (Policy No. 1). Initiatives are wide and varied, particularly in developing economies.

The global energy mix is very different from that in 2000. While total energy consumption is higher (as well as per capita consumption), overall fossil fuel consumption is lower. A number of countries with large energy appetites invest significantly in developing energy alternatives (Policy No. 4). Their motivations are varied; some pursue these initiatives for environmental reasons, some for national security reasons. Some saw the economic potential of developing technologies to sell to other countries. Some start the research process later, after they saw their traditional energy industries declining. The free trade economic system operating over much of the globe spreads successful new technologies quickly.

Other consumption continues to increase (total and per capita), with population growth and standard of living increases outstripping conservation efforts. In an attempt to counter this, a number of industrialized economies have adopted consumption taxes to replace the old-style income taxes (Policy No. 5). This causes a decrease in the growth of consumption, but standard of living measures increase with product quality improvements. Developing economies recognize the benefits later than industrialized economies and are just beginning to consider the consumption tax alternative.

To capitalize further on the change in consumption taxes, many governments eliminate subsidies for extractive industries and farming (Policy No. 6). In addition, subsidies for alternative energy development, technology conversion, environmental–industry development, and ecosystem preservation and restoration (associated with the eco-tourism industry) increase (Policy No. 7).

The globalization of trade and communication has reduced the cultural diversity of world populations. In this process, some knowledge of the unique benefits of local ecosystems has been lost.

Local decision-making, with a move toward a proactive focus, has improved the local and regional regulating services of ecosystems. The focus on energy alternatives and the free spread of new information, technologies and policies creates a rudimentary global system for problems like climate change. The significant decrease in fossil fuel use also helps to tackle the climate issue.

Unfortunately, a new analysis of the trends and conditions of ecosystems and their services shows that society has not achieved long-term sustainability. A graph of the level of ecosystem services still shows a downward slope for several of the services. The rate of decline is not as severe as in 2000, so it can be concluded that the system for managing ecosystems in 2050 is more stable than it was in 2000, but there is still work to do.

NOTES

1 The United States and Australia each negotiated and signed the Kyoto Climate Change accord but did not ratify the treaty demonstrating the difficulty democracies can have in international affairs. In addition, the US departure from the Anti-ballistic Missile Treaty demonstrates that treaties are not ironclad perpetual deals.
2 In certain cases, even reactive management would be an improvement to the current state of play regarding environmental issues. Reactive management means that decision-makers are at least recognizing and attempting to solve problems.
3 The MA includes sub-categories of human well-being, including material well-being, health, security, social relations, and freedom of choice (MA, 2005a: p. 9).
4 Some US environmentalists are now opposing immigration to the US because US residents consume too much, and new residents significantly increase their consumption patterns once arriving in the US.
5 The MA cited 'experimentation with monitoring and analysis to learn more sustainable management methods' as one way of achieving 'active adaptive ecosystem management' (MA, 2005a: p. 16). Cell phones are one type of technology that could assist in the monitoring of ecosystems.
6 In the Philippines, cell phones made possible two 'people power' revolutions that toppled sitting presidents by convening protesters in large numbers.
7 See www.fairtax.org for one version of this proposal.
8 The King II report on corporate governance in South Africa suggested such a requirement. The JSE encourages compliance with all of the King II report, but does not require adherence to all of its recommendations, including the GRI reporting requirement.

REFERENCES

May, P., F. Neto, V. Denardin and W. Loureiro (2002), 'Using fiscal instruments to encourage conservation: Municipal responses to the "ecological" value-added tax in Paraná and Minas Gerais, Brazil', in S. Pagiola, J. Bishop and N. Landell-Mills (eds), *Selling Forest Environmental Services: Market-Based Mechanisms for Conservation and Development*, London: Earthscan.

Millennium Ecosystem Assessment (MA) (2005a), *Ecosystems and Human Well-Being: Scenarios, Volume 2*, Washington, DC: Island Press.

Millennium Ecosystem Assessment (MA) (2005b), *Ecosystems and Human Well-being: Synthesis*, Washington, DC: Island Press.

Nissanoff, D. (2006), *FutureShop: How the New Auction Culture Will Revolutionize the Way We Buy, Sell, and Get the Things We Really Want*, New York: Penguin Press.

Oberman, J. (2005), 'SMS mobilizes to demobilize rainforest destruction', *Personal Democracy Forum*, 2 November, online at: www.personaldemocracy.com/node/756.

8. The implications of interdisciplinary scientific assessments for environmental governance

Richard B. Norgaard*

Through the Millennium Ecosystem Assessment, some 1360 scientists combined their different types of specialized knowledge to take a global look at the effects on human well-being of human-driven ecosystem change. Beyond an impressively detailed, well-organized, and beautifully written report on the human predicament, two very important things emerged. First, many scientists shifted their perspective from a disciplinary to an interdisciplinary one, with many shifting from a local to a global perspective as well. This change allowed scientists to see phenomena which they had previously viewed narrowly – from the indirect drivers of human population growth, urbanization and globalization, and rising material consumption to the direct drivers of land conversion, nutrient loading, and invasive species – and now see them in aggregate as pervasive, global interactive drivers of ecosystem change. Second, many scientists involved in the Millennium Assessment acquired the ability to work collectively to develop deeper, richer understandings of the complex ways that human and ecological systems interact across ecosystems and geographic scales. A large number of individual scientists broadened their perspectives, and a significant number of individual scientists also learned how to think together to combine their strengths with each other. These two outcomes of the Millennium Assessment have significant policy implications.

The primary policy implication of the first outcome, the broadening of perspective among a larger number of scientists, helps move the global environment and development policy debate toward global controls on critical drivers. Traditionally, ecosystem management has been largely reactive, acting only as problems emerge, or preventive, by taking steps to conserve ecosystems by setting aside protected areas. Global accords have largely been in response to outcomes, rather than drivers, in different types of ecosystems including desertification in arid regions and deforestation in the wet tropics. The Montreal Protocol and recent efforts to curb greenhouse gases address global problems with fairly clear direct drivers. The Millennium Assessment shifts ecosystem and economic development policy into a relatively new and contentious area by also highlighting indirect drivers. For example, most people see that the increase in numbers of invasive species is linked to increasing levels of globalization, but the conventional policy response has been to think about ways of strengthening particular ways of reducing particular types of invasions rather than to question globalization, the indirect driver of invasions and ecosystem change. Similarly, the Millennium Assessment begins to address the problem of 'overconsumption' by rich people, but there is relatively little literature to assess on this topic, let alone existing policies whose effectiveness can be assessed. It seems likely that subsequent assessments will improve upon the framework of the Millennium Assessment with its emphasis on both indirect and direct drivers and continue to move the global environment and development policy debate toward both global and systemic management.

This chapter, however, focuses on the implications of the second outcome, the emergence of a collective ability among many of the participants to combine their respective strengths to analyze complex problems across ecosystems and geographic scales more deeply. This ability is not something that can be reduced to basic principles and written up so that other scientists and the public at large can accept it as a basis for policy. Rather, it is an acquired skill of those who were working within the group. For example, with respect to ecology and economics, two key disciplines in the Assessment, the understanding reached does not simply build up from well established basic economics and basic ecology. An economist who learns a little ecology, or an ecologist who learns a little economics, cannot simply see that the arguments and findings of the assessment are logically tight, empirically well grounded, and should be accepted by all as a basis for policy. In short, the Assessment demonstrates that understanding the complexity and critical importance of our systemic interactions with the environment can best be understood through a collective, discursive process of scientists learning together and then applying their newly acquired ability for collective thinking to broad clusters of issues. As a demonstration that scientists with widely different disciplinary expertise and geographically dispersed experiential knowledge can combine their strengths into a new collective ability to understanding the whole, the Millennium Ecosystem Assessment was a resounding success.

Within this overall context of success, the scientists attempted to review the ecosystem policy and management literature and to derive general policy and management responses. Assessing responses, however, proved much more difficult than assessing current conditions and trends or projecting future scenarios. One reason for this difficulty was that the scientists who developed the ability to think together were working on conditions and trends, scenarios and sub-global assessments. The deeper understanding that they were able to generate amongst themselves did not reduce to transferable general principles for the other teams of scientists assessing policy responses. Further, the policy literature that was assessed largely assumed that generalized principles actually did exist, yet it was also clear that this policy literature rarely engaged the degree of complexity and interactions over different ecosystems and scales that were addressed in the Millennium Assessment. Additionally, to the extent that some of the understanding about appropriate responses could be transferred, the transfer is difficult because the existing 'receivers' of transferable principles in environment and development agencies to whom Millennium Assessment scientists were trying to communicate overall are trained in separate sciences, reside in separate agencies, and work on specific aspects of interconnected problems. Integrated principles are not easily received by scientists, managers and policy makers whose institutional environment reinforces the fragmentation of science or to those whose working environments do not allow them to actually implement integrated policy and management principles.

For these reasons, scientists and society as a whole must radically rethink how science relates to governance and begin to restructure science and governance in accordance not simply with new understandings but with the processes by which new understandings of complex systems are being acquired. This chapter helps lay the groundwork for that radical rethinking by presenting an analysis of the Millennium Assessment process and how its approach does not conform to historic beliefs about science and about how science informs governance.

Most people working in both the fields of environment and of development broadly understand that reductionism is failing us. Reality cannot be reduced into separate parts that are then studied by separate scientists whose findings are implemented by separate agencies. The critiques of reductionism and of the structure of governance are not new. The Assessment's process, however, gives new substance to the critique. The success of this process lends legitimacy to collective approaches to understanding complexity and gives new insights into how science and governance can be reconfigured.

Since the complexity of social and natural system interactions cannot be reduced to simple principles to inform the exercise of governance by separate bureaucratic agencies, collective processes of learning need to be fully and broadly integrated into society's understanding of governance. This chapter complements the argument for a significant transition toward discursive or deliberative democracies that go beyond greater citizen participation to build shared values, and emphasizes the construction of knowledge within science and across the body politic (Dryzek, 1990 and 2000; Ozawa, 1991). This is not an argument against reductionist science or specialized management agencies, but rather that their relative importance in our understanding of science and governance should be diminished. Collective learning processes, such as the Millennium Assessment process, need to be dramatically increased. Most importantly, these methods need to be institutionalized in a way that they are ongoing and able to inform new, more integrated governance.

To some extent, the processes of science and policy-making have been moving steadily in the direction of more deliberation across disciplines, within inter-agency task forces, and through citizen participation. At the same time, we are in a period of great backlash (Ehrlich and Ehrlich, 1996) where special interests are using older, narrower beliefs about science and governance (Gough, 2003) to attack the new, not to replace them with the old, but rather to replace rational governance with raw power politics (Gelbspan, 1997; 2004). Thus, it is important to step back and see the big picture before developing recommendations about science and environmental governance.

This is a radical argument. It does not argue for scattered adjustments out on the branches of governance today. Rather, this analysis moves toward the roots of the tree by focusing on basic questions about science and governance. The first part of this chapter is a brief exploration of historic beliefs about science and discussion of how these beliefs are embedded in the structure and processes of environmental governance today. In Section II, an overview of the MA is provided followed by an explanation of how its portrayal of the complications of reality and the process of coming to this portrayal illustrate a promising new way of doing science. This section also shows how the Millennium Assessment process compounds current confusions about science and governance. In Section III, it is argued that many of the reasons ecosystems are changing in undesirable ways can be traced to the historic structure and processes of science and progressive governance. In Section IV, the process by which the participants in the MA acquired their understanding of the complexity of ecosystem change is elaborated on, noting how this is quite different from the historic beliefs. In Section V, links are made between the

Assessment's process and its implications for governance. Lastly, some practical steps are suggested to initiate the needed changes.

1. OUTDATED UNDERSTANDINGS OF SCIENCE EMBEDDED IN PROGRESSIVE GOVERNANCE

Prior to the nineteenth century, knowledge fell into two broad categories: natural philosophy and moral philosophy. During the nineteenth century, thinking became more specialized, with the separate disciplines of physics, chemistry, geology and biology arising in the natural sciences, and economics, sociology, and so on defining the social sciences. In effect, there was a transition from philosophy to what we call science today. Dividing, reducing or structuring the world into distinct separate realms of learning and research was key to the early success of science. The structure of science became mirrored in progressive governance as scientists began to play an increasing role in government during the end of the nineteenth century to further progress. The same divided-world format of the scholarly disciplines structured agencies. Thus in the United States, for example, there were schools of agriculture and a Department of Agriculture, schools of forestry and a Forest Service, departments of geology and a Geological Survey, departments of fish and wildlife and the Fish and Wildlife Service. Dam builders and irrigation engineers went into the Bureau of Reclamation, sanitary engineers into the Public Health Service. This division made a great deal of sense at the time, but it also meant that foresters, dam builders, fisheries biologists, and public health experts were intellectually and professionally pretty much out of touch with each other once they picked an area of study in college.

Western people early adopted the belief that science is about robust predictions. Understanding the mechanics of the solar system preceded successful thinking about the complexities of life. Newton preceded Darwin. As a consequence, a modern belief arose that good science was predictive, even precisely so. The solar system has a few small wobbles that turn out to be very important to life on earth over long periods which have taken a while to explain precisely, but overall the system is relatively simple and is now well understood. Following this belief about the nature of science, the US government has argued in the courts that climate science lacks legitimacy because its predictions are vague. Had the life sciences of complex systems blossomed before the physics of relatively simple systems, less emphasis would probably be put on prediction as a characteristic of science. The American pragmatist-philosopher John Dewey also argues that the 'quest for certainty' lies deep in a desire of the western psyche to rationalize action (1929). In any case, it is now known that much good science is not predictive in the ways originally thought for science. Nevertheless, business-as-usual interests can still place the burden on environmental and development agencies of being able to fully predict the outcomes of a policy change before initiating it, even while no one can predict the outcomes of staying the present course.

Newton's success with the laws of motion and the subsequent development by chemists and physicists of the laws of thermodynamics also embedded the hope that everything in nature could be eventually explained using a relatively small number of universally true principles. While the life sciences are increasingly being held together by evolutionary theory, no universal principles comparable to the laws of physics have arisen in biology. The social sciences also have nothing comparable. The success of the physical sciences, however, still influences the public's understanding of the nature of science. Principles for ecosystem management that can be applied in different combinations across the diverse lands managed by federal governments are still naively sought.

Historic beliefs about science and progressive governance have brought us to a difficult situation today. The situation is vividly illustrated by the dilemmas of federal land management in the western United States (Norgaard, 2003). Progressive agencies tried to manage forests and rangelands scientifically before the science was adequate. Due to bad science, large predators were hunted excessively, rangeland was over-grazed, too many dams were built given the flow of the Colorado River and the value of fish on the Columbia and Snake Rivers, combustible material accumulated in forests due to excessive fire suppression, and inappropriate logging and road building left fish-spawning beds covered with silt. These were not simply failures of imperfect or incomplete knowledge, but were also due to the selective focus of each of the agencies and the specialized scientists staffing them. Interagency planning teams helped broaden the process, as did public input into environmental assessments, but, in the end, planning teams and the citizen participants did not have the management authority to fix the process. Problems also arise when directives from Washington, DC did not fit the contexts of particular places. Here is the conflict between universal principles and the differences in people, economies and ecosystems in particular contexts. These gaps between how good science was supposed to result in good land management and the actual outcomes, given how specialized science and agencies actually worked, opened the possibilities for power politics: for timber companies, local communities and urban-based environmentalists to argue for their preferred outcomes.

This partial explanation for why ecosystems are mismanaged – beliefs about reality, science and the design of institutions – applies as well to international difficulties with environment and development. Global environmental accords – the conventions on climate change, biodiversity, desertification and wetlands – and the agencies that work to implement them are narrowly focused, rarely able to join their arguments when their concerns are complementary, and sometimes work at odds with each other. Environmental accords and their respective implementing agencies certainly work at odds with accords and agencies addressing economic development. The call for sustainable development certainly has brought greater realization to the need to bring development economists and conservation biologists together, for example, but ultimately single-focus agencies are budgeted to pursue their focus. They are quick to return to their narrow mandate when efforts to cooperate with other agencies to solve the complex nature of problems prove more difficult than they expected. Of course, their expectations, like the founding of the agencies themselves, arose under the premises about the nature of reality and of science as well.

The implementation of global environmental accords at the local level has also proven difficult, just as federal regulations have been in the United States, because the global accords cannot take into account all of the local differences that exist at the time, let alone those that arise over time. There will always be conflicts between global interests in saving elephants, for example, and the local interests of not having elephants stampeding through villages. The responsibilities of sovereign nations both to their own people and to the international environmental accords to which they have agreed, however, are especially difficult to reconcile when flexibility is constrained. The central issue is how to expand flexibility in a scientifically legitimate way. The Millennium Assessment process demonstrated that it is possible to come to more complex, contextualized understandings.

This is not to say that the tensions between specialized agencies and solving complex problems or the tensions between global and local interests can magically disappear. The ability to respond to problems, however, has been excessively weighted towards continued specialization and global rules because of the dominance of earlier mindsets about the nature of reality and science.

II. THE PROCESS OF ACHIEVING COLLECTIVE UNDERSTANDING

Scientists participating in the Millennium Assessment developed a systemic understanding of ecosystem change across spatial scales using a new style of doing science. This new approach reflects the underlying complexity of the interactions between social and environmental systems. Familiarity among participating scientists with the patterns of thinking and established knowledge in many different disciplines, as well as their cultural and experiential knowledge, is necessary to make the process work toward understanding the whole system. There is no overarching conceptual model that links the separate, multiple disciplinary knowledges together, let alone the cultural and experiential knowledge. For the Millennium Assessment, simulation models were developed to help explore very general future scenarios, but these models used only a small fraction of the knowledge of the Assessment's participants. The systemic understanding had to be built and was held together through shared learning among the participants. Ecologists, economists, sociologists, hydrologists and silviculturalists all had to learn to think a little more like each other. Each participant stretched their thinking in several new directions and learned how to link processes and particular facts into a bigger picture. No one learned how to think like all of the participants together; no one absorbed all of the facts known to everyone else. No one was able to put that much effort into the process. Sufficient linkages, however, were made in understanding among the participants so that different combinations within the overall group could collectively explore various intricacies of different aspects of the complex interactions between social and ecological systems. A few participants also began to see this more complete picture at another level of abstraction. In short, the Millennium Assessment connected the multiple disciplinary perspectives and understandings largely through an intensive shared learning process.

This shared learning occurred through the many interactions of its 700 participating scientists in meetings around the world, email exchanges and international conference calls. These intellectual exchanges were all structured around writing particular parts of the multiple chapters underlying the synthesis document, but insights from many of the conversations never fit into the final document. Not all participants learned how to do this; indeed some refused and dropped out of the process early. On the other hand, those who saw the process through to the end and worked on the synthesis documents had a sense of shared success, a touch of elation as they explored the interconnections between issues together. A number of the scientists were eventually able to think together in an integrated manner. Those who reached this collective interactive understanding had a new sense of accomplishment. This is one of the Assessment's major accomplishments, for it demonstrated that it is possible for scientists from diverse positions to come together, address complex problems, and agree that different answers were appropriate for different situations. It is for this reason that this chapter emphasizes how the process built a team that could address complex, contextually dependent questions rather than a team that could reveal general findings, especially appropriate responses for development and ecosystem governance.

III. EPISTEMOLOGICAL CHALLENGES

The Millennium Assessment was not a simple review of the data and literature on ecosystem change and its implications for human well-being compiled on a global scale. Rather, the participating scientists had to undertake new thinking that required considerable learning. Most importantly, much of that new thinking had to be done collaboratively. There are many reasons why the participating scientists had to work as hard as they did.

The first challenge arose in the assessment of current conditions within broad areas such as drylands, wetlands, coastal zones and forests, which makes up much of the first volume of the Assessment. The review process took as much time and effort as it did because the scientists were assessing the significance of individual studies about specific areas. Each individual study emphasized particular aspects of particular ecological and social systems in particular places. The individual studies were undertaken at particular scales with particular drivers of change being more important than others, depending on the case study. The analyses being reviewed were conducted independently and did not measure the same outcomes, address the same driving forces, or use the same underlying descriptive model. Ecological studies emphasized particular species or processes; few ecologists paid much attention to social system dynamics. Similarly, the studies by social scientists that emphasized people were light on ecological concerns. Thus the information base being assessed had considerable disconnected information for which connections had to be made by the assessors. The situation was additionally problematic since some of the conclusions of the individual studies seemed contradictory because the studies had used different models or emphasized different variables.

There was not, however, simply a problem of getting assumptions and variables matched and working together. The actual information used in the separate analyses reviewed had real differences because ecological systems and social systems, even within the general ecosystem categories of the assessment, were different. Hydrologists, for example, came to the Millennium Assessment familiar with the findings of studies in their own regions with respect to the relationships between upstream vegetation and downstream flows. They were prepared to extend these findings to other areas. They were surprised to discover, through interactions with members of their chapter's assessment team and through external reviews, that very different relations seemed to hold in environments that were only slightly different. Social scientists tended to be better prepared for the subtleties of cultural, social, economic and political context, but their prior awareness did not make it easier to synthesize these rich contextual understandings into a global assessment. Participating scientists also learned that the definition of a word is important beyond its scientific correctness: the definitions of words are embedded in the establishment and funding of institutions. The definitions of hunger, water quality and desertification all have serious political and economic ramifications.

The scientists were trying to get a 'snapshot' of the current condition of social and ecological system interaction, but the drivers that had led to current conditions were different in different regions. Some scientists could see how the current conditions of social and ecological system interactions may be comparable in two studies, while other scientists were more impressed by the fact that the driving forces of ecosystem change, the 'arrows', were shooting in different directions. From this incomplete and contradictory information base, the scientists were given the task of assessing the significance of an aggregate of particular studies for describing the relations between people and nature within broadly generalized types of systems.

Some ecologists came to the Assessment aware that local ecosystem dynamics can be understood as operating within a hierarchy of ecosystems, with some things being very local, others regional, and still others global. Within this framing, there are short- and long-term phenomena, sometimes seen in cyclic terms, which inform the temporal understanding of ecologists. This hierarchical framing of how things happen over space and time within ecosystems needed to be expanded by including social systems. Unfortunately the ecologists who best understood this productive framing did not have a very sophisticated understanding of social systems, and few social scientists thought of space and time in this way at all. Economists, on the other hand, have very strong ideas about how economic development takes place over time, and ecological phenomena frequently had to be reinterpreted in order to fit with economists' temporal way of thinking.

Working together, scientists were able to realize that the changes they were seeing at their field sites were also occurring globally. Of course most already knew this, but being with scientists who were personally experiencing the changes first hand around the globe shifted their understanding from abstract knowledge to personal knowledge. Particular instances of ecosystem change, indeed the emergence of whole new ecosystems, with which individual scientists were familiar, merged into a larger understanding of how the globe is rapidly changing overall.

The most significant challenge, however, was that most of the case studies from the ecological literature had too few links to economic systems, let alone policy or cultural factors, while most of the case studies in the social science literature had too few links to ecological phenomena. And so the contesting framings of space and time were embedded with the practical problem that the needed linkages between social and natural systems from the respective academic literatures simply were rarely available in the appropriate form. The nature of disciplinary thinking and their formal models is that they simplify or ignore systemic connections beyond the focus of the discipline, so pieces of information taken from the separate disciplines do not fit together by the nature of the way science is structured. Participating scientists had to take information gleaned through the particular ways of analysis available in the literature and imperfectly 'attach' it to a new way of thinking to assess current conditions.

In summary, undertaking the assessment of current conditions alone required the scientists who participated most successfully to adjust their way of thinking significantly throughout the Assessment process. This adjustment required a change in how they thought about systems and forced them to find ways to attach imperfect data to interpret the conditions. Most of the discussions between scientists working on the Assessment were centered on how to resolve these difficulties. As a given team of scientists settled in on one way of resolving the difficulties, they would discover that other teams of scientists were choosing other ways to resolve them. The ways of thinking may have been sufficiently compatible, but the points being emphasized in the different parts of the Assessment were not proving complementary. Regardless, some resolution had to take place between teams.

The process of reframing and collaboratively thinking through a problem together worked especially well among those doing sub-global assessments. These assessments focused on a particular area and problem and then moved to larger framings to identify the regional and global drivers. By focusing on a specific area, these team members were less plagued by discrepancies in studies between areas and the challenge of making statements across economic and social systems. The scientists doing the sub-global assessments also tended to be younger, more interdisciplinary, and hence more adept at changing frameworks compared to the scientists doing the global assessments where stature in their fields weighed more heavily in the selection process. Participants in the global assessments rarely knew each other before they were put together in teams, whereas with the sub-global assessments, there was often a core of the participants who were already working together who then selected others to work with them. This reduced the time required to build trust and learn how to communicate between participants.

Though the global and sub-global working groups were different, the important thing is that a significant number of the scientists participating made these shifts in thinking. To a large extent, this change entailed becoming comfortable with irresolvable ambiguity and letting it be. This does not mean ambiguity was sealed off and forgotten; in fact, it was quite the contrary. Scientists who accepted ambiguity were able to be more active in the whole process of working toward consistency where possible between the different parts of the Assessment. These scientists were also better prepared to participate in the synthesis of the multiple chapters into shorter, more general documents for the public and different types of decision-makers. These were also the scientists who were most enthusiastic about the Millennium Assessment because it helped them see the complex interactions of people and the environment in a deeper way. Hence there was considerable excitement among these individuals as the process came to a close.

The excitement is not well portrayed in the final documents. The sense of discovery was palpable as eclectic thinkers discussed the complexities of our interactions with the environment, but this was an experience of the participants, not a conclusion that translated into text. It was a shared realization that while what any one person knew was limited, there could be a deeper understanding of complex problems by thinking them through collaboratively.

Those who assessed the policy response options did not experience a comparable sense of rising excitement as they came to the end of their assessment. Indeed, for some, myself included, there was a tendency to descend into a modest despair. The scientists assessing response options, mostly social scientists, were not seeing the parts and the whole in a new way, let alone across spatial and temporal scales. They were not developing policy responses to fit the hierarchical understanding of the other participants. They did not seem to have the opportunity, or even the freedom, for analogous learning and flexibility in their thinking. Surely one problem was that the Millennium Assessment schedule did not allow a sufficient number of those assessing response options to participate in the discovery of the nature of the problem and then think of responses. Rather, they had to identify responses starting at the same time as those assessing conditions and trends and scenarios in order to meet the demands of the four-year schedule. There are a variety of other reasons why the sense of accomplishment was less, but the primary reason in my judgment is that the literature on policy responses and institutions being assessed was rooted in the earlier, narrower concepts of the nature of systems and science.

IV. TOWARD RECONFIGURING SCIENCE AND GOVERNANCE

It is difficult for an individual scholar to argue for particular changes in science and governance after elaborating on the virtues of collective understanding. Surely this is a complex problem that needs to be attended to by a diverse group of scholars. Further, I have authored a book arguing for deliberate experimentation and acceptance of what proves fit rather than striving for optimal 'intelligent design' by human reason. Thus, the following proposals are humbly put forward as points for discussion and elaboration rather than as answers.

Research and Education on Deliberative Processes of Knowing

New ways of relating science to governance cannot simply be grafted onto the old philosophical underpinnings of science and governance. This analysis has emphasized the mismatch between our emerging understandings of the complexity of reality, the ways scientists have come to understand this complexity, and the way science connects to politics, policy and management. Improvements in development and environmental governance must be based on a widespread understanding and acceptance of the richer understandings existing among a few scientists and policy makers. This shift in understanding could come about quite quickly because it reduces the elite nature of scientific understanding, which is a source of tension, even divisiveness, in societies espousing the egalitarian ideal of one person, one vote. Deliberative processes open up scientific understanding to experiential knowledge through public participation.

More research is needed on how scientists collaborate and how the processes of collective understanding can be improved (Norgaard and Baer, 2005a and b; Lele and Norgaard, 2005). The process can probably be accelerated and made less expensive than it has been for climate science and for ecosystem change. Better ways of thinking about complex phenomena across space and time need to be developed in order to design better institutions of governance to give us better feedback on how our actions play out locally and globally. Clear thinking is needed on how to avoid the pitfalls of arriving at a false consensus in the shared learning process between scientists. The richer understandings are needed within the community of scientists as well as among the public. More complete understandings of how science works need to become a part of the training of scientists. This richer understanding also needs to be taught to some degree at every level of education, and deliberate efforts need to be made to extend this understanding to the public at large.

In the United States, and to lesser extents in other parts of the world, the need to enrich the understanding of how science works comes at a time when the nature of reality, how scientists understand, and the role of science in governance are being challenged by some with fundamentalist beliefs. This is a period when scientists are challenged to explain science regardless of the challenges of understanding the complexities of ecosystem change. There is considerable danger that because of the multiple assaults on science and assaults on the roles scientists play in governance, scientists will simply defend the earlier, simpler beliefs about how scientists understand. Even if the case for a richer understanding of reality, science and governance can be brought into the discourse of scientists, they are likely to argue that the seriousness of the current situation requires them to take the safer route of defending earlier understandings previously accepted by the public.

Increase Publicly-driven and Publicly-funded Assessment Science

Good governance needs to be based on the big picture. Deliberate efforts need to be made to ensure that the scientific community generates the broad understanding needed. Greater public funding of science is needed to bring together significant numbers of scientists from diverse disciplines and user groups to assess energy futures, the potential and pitfalls of genetic engineering, the likely impacts of nanotechnology, and other big issues before us. Policy makers need to help frame and participate in these studies in order to understand how they themselves can work with complex issues. This increased effort should not come at the expense of existing ways of doing science.

Promote Deliberative Democracy

Citizen participation in development policy and environmental management increased dramatically beginning in the 1970s, and was given another boost in connection with efforts to design and implement sustainable development. A rapid rise of non-governmental organizations (NGOs) occurred during the 1980s that helped facilitate public expression. Lay knowledge is being combined with scientific knowledge (Stern, 2005) and indigenous knowledge into scientific deliberation and planning processes (see, for example: Emery, 2000). Participation, however, has also been adversarial more frequently than a contribution to a joint learning process, especially in the case of environmental impact analyses in the United States. Participation has been used disingenuously by development agencies and NGOs alike, and more recently by corporatist NGOs as well. With the rise of globalization, popular participation often has turned to protest. One of the reasons for the mixed outcomes of participation is that the roles of deliberative processes in science and their extension to citizens' participation and our concept of democracy and an informed electorate need to be broadly accepted.

Some political philosophers stress the advantages of a more deliberative democracy (Dryzek, 1990 and 2000; Fishkin, 1991); others emphasize stronger community bases for democracy (Etzioni, 1993; Sandel, 1996). Though their philosophical roots are somewhat different, both deliberative democracy and communitarianism stress the interdependence between our private and public lives. Both address the failure of relying on expertise apart from the polity, the intertwining of values and understanding of reality, and how common understanding of this intertwining can be built through democratic processes. The deliberative learning and trust-building processes of the Millennium Assessment are consistent with deliberative democracy and communitarianism. Participation in the Assessment, however, was almost limited to scientists engaged in research or in the policy process. The challenge will be to implant the learning, trust-building, and community-forming processes into policy processes themselves. Simultaneously, the same learning, trust-building, and community-forming processes need to be built into our understandings and practices of democracy to politically address the complexities of development and environment.

Agency Responsibility in a Complex World

At whatever scale environmental and development problems are considered, the problems are complex. They are not as clearly defined as the responsibilities of the agencies we have set up to work on them. One problem is that there are clear gaps in the existing structure such that responsibility cannot even be exercised. This seems especially true with respect to the ability to take responsibility for integration at the biome to continental scale (see Rajan's discussion in this volume). Clearly, new agencies are needed to fill in some of the biggest gaps, but the far more challenging issue will be to define responsibility and effective agency behavior in the complex, heterogeneous milieu of environment and development. Heretofore, responsible behavior has been judged based on its consistency with respect to the agency's mandate, objective scientific knowledge, and across applications to particular situations. A major quandary among the Millennium Assessment scientists assessing appropriate policy and management responses centered on the problems of the difficulties of being consistent across applications while acknowledging the particulars of local ecosystems and cultures. This analysis raises the additional but related problems with using scientific criteria to judge responsible behavior when the understanding of complex systems can best be collectively reached through a discursive process rather than deduced from basic scientific principles. Clearly one of the policy implications of the Assessment process is that a sense of understanding and trust will have to be developed so that agencies can have more leeway in how they respond to particular situations themselves and how they combine their efforts with other agencies. For the US, this means a move away from distrust of and litigious relationships with environmental and developmental agencies. Of course, the agencies will also have to earn this new trust. Fortunately, the proposed shift toward discursive science and democracy complements the shift in understanding and building of community and trust that is necessary.

V. CONCLUSIONS

There is no way around the difficulties of having to work on particular aspects of complex systems, and as individuals this is typically about all that we can do. At the same time, some scientists and those who implement science through policy and management must coordinate with others so that unwanted systemic effects are avoided. Similarly, there is no way around the difficulties of working across scales, from the local to the global and back again, without dealing with contradictions between grand global goals and locally lived lives. Dilemmas and ambiguities will always be with us. It is the nature of real social and ecological systems that everything is connected to everything else in very complex ways.

The problem is that earlier, narrow concepts about the nature of science still dominate and have structured our social organization such that our efforts to coordinate our understanding and adapt it to the problems at hand are always seriously constrained. We accepted the concept of private property and put the burden of proof on those who sought the public good. We allowed science to fragment into disciplines and gave the disciplines the primary responsibility in universities to judge the styles and quality of scientific work. We established agencies to promote particular ends using particular parts of science. In short, we built a world around the belief that complex systems could be divided into parts. These separate institutions and agencies channel the bulk of the resources not only of science and how it relates to economic development but also how it relates to biological conservation and speaking for the poor. Similarly, we have instituted international agreements and designed international institutions as if solutions could be global. We try to work through our imperfect organizational structures to solve the problems that result in part from fragmentation and false beliefs about the broad applicability of economic, social and environmental mandates. We know what the problems are. Nevertheless, both how hard we try and the acceptance of what we learn are still limited by the dominance of earlier beliefs about the nature of social and environmental systems, about the nature of science, and hence about the nature of the organizational structure built around those beliefs.

The Millennium Ecosystem Assessment offered an unusual opportunity for a significant number of scientists to work outside the conventional fragmented structures of science and its implementation. The work was difficult and time consuming, but most of the participants pursued it until the end. The Assessment process went a long way toward developing the capability among the participants to comprehensively address the complexity of social and ecological system interactions at multiple scales. It also went a long way toward developing the trust necessary to facilitate treating apparently similar situations, such as balancing economic needs with pollution control or species conservation, differently in different situations.

The primary political and policy implication of the Millennium Assessment is that we need to substantially increase our understanding of how we collectively assess complex problems and then extend this understanding to all of the scientific community, policy makers and politicians, and the public at large. Then we need to reconstruct our institutions of governance to match our new richer understanding of the complexity of our interactions with our environment and the collective processes by which we can understand this complexity.

The new structure needs to exist in parallel to the existing structure that connects scientists in universities and research institutes to implementing agencies, but the processes within that structure need to emulate the systemic learning and trust building of the Millennium Assessment rather than the fragmented and fragmenting approach of the existing connections between science and implementation.

This will require that scientists themselves acknowledge that there are multiple ways of doing science, including ways that are more open to indigenous and experiential knowledge. It will also require that those economic interests that benefit from the existing structure of knowledge fragmentation and scale specificity and its implementation acknowledge that the future of their children and the economic system that will support them will require such a countervailing approach with a sufficiently supportive structure.

In short, this is a radical proposal for substantial change. The proposal stems from the process of the Millennium Assessment rather than its conclusions. Scientists working on global environmental and developmental problems typically find the conclusions of the Assessment neither surprising nor even very novel. What is both novel and surprising is how well the participants in the process shifted their understanding. The political and policy implications are in the assessment process itself.

NOTE

* This chapter builds on earlier work under an NSF grant on understanding biocomplexity (NSF SES-0119875). As a participant-observer of the Millennium Ecosystem Assessment, I was in conversation with many of those who established the underlying framework of the assessment in the early months. In the final fifteen months, I served as a chapter review editor, both helping the authors respond to the review comments of other participants and another 600 external reviewers and assuring that they responded adequately to maintain the scientific integrity of the report. I was a review editor on 10 of the 60 original chapters. In the process, I also ended up becoming an author on one of the chapters and participated as a chapter review editor on most of the syntheses chapters.

While my PhD is in neoclassical economics, I have spent three decades working with natural scientists on ecological issues in developed and developing countries.

REFERENCES

Dewey, J. (1929), *The Quest for Certainty: A Study of the Relation of Knowledge and Action*, New York: G.P. Putnam's Sons.

Dryzek, J.S. (1990), *Discursive Democracy: Politics, Policy, and Political Science*, Cambridge, UK: Cambridge University Press.

Dryzek, J.S. (2000), *Deliberative Democracy and Beyond: Liberals, Critics, Contestations*, Oxford, UK: Oxford University Press.

Ehrlich, P.R. and A.H. Ehrlich (1996), *Betrayal of Science and Reason: How Anti-Environmental Rhetoric Threatens Our Future*, Washington, DC: Island Press.

Emery, A.R. (2000), *Guidelines: Integrating Indigenous Knowledge in Project Planning and Implementation*, Washington, DC: World Bank and Hull, Canada: Canadian International Development Agency.

Etzioni, A. (1993), *The Spirit of Community: The Reinvention of American Society*, New York: Simon and Schuster.

Fishkin, J.S. (1991), *Democracy and Deliberation: New Directions for Democratic Reform*, New Haven, CT: Yale University Press.

Gelbspan, R. (1997), *The Heat is On: The High Stakes Battle Over Earth's Threatened Climate*, Reading, MA: Addison-Wesley.

Gelbspan, Ross (2004), *Boiling Point*, New York: Basic Books.

Gough, M. (2003), *Politicizing Science: The Alchemy of Policymaking*, Stanford, CA: Hoover Institution Press.

Lele, S. and R.B. Norgaard (2005), 'Practicing interdisciplinarity', *BioScience*, **55** (11), 967–75.

Norgaard, R.B. (2003), 'Which we, whose wisdom, whither the mountain west?', *BioScience*, **53** (3), 278–81.

Norgaard, R.B. and P. Baer (2005a), 'Collectively seeing complex systems: The nature of the problem', *BioScience*, **55** (11), 953–60.

Norgaard, R.B. and P. Baer (2005b), 'Collectively seeing climate change: The limits of formal models', *BioScience*, **55** (11), 961–66.

Ozawa, C.P. (1991), *Recasting Science: Consensual Procedures in Public Policy Making*, Boulder, CO: Westview Press.

Sandel, Michael J. (1996), *Democracy's Discontent: America in Search of a Public Philosophy*, Cambridge, MA: Harvard University Press.

Stern, P. (2005), 'Deliberative methods for understanding environmental systems', *BioScience*, **55** (11), 976–82.

9. Looking back and ahead in ecosystems: Reflections on the lessons of the Millennium Ecosystem Assessment

Anthony C. Janetos

1. INTRODUCTION

The Millennium Ecosystem Assessment represents the considered findings of nearly 1400 scientists from around the world, who were asked to evaluate the state of knowledge in the trends and conditions of ecosystems and the services they provide, provide some indication of potential future conditions, and evaluate the successes and failures of possible policy responses. They have produced a growing number of assessment volumes, including the publication of the original intellectual framework, an overall synthesis document, four topical volumes, and a variety of special products for the multilateral environmental conventions.

There are many novel features of the Millennium Ecosystem Assessment and many others that are familiar to participants of other scientific assessments that have been performed over the past decades to address significant environmental issues, especially the scientific assessments of the Intergovernmental Panel on Climate Change. This chapter focuses on several topics that merit additional thought:

- the credibility and legitimacy of the Assessment
- the political targets/clientele of the Assessment
- what those actors are being asked to do
- the ability of the Assessment to evaluate policy choices
- the long-term sustainability of the Assessment
- recommendations for policy actions

2. CREDIBILITY AND LEGITIMACY OF THE MILLENNIUM ECOSYSTEM ASSESSMENT

The scientific credibility and political legitimacy of the Assessment are crucial to any possible success it might have in informing and influencing policy debates over the management of natural resources and ecosystems. If either were to be compromised in any significant way, its overall impact would certainly suffer. This is not to say that one can expect that there will be no mistakes in the Assessment. There will be people backing particular interests who will question some aspects of participation in the Assessment (more on this below). But the processes by which participants are chosen, by which their work has been reviewed, and by which their products are both solicited and received must have sufficient transparency, be sensitive to a large number of competing issues, and be done with sufficient skill and expertise to lend substance to the overall results.

Nearly 1400 scientists participated in the Assessment, with another large number asked to provide independent reviews. The participation was purposely balanced between natural and social scientists at the coordinating lead author level, and this was also additionally attempted at lower levels of contributors and authors. The Board and Assessment Panel were also carefully balanced in terms of academic and experiential backgrounds. In this respect, the Assessment was more carefully constructed for intellectual balance than many other scientific assessments. In part, this simply recognizes that managing ecosystems is inherently an interdisciplinary enterprise. But the Assessment has gone substantially farther to acknowledge this interdisciplinary nature than any other international scientific assessment.

In addition, the Assessment purposely attempted to balance the geographical distribution of scientists who were involved, especially the balance between the developed and less developed countries. As with other international scientific assessments, this was carefully done, but it remains an enormously difficult task. It may well have been a more difficult task for the Assessment because of the importance of experiential knowledge and direct research experience with local systems in ecological science.

The process for scientific review of the results of the various chapter teams built on the model of the IPCC, and was implemented well. The chapters went through several rounds of expert and governmental review, and review monitors were required to sign off on the responses to reviews before the chapters were accepted for publication. Authors were clearly responsible for the intellectual content of the chapters.

The concept of ecosystem services encompasses both goods with prices traded in markets and unpriced services that for the most part are not traded. Being able to conduct the assessment required experts in the various disciplines to learn to communicate with each other and to make joint judgments about concepts and the strength of knowledge in different topical areas (see Norgaard's review in this volume for more information). This is not an easy task, and although the Assessment went a long way towards fostering the kind of interdisciplinary discussions that will be productive, the process will need to continue in order to mature. There were certainly difficulties, many of which may have been due to problems in interdisciplinary communication. Early joint meetings of the scenarios and conditions and trends groups, for example, were clearly characterized by misleading confidence in model simulations, for example statements by some economists at the beginning of the process that one could make excellent predictions of GDP over 20- and 30-year timeframes, but could not do so for net primary productivity. These kinds of difficulties are inherent in the beginning stages of most interdisciplinary collaborations, and one key characteristic of the Assessment and its review process is that it was successful at preventing such obviously nonsensical statements from being published. In my view, this is due both to the repeated interactions of the various author teams, which served to improve interdisciplinary and personal communications, and review by outside experts, and especially by the review monitors.

The bigger challenge for the Assessment was ensuring that the authors who were involved were indeed the world's experts. For the most part, the Assessment was successful in this respect. But the world of scientific assessments has become extraordinarily crowded, and there is now competition to get the best people involved, keep them involved, and at the same time ensure that they can continue to produce excellent research. This will be a longer-term strategic challenge for follow-on activities, such as the launching of national-level integrated ecosystem assessments.

Invitations for participation were sent out broadly, and there appeared to be reasonable financial support available for travel and expenses that would allow scientists from many countries (or from non-prosperous institutions within prosperous countries) to participate as much as they were able. As has become the practice in international scientific assessments, the Assessment did not provide salary relief for its participants, which did create issues for some participants, but did not seriously compromise its overall legitimacy. The resulting scientific credibility of the Assessment will be a major strength of the process, and its policy clients will recognize this. Some questions might be raised about national and regional representation of different countries' scientific communities in the Assessment, and about the ability of national governments to participate. For the most part, any issues about the participation of different countries' scientists can be answered effectively.

The one caveat about this last statement is that the unpaid time of participation in the Assessment process may have hindered full participation from some non-governmental organization experts, since many operate on a pay-as-you-go basis. Non-governmental organizations support a larger community of experts with respect to ecosystems, natural resources and conservation than on such issues as climate change and stratospheric ozone depletion. Ensuring that this expertise can be fully utilized will be an issue that needs to be addressed in future follow-on activities of the Assessment.

3. POLITICAL TARGETS AND CLIENTELE OF THE MILLENNIUM ECOSYSTEM ASSESSMENT

The main political targets/clientele of the Assessment were the international environmental conventions, specifically the Conventions on Biological Diversity Wetlands Desertification and Migratory Species, and to a lesser extent, the Framework Convention on Climate Change. The UN Environment Programme and to some extent the UN Development Programme were also viewed to be important clients, as well as the UN Commission on Sustainable Development, and other UN-based activities. Indeed, some of the earliest targets for obtaining political backing for the Assessment were the ecologically-oriented conventions, UNEP, FAO, UNDP, and associated working groups among those agencies. When Kofi Annan called for the creation of the MA in the UN General Assembly, the consensus among most of the Assessment participants was that this was a major milestone for the political foundation of the overall process. The participation of key individuals from the World Bank and from FAO was viewed to be critical towards ensuring that those institutions were fully informed stakeholders in the Assessment.

A key difference between the Millennium Ecosystem Assessment and other scientific assessments was the attempted inclusion of representatives from the business community at the very beginning of the process. In particular, the Millennium Ecosystem Assessment's Board of Directors made a concerted and successful effort to get representatives of very large corporations to participate, based on the argument that increased knowledge about ecosystem goods and services was a key ingredient in corporate strategies for sustainability, and that there was no a priori reason to expect that the corporate community was necessarily going to be viewed as being in an adversarial role, as has been the case in climate change issues, for example. The involvement of the corporate community was more successful at the Board level than at the participant level. Natural resource consultants were involved in a few chapters, and there were business participants in the drafting team for the synthesis report that was targeted at the business community, but the overall numbers were extremely small compared to academic scientists. This is to be expected to some degree, but it is an area in which future scientific assessments would probably be wise to make an even greater investment of time and effort. Such an effort would be worthwhile particularly if arguments about ecosystem services are going to be pursued. If these services have values that can be translated into economic terms, then the involvement of appropriate expertise and experience from the private sector will provide a valuable contribution.

The involvement of national governments has been more problematic for the Assessment. Early in the process, the organizers recognized that national governments needed to be involved in some way. For an assessment that was not called for as an intrinsic part of any of the pre-existing multilateral environmental agreements, what that involvement might be was more difficult to articulate. The effort to involve national governments has had three components: one is the involvement of scientists supported by their national institutions. As reviewed above, the international participation of scientists was broad, but as is typical in the natural sciences, relatively few of the participating scientists had direct, tangible ties to their country's policy-making apparatus. This effort has turned out to be a relatively weak approach to ensuring strong national participation. Of particular concern in this respect has been the relative inability of the Assessment to generate support for its scientific participants within the national governments of many of its participants. Other international scientific assessments, and in particular the IPCC and Stratospheric Ozone Assessments, have been quite successful in getting the scientific funding community to share in the actual costs of participation for many scientists around the world. The Assessment was striking in its relative inability to help generate such funding. Some of this may have been due to its relative strength in securing private foundation and intergovernmental funding. But many national governments, and in particular the US, did not provide the additional support that many, if not most, scientists need in order to participate in such activities.

The second component of ensuring national participation was the selection and conducting of the sub-global assessments. The rationale for the sub-global assessments was primarily that there was a need to investigate how issues concerning the sustainability of ecosystem services were playing out in many different parts of the world. In some cases, this fitted nicely with existing national agendas. For example, the Chinese government viewed an ecosystem assessment of the western provinces as being enormously important to its plans to boost economic productivity in the west in environmentally sustainable ways. In other countries, it was not so clear that there was a strong national commitment to the sub-global assessments, which led to dramatic disparities such as in funding levels and the level of expertise applied to the assessments. There were many questions early in the process about how sub-global assessments were to be proposed, whether there was funding available, how they were meant to fit within the context of the rest of the Assessment, and so on, which probably reduced the total number of sub-global assessments. The sub-global assessments turned out to be a reasonable way of involving national governments in the Assessment process, but only when the assessment agenda and national agendas already coincided. At present there is no evidence that this method of involvement has been or was successful in changing national policy trajectories.

The third component of ensuring national participation was the involvement of national governments in the peer review process. In this respect, the Assessment operated like the IPCC in inviting comments from national governments during peer review, in addition to inviting comments from other experts in the appropriate fields. There are two levels at which to address this issue in the review process: the hierarchical level of governmental participation and the level of experience/expertise and familiarity with the Assessment process of the government reviewers.

Compared to other governmental reviews in international scientific assessments, the Assessment process was handled at a much lower bureaucratic level, and resulted in less elicitation of experts' views than might have been expected. In other processes, the governments often attempt to solicit the reviews of the external academic expert community in addition to the views of their own internal scientific communities. The other major international scientific assessments also demand the attention of senior science policy officials within national governments, and a well-organized and quite public process of soliciting comments for inclusion in the official response. The Assessment had very limited success in achieving similar levels of engagement of governments in the review process. For the most part, senior science policymakers were not heavily involved, and opinions outside the government were not solicited in a broad, open, public manner. The resulting reviews were often misinformed as to the nature, intent and procedures of the Assessment, and were less helpful as a rule than such reviews have been in other scientific assessments.

One of the main challenges that the MA has faced in terms of its institutional visibility is that much of the research it summarizes is context-specific. Many of the examples and data on ecosystems and their services are quite case specific, and although the Assessment attempted to be as broad and encompassing as possible, it was impossible to use case examples from every nation. One is left to wonder, for example, how relevant the United States feels that the emphasis on desertification, deforestation and global trends in agricultural expansion are for its own national interests, where none of these issues are pre-eminent domestic environmental challenges.

A second challenge for the Assessment in reaching its intended audiences is that the issues it analyzed are so interconnected with other energy, environmental and development issues that there is not a single home, even within national governments, for actions to be taken. For example, to take action to address just some of the issues raised regarding improving the sustainability of agriculture might involve every ministry from Agriculture to Environment to Finance. While this level of interconnectedness reflects the real-world complexity of ecosystem services, it can also mean that there is no institution whose interests are so strongly connected to the issues raised in the Assessment that it would automatically benefit or be threatened by the findings and therefore would become a focal point for policy actions.

A particular case in point has been the low level of visibility of the Assessment process and findings in the United States. In striking contrast to the IPCC or the Ozone Assessment, the major science agencies in the United States involved in the Assessment have only been participants at low staff levels. The coordination of the United States government reviews was handled by a combination of the State Department and the Office of Science and Technology Policy, but unlike the IPCC, the US review involved government scientists almost exclusively, and was handled entirely at a staff level with very little involvement or notice of senior officials. The domestic natural resource policy apparatus of the United States government has been almost entirely disconnected from the Assessment process, and the international natural resource policy apparatus only slightly less so. This has resulted in very low visibility for

Millennium Ecosystem Assessment products and procedures within the US, with several of the major science agencies almost entirely unaware of the Assessment at high levels, and with major policy agencies only slightly aware. There is some growing understanding of the Assessment in a few of the relevant committees in Congress, but without a strong institutional home within an agency or a prominent non-advocacy non-governmental organization, the Assessment will have a difficult time ensuring that its messages are heard effectively within the United States public policy making or funding apparatus.

It is difficult to obtain sufficient insight into the policy or bureaucratic institutions of other large countries to be confident about whether the low visibility within the national government of the US is also true in other places. It is generally not true in China, but that may be because of the large Chinese investment in money, staff and prestige in conducting their own sub-global assessment of the western provinces. However, the Assessment has become a visible political influence in the scientific establishments of the United Kingdom and France. Its visibility in other western European countries, South America and many places in Asia remains to be seen.

The buy-in of scientific establishments is an important issue for the political vitality and effectiveness of the Assessment's messages to governments. There have traditionally been two sorts of influences that other international scientific assessments have had. One is direct influence on current policy debates and discussions. The other is the longer-range influence on research agendas, which then can influence political and policy discussions on relevant issues. Other international scientific assessments have generally had both types of influences, even though they, like the Assessment, have generally enforced a rather strict internal rule of being relevant to policy without being prescriptive about policy. In the climate debate, for example, the analyses by IPCC of emissions scenarios on the consequences of climate change or the role of aerosols in climate forcing, have an enormous impact on the tenor and often on the specifics of the policy debate. IPCC and the ozone assessment also have large effects on the future research agendas of participating agencies and countries, even though they have not always featured chapters focusing specifically on research needs. The sections of these assessments focused on major uncertainties are widely understood to form the agenda for the next round of research funding by the relevant scientific communities.

The Assessment, with only a few exceptions, may be more successful in influencing research agendas than in influencing public policy agendas. This is primarily due to the scientific prominence of the individuals that were involved, especially in its leadership positions, and to the relative inability of the Assessment to gain visibility within national governments. But influencing scientific research agendas in this way, and establishing that the sort of interdisciplinary analyses that the Assessment sponsored are indeed helpful, may yet lead to policy impacts at the sub-global, or even sub-national, level as more scientists attempt similar tasks.

4. WHAT ARE POLITICAL ACTORS BEING ASKED TO DO?

Part of the Millennium Ecosystem Assessment's initial lack of visibility in public policy may result from the general lack of clarity about what the various political actors and clienteles are actually being asked to do. The scientific part of the Assessment is primarily about defining what the issues are with respect to the status, trends and sustainability of ecosystem services and why various policy communities ought to care about the results. This step is necessary and enormously valuable, but it does raise the question of what the various actors should then do about the findings and results.

One answer to this question is to sponsor more and better research to understand the variety of issues raised by the Assessment. For example, it has proven enormously difficult to demonstrate clearly that reductions in ecosystem services are also associated with reductions in human well-being on local or regional scales, although there are good reasons for thinking that this might be so. Some of the difficulty stems from the obvious – in prosperous regions and countries, there can be substitutes, such as water treatment plants, to account for the ecosystem services that are lost, in this case the loss in water purification services due to losses in wetland extent. Some difficulty also stems, though, from not understanding the specific relationships between services and well-being, not having kept track of the appropriate data, or not having a common set of definitions of well-being and services that would enable the discussion to be held clearly in the first place. Political actors could undertake steps to do research to understand conditions and trends better, or affect the proximate or ultimate drivers of changes in these conditions, but specific options for action are difficult to define.

At least some of the sub-global assessments have a clearer view of what their political and policy constituencies are really asking, and therefore what the Assessment findings are in effect asking them to do. In the western China sub-global assessment, for example, there is a very clear policy dimension, which all the participants understood – the intent of the Chinese government to boost economic productivity in the western provinces. However, the Chinese government is aware that the environments there are arid and fragile, and they need to figure out how to make economic gains, while protecting the natural resources on which the gains ultimately depend. The southern Africa regional assessment is also looking at fundamental issues of the sustainability of ecosystem services in a part of the world that is struggling with its own environmental sustainability. This context is also well understood by all the participants.

The more global parts of the Assessment are limited to some extent because they have such broad mandates. For example, it is difficult to know what lessons to draw with respect to all forest ecosystems for all actors involved even though forests are enormously well-studied, and there are well-known issues in the forest sector of importance to public policy.

The final difficulty with being clear about what the different potential political actors should do relates to the initial strategy chosen by the Assessment to focus on the multilateral environmental conventions as key target audiences. This choice has had strengths, such as enabling the Assessment to avoid any strong opinions and points of view of any one country that could have adversely impacted the Assessment's objectivity or affected the topics on which it chose to do analysis. For example, because the Assessment did not have to deal specifically with the topic of 'avoided deforestation' or deal with the policy ramifications of forest definitions as they affect the implementation of Kyoto, it actually had more political space within which to review and assess the state of our knowledge on changes in forested ecosystems. The Assessment's broad interdisciplinary approach also enabled it to address multiple issues associated with changes in forested ecosystems, such as measuring changes in both the regulatory services of carbon sequestration and the provisioning services of timber production.

This political freedom and broadness of areas for discussion comes at one important cost. The conventions are not themselves potent political actors. It is the countries that have the political power and authority to make public policy decisions over their own resources. For example, the Convention on Biodiversity explicitly recognizes each country's sovereign control over its own natural resources, and to date has very few, if any, binding protocols regarding the reporting of the conditions and trends in those resources, or services they provide.

The mismatch between main political audiences and authority to act for the Assessment raises the potential for individual countries to ignore its findings, unless specific actions are taken to ensure that the findings are confronted. For example, one of the main findings of the Assessment is that the capacity of many ecosystems to continue to provide services has been severely eroded over the past 50 years. Some ecosystems, like open oceans, are already showing signs of the loss of biological productivity in the form of reduced fisheries yields. Other systems, like many agro-ecosystems, continue to show high levels of services, but are undermining their capacity to continue to be productive in the future. Yet these findings do not easily translate into particular case studies or implications for particular countries, and therefore those countries that have the capacity to act do not always have the incentive to do so.

5. ABILITY OF THE MILLENNIUM ECOSYSTEM ASSESSMENT TO EVALUATE POLICY CHOICES

A major conceptual advance of the Assessment is that there are tradeoffs in ecosystem services, and a major finding is that for many decades, there have been significant enhancements of provisioning services (for example production of food) at the expense of regulating services (for example the sequestration of carbon). The Assessment shows that such large tradeoffs on global or regional scales can already be documented. It also shows that there are good conceptual grounds for believing that such tradeoffs also exist at more local scales, and these too have occasionally been documented in case studies in different parts of the world. A further finding of the Assessment is that there are tangible, albeit sometimes indirect, links between the wealth of countries and the degree to which those countries benefit from ecosystem services, although this is harder to document.

A challenge for the Assessment, however, is that once having established that tradeoffs are a major issue and that the supply of some services is declining, there is relatively little ability to evaluate policy choices that might serve to increase the reliability or the level of ecosystem services. Some of this difficulty is intrinsic to the scientific assessment processes. If in evaluating actual policy choices the Assessment was perceived to be prescriptive as to which policy was favored, the credibility of the Assessment as a scientific document could be decreased. This could be avoided by having the Assessment evaluate actual policy choices from several different perspectives and illustrate the tradeoffs between policies, rather than pick one option over another. However, for a global assessment it is infeasible to evaluate all possible policy choices or even a large fraction of available policy choices. A third challenge is that policy evaluation also has embedded within it some important value judgments and contextual information about particular national situations. It would be quite difficult for the Assessment, as a global process, to reach agreement on such values or to be specific about national situations. Consequentially the Assessment, like other international assessments, is constrained in its ability to evaluate policy choices.

This difficulty for international scientific assessments is by no means a fatal flaw. One of the strengths of the Millennium Ecosystem Assessment, as with other international assessments, is that it can provide a systematic model for how policy evaluations could be carried out by taking the concepts of tradeoffs among ecosystem services into account. Different countries will have different needs for such models, of course, but particularly the less prosperous nations may find that having an internationally accepted model for assessments can help them evaluate their own choices more carefully. Again, the example of the western Chinese sub-global assessment is instructive. That assessment benefited tremendously from the publication of the Millennium Ecosystem Assessment's conceptual framework and from the interactions with those working on other sub-global assessments even as it was being developed and implemented.

Major policy choices that will require international as well as national actions could, in principle, be catalyzed by the Assessment findings. Many of these choices have a critical time dimension: for example, as human population grows to somewhere around 8–9 billion by the middle of the century, best estimates are that food demand will roughly double. This is clearly going to put major stresses on agricultural systems, as marine systems are clearly already undergoing major problems in continuing to provide food. The technology is available for increased food productivity through more efficient use of water and fertilizers, improved crop varieties, and so forth, but the costs of either continuing to intensify current agriculture or to expand the area devoted to agriculture are relatively poorly understood.

The Assessment's method for illuminating the kind of policy choices that might be made is to use illustrative scenarios that identify different development and environmental trajectories. Such scenario-building has a long history in private industry, in national defense planning, and in other scientific assessments. But this approach can only handle a relatively small number of important policy choices, and it is difficult to assign probabilities to any individual scenario. However, as Richard Moss and

Steve Schneider have emphasized within the IPCC, assigning even subjective probabilities to different scenarios is extremely important for their overall utility and credibility. In this respect, the Millennium Ecosystem Assessment has quite similar strengths and weaknesses in its scenarios as does the IPCC.

6. LONG-TERM SUSTAINABILITY AND EFFECTIVENESS OF THE MILLENNIUM ECOSYSTEM ASSESSMENT

The Assessment was begun with the anticipation that it, like the IPCC and the Ozone Assessment, would become a rolling process, that is that it would be repeated at some reasonable interval in order to track changes in the scientific understanding of ecosystem services. During the Assessment, however, it became clear over time that the goal of having a rolling process could not be achieved in the short term. This is no doubt due to many factors, among them difficulty in raising funds from private donors and the competition among many scientific assessments for the best participants. However, it is important for there to be follow-up actions to the Assessment. Three important avenues exist for such follow-ups.

One such avenue is the continued production of special reports that illuminate issues that are especially important to one or another policy community or client. The Assessment has done an excellent job of this already with special reports to the conventions. But there is much more that could be done in this regard. For example, analyzing exactly what kinds of data and models would be necessary to evaluate tradeoffs between agricultural production and other services, evaluating how marine fisheries might recover, or discovering whether invasive species are in fact causing declines in other ecosystem services might be of particular interest to political clientele. These issues have only been able to be pursued at gross levels in the Assessment, while their resolution at regional and national scales will require the sort of analysis that more detailed follow-on work could provide.

A second avenue is the targeted marketing of the findings and conceptual advances themselves. Because the main policy audiences have been the international conventions, and there has thus been relatively low visibility within national audiences, there is a great need for specific activities to enhance national visibility in potential client nations. There are at least two important reasons for doing this. One is to ensure that the findings are highlighted for countries that have the power and willingness to act to sustain ecosystem services. The second is that by engaging policy and scientific audiences directly in these countries, the Assessment may ultimately succeed in building a series of national consensuses that it has been an important and productive process and should become a rolling process. Ensuring that additional individual countries become backers of the Assessment process is absolutely critical. Not only are they the main decision-makers with respect to natural resources and ecosystem services, but they are also the only actors with enough institutional stability to ensure that the work begun so ably by the Assessment can continue.

A third is to concentrate on remedying some of the key scientific issues that appear in the Assessment, so that subsequent regional or global analyses can be done in a more sophisticated manner. This would be entirely analogous to the process that has followed each of the IPCC and stratospheric ozone assessments. It has resulted in both more advanced science, and vastly better foundations for policy making.

7. RECOMMENDATIONS FOR POLICY ACTIONS

The following recommendations for policy actions based on the findings and the process of the Millennium Ecosystem Assessment primarily cover the conduct of scientific assessments that are undertaken in response to policy concerns. The recommendations draw not only on the lessons of the Assessment itself, but on previous experiences in the IPCC, the Global Biodiversity Assessment, the US National Assessment of the Potential Consequences of Climate Variability and Change, and other such processes. They do not focus on the steps in the conduct of scientific assessments, like the need to ensure adequate representation from developing countries, the need for rigorous peer review, and the need for independence of the participants' analyses from particular policy interests and advocates, as these are all areas that are widely understood by all stakeholders as necessary. Rather, they focus on those areas in which additional thought would seem to be wise.

Ensure that National Governments have a Strong Role and Voice in Scientific Assessments

This recommendation is crucial for both long-term and short-term success of international scientific assessments, such as the Millennium Ecosystem

Assessment and the IPCC. The Millennium Ecosystem Assessment planning process recognized early on the importance of having the endorsement of international organizations, such as the various environmental conventions. The same is true for IPCC. The lack of such endorsements, for example from the Biodiversity Convention, was a major factor in ensuring the lack of reception for the Global Biodiversity Assessment. It is key to involve national governments from the beginning for two main reasons, both of which are pragmatic. One is that they are the institutions with the resources available for ensuring that there is continued investment in the assessments and in follow-up work. The international conventions and United Nations agencies are quite constrained financially, so any new assessments, no matter how focused, will almost certainly find that existing commitments and desires on the parts of these institutions make additional investments quite difficult to come by, and nearly impossible to maintain. While national governments may claim to be financially constrained, the fact of the matter is that especially for the OECD countries, there is substantially more flexibility in national budgets. There will be very little tendency to expend those budgets on new scientific assessments, however, unless the countries are involved themselves from the very beginning.

Secondly, the international conventions recognize that only individual nations have the authority to make real policy changes. This may seem obvious, but a process that does not involve the individual countries independently of convention secretariats will be much easier to dismiss. There is of course no guarantee that involving individual countries will ensure that there is policy and/or financial follow-up – but without that involvement, the countries are nearly completely free either to dismiss the findings of the assessment, to ignore them, or not to consider any serious policy measures. A corollary of this point is that while it may be necessary to have a foreign ministry as an official point of contact, there also need to be national contacts within ministries/agencies that are going to be the critical ones for follow-up activities and for funding. To use an IPCC example within the United States, although the State Department is the official point of contact for the international environmental conventions, it is a weak player in domestic policy, and the science agencies, Department of Energy and Environmental Protection Agency are far more important for domestic responses, not to mention the White House itself and its own agencies. An international scientific assessment process that primarily engaged the State Department without engaging relevant individual agencies would have little or no short- or long-term impact.

Ensure that New Scientific Assessments are Undertaken as part of an Ongoing Process

There are several reasons for this recommendation. One is that environmental problems that can be solved quickly within a few years are poor candidates for garnering enough political and financial resources for a major scientific assessment. Another is simply pragmatic. The landscape for scientific assessments is crowded. In order to ensure that one can motivate the best scientists around the world to participate, there needs to be some assurance not only that their efforts will lead to something important, but that their participation is necessary for a longer-term improvement in the use of science for informing environmental policy. There is very little appetite in many of the best potential scientific participants for one-shot efforts with the level of commitment and amount of work involved in international assessments. In the case of the Millennium Ecosystem Assessment, for example, many participants were originally entrained in the Assessment in part because they felt some assurance that there would be an ongoing assessment process. The failure of the Assessment to achieve this is problematic and may hinder the recruitment of some of these same key individuals for future efforts.

The IPCC, for example, has been very successful in ensuring that there is an ongoing process. In fact, although there is a nominal repeat time of five years for IPCC assessments, the reality has become that there is either one of the main assessments, planning for the assessments, or special reports requested by the IPCC Board going on nearly all the time. The IPCC has thus become a nearly continuous stream of activity, involving many different scientific and engineering communities. In some cases, it may not be possible in the end to ensure that a scientific assessment does become part of a longer-term process. A useful example here is the United States National Assessment of climate impacts, which failed to achieve any measure of continuity, although it had done a tremendous amount of engagement of stakeholders at national, state and local levels of government (and in some cases, the private sector). The failure to achieve any continuity has alienated some key Congressional audiences (while delighting others), but has also ensured that when national climate impact assessments begin again, they will not only have to start from scratch, but will have to make significant efforts to convince potential participants that they will be heard.

Ensure that there is Substantive Participation from the Private Sector

The Millennium Ecosystem Assessment has improved over the performance of the IPCC in this regard, having had representatives from the private sector on its Board, and thus having conduits for private sector involvement in setting follow-up policy. But even within the Millennium Ecosystem Assessment, there was very little private sector involvement in funding (other than from philanthropic foundations), and very few participants involved from the private sector. Even though there was a synthesis product targeted at the private sector, it was not planned and done until very late in the Assessment process. Some of the business participants have lamented this fact, asserting that more could have been done to get their sectors more fully involved.

There are, however, substantial difficulties in achieving this recommendation. For some environmental issues, like climate, there is often a perception that many elements of the private sector are in an adversarial relationship with other participants in the assessment, especially with governmental and non-governmental organization participants. Another potential problem is that unlike the university-based scientific community, many firms do not have the personnel or financial flexibility to assign senior scientific personnel to participate in a broad-ranging international assessment. Businesses simply have other things that they must do to enhance value for their company, and the potential benefits from an assessment are diffuse or far off in the future. There is also a challenge because there is no single business entity that one can go to for participation. Depending on the topic of the assessment, there may be particular economic sectors that one would like to see involved. For something as broad in scope as the Millennium Ecosystem Assessment, it is not immediately clear which sectors should be involved and whether trade associations are appropriate entities for involvement. Because of this, one is then left trying to secure the participation of individual key, often very large, firms. This strategy works for partial representation of the private sector, but the character of that representation is then somewhat questionable for the broader community.

Devote Resources to Analysis of Different Potential Policy Choices without Necessarily being Prescriptive

There could be a combination of post hoc analyses, analyses of current potential choices, and analyses of the potential consequences of different policy choices in the future explicated through scenario analysis. Post hoc analyses are in principle the easiest to do in a defensible way, but in practice even these have been quite difficult. The Global Biodiversity Assessment, for example, had great difficulty in persuading authors to address the question of which conservation policies and practices had had the greatest (or least) effects. In part, this was because many of the authors had professional ties to institutions that were attempting to implement very different kinds of policies, and they were reluctant to address the question overall.

The Millennium Ecosystem Assessment did not attempt very much analysis of these sorts of questions. Analyses that are immediately relevant to current policy choices are extraordinarily difficult to gain support for in a non-partisan, non-prescriptive environment. Analyses could be structured, however, to investigate the scientific dimensions of different policy choices. Interesting examples of this tactic come from the IPCC where special reports and individual analyses have attempted to examine the scientific underpinning of policy choices while stopping short of recommending one or another particular policy decision. However, the IPCC experience has also shown that such analyses can be misinterpreted by particular interest groups. For example, the IPCC Special Report on Land-Cover, Land-Use, and Land-Use Change is clear that forests and other types of ecosystems can in fact be used as carbon sinks, that the measurement issues are tractable without major advances in methods and technologies, that the accounting issues are resolvable, and that the size of the sink is large enough that it could be augmented usefully. Each of these conclusions has on occasion been misrepresented by governments and non-governmental organizations that have previously expressed strong positions on whether biological sinks should be part of an overall policy response to climate change.

Regardless of the difficulties, however, it is clear that scientific assessments can and should do more to illuminate current policy choices. For ecosystem management, this may be particularly important since many of the current policy choices about how land is used effectively foreclose future options. The opportunity costs in this area can be very high, and thus careful analyses are needed now.

Finally, the scenario analyses should continue to be part of future scientific assessments. There are many complications associated with scenarios, but they are useful tools for depicting the possible consequences of different strategic decisions about policies. For them to be even more useful tools, however, they should highlight the differences that specific choices can make. How, for example, might actually associating monetary values with ecosystem services lead to different outcomes with respect to conserving biodiversity? Or managing water? There are now beginning to be examples in the energy literature in which the revenue for utilities in the future is as much a function of how much carbon they are able to sequester in geological formations as it is a function of how much electricity they sell, depending on what assumptions are made about the cost of trading carbon. Similarly explicit studies would be extremely interesting for the follow-on steps of the Millennium Ecosystem Assessment and for future scientific assessments.

8. AUTHOR'S DISCLOSURES

In the spirit of full disclosure, the author notes here that he was involved in the Millennium Ecosystem Assessment process from near its beginning. He was part of a group of United States' scientists and federal agency personnel that was originally consulted by the World Resources Institute about the potential need for the Assessment. While a staff member of WRI, he helped promote the establishment of the Millennium Ecosystem Assessment. He participated in the Assessment as a coordinating lead author on one chapter (Chapter 28, 'Synthesis', in the *Conditions and Trends* volume), as a contributor in others, and as a principal investigator on a study of rapid changes in global land-cover that was commissioned and partially supported by the MA. The author does not claim to be completely dispassionate about the Assessment, and hopes that his observations and opinions will nevertheless be viewed as reasonably objective.

10. Policy actions to help move us towards ecosystem security

Karin M. Krchnak[1]

I. INTRODUCTION

The Millennium Ecosystem Assessment's greatest strength is the clarity with which it describes the range of ecosystem goods and services and their past and present degradation. Where the Assessment falls short is on recommending policy responses that decision-makers need to take to ensure that the world will not face the same decline in ecosystem health in the coming decades that it has over the last half of the twentieth century.

The Millennium Assessment offers startling statistics about the rapid rate of change in the world environment from water shortages to climate change to the collapse of fisheries. The problem this poses is threefold: (1) how do governments move toward integrated decision-making to slow and avert these changes and losses? (2) how do the broad range of stakeholders that are not governments participate in and contribute towards improved management of natural resources? and (3) how can new ways of thinking and innovative approaches to managing resources be encouraged to ensure the integrity of these critical environmental services in the long term?

At the international level, governments have discussed for some time the need for integrated decision-making. This verbal commitment often breaks down, however, when it comes to action. Priority-setting poses countless difficulties when political agencies are rewarded for maintaining their own turf. The Millennium Assessment in some sense provides too much information on the status of ecosystem services to the point that decision-makers may be unable to decide on how best to meet the needs of humans sustainably. For example, the Assessment notes with regard to halving the population living on less than $1 per day that 'reaching this target in the Middle East/North Africa and in sub-Saharan Africa is unlikely…' (MA, 2005c). This does not create a sense of optimism. However, the Assessment does do well in describing the trade-offs[2] between ecosystem services and human well-being. This should be useful for policy-makers as long as decision-making moves toward an integrated approach (MA, 2005a; 2005d). If sectoral decision-making remains the norm, it will be next to impossible for policy-makers to consider the full impact of trade-offs.

To move toward integrated decision-making, governments, as a whole and individually, may undertake a number of policy options, including: (1) reforming the Commission on Sustainable Development; (2) addressing population growth and consumption issues; and (3) developing integrated governance systems for ecosystem security.

Experts agree that effective management of resources can mitigate the impending threat of scarcity. Resource managers often face multifaceted challenges posed by a variety of sectors and users. Governance systems that mandate open, transparent and accountable decision-making can help ensure that decisions are based on sound science, take into account the interests of all stakeholders, and effectively allocate resources for current and future uses (MA, 2005a).[3]

The natural environment, communities and governments change over time, and thus a key element of any form of integrated resource management approach is that it be iterative. More importantly, the dynamic learning process should involve stakeholders in a meaningful way with participation, not simply consultation (MA, 2005a).[4] More open and transparent processes for involvement in decision-making processes can help identify the appropriate goals for allocation of resource use. In 1992, over 170 governments affirmed the importance of public access to information, participation and justice in decision-making in Principle 10 of the Rio Declaration on Environment and Development. Ten years later, governments reaffirmed this commitment at the World Summit on Sustainable Development. Although many governments have developed laws relating to Principle 10, implementation of these rights has been weak. Some policy options for the global community in increasing the role of non-governmental actors, from environmental groups to business, include: (1) adopting the major group model by international institutions and processes and (2) making gender equality a reality.

Startling statistics should make action urgent, but without additional information on causes, it may be difficult for policy-makers to know what response should be selected (MA, 2005b).[5] As the Millennium Assessment points out, particularly in developing countries data is often lacking (MA, 2005a).[6] In many cases, resource management decisions must be made in the absence of comprehensive information. This is where new ways of thinking and innovative approaches to resource management become important. Environmental and conservation groups could join forces with poverty, development, human rights, and other activist groups to help policy-makers understand the links, for example, between human health and polluted water and take actions that will bring multiple benefits – conserve watersheds through creating a protected area which leads to a decrease in water-related diseases among the nearby population.

As part of encouraging new ways of thinking on resource management, it is imperative that conservation and environmental groups take into account the human dimensions of ecosystem degradation. Millennium Development Goal (MDG) No.7 – Ensure Environmental Sustainability – and specifically 'integrate the principles of sustainable development into country policies and programmes; reverse loss of environmental resources' could be said to be the weakest of all the goals and also the easiest for countries to overlook (for more on links between the Assessment and MDG 7, see Schmidt-Traub and Cho's chapter in this volume). Conservation and environmental groups have the challenge of convincing policy-makers that protecting 'environmental resources'[7] will promote development. The Millennium Assessment begins this process by underlining the importance of ecosystem services.

Potential policy responses to encourage improved natural resource management include (1) recognizing adaptive management in policy frameworks and (2) establishing payment for ecosystem service (PES) fee systems. These are just two possibilities; much work remains to be done to find new and creative ways to slow and reverse the rapid loss of ecosystems. Ultimately, institutional and political focus on improving coordination, inclusion, and participation can help build the trust, experience and confidence necessary to move toward sustainable development.

II. INTEGRATING RESOURCE MANAGEMENT

To help move toward integrated decision-making, governments, as a whole and individually, may undertake a number of policy options, including: (1) reforming the Commission on Sustainable Development; (2) addressing population growth and consumption issues; and (3) developing integrated governance systems for ecosystem security. The first focuses on changes in governance at the international level to help guide countries in integrating resource management at the national and local levels. By developing policies that address population and consumption issues, the second recommendation, governments can obtain a more realistic assessment of rates of change, where change is occurring, and what policies and practices need to be put into place to be able to provide services to a growing population. The third policy option focuses on the design of comprehensive systems of governance for ecosystem security.[8] This last one is particularly challenging since developing systems of good governance requires putting into place sound political, social, economic and administrative systems.

1. Reform of the Commission on Sustainable Development

The United Nations Commission on Sustainable Development (CSD) was created to ensure effective follow-up of the United Nations Conference on Environment and Development (UNCED). The CSD was established as '*the* institution to monitor and report on implementation of the Earth Summit agreements at the local, national, regional and international levels' (UN Resolution, 1993).[9] Although world leaders adopted Agenda 21 in Rio, a global plan of action for sustainable development, implementation of the plan's over 2500 recommendations has been weak. To some extent, this is understandable; it is not realistic to expect that nations will have created the enabling legislation, the institutions and the systems to implement all of Agenda 21's recommendations when national leaders are often faced with conflicting demands. However, regardless of what level of progress individual nations make in achieving sustainable development, it remains unclear why nations have failed to support the CSD as an institution, improving it each year to ensure that it can meet its objectives.

Despite its problems, the CSD is potentially the ideal institution for helping governments move toward integrated decision-making. It offers an established structure through which dialogue on the trade-offs between ecosystem services and human well-being could occur. Some recommendations for 'reform' of the CSD include focusing on the review functions of the CSD, using the Millennium Assessment and other future assessments as baselines for discussion around ecosystem health, and structuring CSD sessions to encourage dialogue among all participants on how to integrate resource management into practice. Since ecosystems do not typically follow political boundaries, it is important that the international level be used as a platform through which to define integrated management and how it can be achieved. Regional dialogues are also important for this, but the international level provides the broadest sharing of experiences in how different regions of the globe approach integrated resource management.

The 13th Session of the CSD (April 2005) was considered by many to be a setback to the process, barely maintaining the status quo after Johannesburg. This outcome was a disappointment considering the changes that were introduced at the 11th Session of the CSD. At the 11th Session, the CSD agreed to work in seven two-year cycles, with each cycle focusing on selected thematic clusters of issues – CSD 12 and 13 focused on water, sanitation and human settlements. Moreover, delegates decided not to negotiate text at each meeting. 'Policy' sessions are to be held every other year to address means for implementation on every relevant issue, action and commitment for each topical cycle. 'Review' sessions are also to be held for each cycle. Linkages to crosscutting issues (such as poverty eradication, changing unsustainable patterns of consumption and production, and gender) are also to be addressed for every cycle. This discussion about cross-cutting issues offers a way to evaluate trade-offs between ecosystem services and human well-being; unfortunately, the CSD sessions so far have not been structured in a way to encourage discussion across these issues.

In addition to these changes, the 11th Session of the CSD reaffirmed that voluntary multi-stakeholder partnerships are a complement to, not a substitute for, intergovernmental commitments. A set of criteria and guidelines for partnerships were agreed to at the 11th Session, and the CSD was determined to hold the role of overseeing the partnerships (CSD, 2003).[10] The 11th Session embraced a new element in that a Partnerships Fair was held as well as a Learning Centre aimed at building the capacity of those attending the CSD as well as other interested persons.

Where did the breakdown occur to the point that some have described the 13th Session of the CSD as 'taking steps back in sustainable development'? There could be considerable debate over this, but one element stands out – the international community has lost sight of the original purpose of the CSD to 'monitor and report on implementation of the Earth Summit agreements at the local, national, regional and international levels'. One person has described the CSD as 'a hodge-podge of rhetoric and recycling of ideas and phrases from the past ten years'.[11]

Even in the new approach of having a 'Review' year, most delegates fell into the pattern of preparing statements and recycling language. There was little actual *review* of the conditions and trends in the 12th Session. If one visits the website of the CSD for the 12th Session, for example, there is no chart or graph to indicate how countries are doing in meeting their Agenda 21 and the Johannesburg Plan on Implementation (JPOI) commitments. Nonetheless, the review year was very useful in that much of the experience gained through the implementation of water, sanitation and human settlements projects and programs was shared in the meetings and related events.

If we accept, as the Millennium Assessment points out, that 'national governments are likely to maintain their central role in coordinating and regulating most aspects of socioeconomic development, including societies' relationships with ecosystems', then it is important that some institution be charged with tracking how countries are doing in socioeconomic development. Rather than discuss the creation of new institutions,[12] the focus should be on bringing the CSD back to its original purpose and improving the way in which governments and civil society engage in it (MA, 2005a; 2005c). The CSD sessions have an extensive body of expert and policy-maker participants, and it is disheartening to think that such a group cannot be used to advance the CSD and implementation of global sustainable development commitments. The CSD, if structured more effectively, offers a structure through which to engage multiple stakeholders (for example, governments and civil society) on defining integrated resource management and what approaches are effective.

There clearly are data gaps that make it difficult for countries and institutions like the CSD to accurately assess progress in implementing governmental commitments, but tools such as the Millennium Assessment can help to fill this void. 'By identifying gaps in data and information that prevent policy-relevant questions from being answered, the assessment can help to guide research and monitoring that may allow those questions to be answered in future assessments' (MA, 2005a). The Millennium Assessment and future assessments should be embraced as part of the CSD agenda because they provide the most current baseline of information on the state of ecosystems.

What the CSD as an institution can do now is gather governments and civil society together to analyze which policies are working and which are not in achieving sustainable development – the 'Policy' sessions of the CSD should include a *review* of policies and the obstacles governments face in working toward sustainable development. No new text should be negotiated except to the extent that governments, with input from civil society, may add targets, timetables, policy actions, funding commitments, and so on to those already agreed upon before. Moreover, CSD sessions should provide an opportunity to discuss the institutions needed to track, monitor and assist in the implementation of international agreements. For example, the 13th Session of the CSD should have focused on the institution(s) which will track the progress of countries in developing and implementing integrated water resource management plans (IWRM) as called for under the Johannesburg Plan of Implementation (JPOI).

As an example, the 12th and 13th Sessions of the CSD could have been organized along the following elements.[13]

- Governments report (via online surveys) on their progress in implementing Agenda 21 and the commitments made at the World Summit on Sustainable Development relating to water, sanitation and human settlements.
- The Millennium Assessment and similar research products are made available to CSD delegates before sessions to better understand the state of freshwater and related ecosystems.
- Plenary sessions are used to set out the major questions on integrated water resource management including how to handle water resources across countries and regions, particularly integration with coastal zone administration. In addition, the cross-cutting issues of the CSD are used to structure plenary sessions around such issues as gender equality and water scarcity as well as effective means of production and consumption.
- Governments agree upon new commitments that relate to water, sanitation and human settlements that *add* to those previously made to take effect once existing commitments have been met.
- Governments and the Major Groups[14] divide into smaller sessions to discuss policy options with the end result not being agreement on text, but rather a learning process whereby governments and civil society share obstacles to policy actions. There is discussion around reasons why policies, even when implemented, fail to achieve their original objectives (for example, policy incoherence across scales with different policies working against each other) and brainstorming on how to overcome these obstacles.
- Governments and the Major Groups share lessons learned on partnerships relating to water, sanitation and human settlements. All background information on the partnerships is placed on the CSD website, enabling the meetings to be facilitated dialogues around key questions.
- The CSD and supporting institutions enlarge the Learning Centre to allow for more training programs on methodologies, software, tool-kits related to water, sanitation and human settlements.

The CSD should not be used for negotiating different forms of the same text. Rather, the CSD sessions should be practical meetings in which the broad range of experts dedicated to sustainable development discuss the status of and challenges to implementation of Agenda 21, the JPOI and other global commitments. The 13th Session of CSD could have made progress if, for example, participants broke into groups to deliberate on the topics within the User-Friendly Matrix of the Chair's IPM Summary (CSD, 2005). Facilitated dialogues could have focused on, for example, how does one strengthen river basin organizations? Who has developed gender-sensitive indicators and how can they be used by governments, international financial institutions, civil society, and so on? What types of incentives do and do not work for adopting water-saving technologies? How does a country incorporate sanitation into integrated water resource management plans? How do countries develop the legal and regulatory systems for river basin commissions to function effectively within countries and across national boundaries?

It would be useful if organizations like the CSD and future Millennium Assessment-type evaluations were to integrate matrices of policy responses to help move the dialogue to focus on the most effective policies and the necessary conditions for their success. The Millennium Assessment policy matrix, for example, could be revised to coordinate more closely with the JPOI policy commitments.

2. Addressing Population Growth and Consumption Issues

One strength of the Millennium Assessment is its acknowledgement of the role of population growth in the decline of ecosystems (MA, 2005a; 2005b).[15] The Assessment could have been improved, however, if it had recognized that the projected population growth – approximately 9 billion people by 2050 – will create serious problems for ecosystem services long into the future (MA, 2005c).[16] Unfortunately, political sensitivities typically hold leaders back from facing population issues. Not since the 1994 International Conference on Population and Development (ICPD) have world leaders sought to address population growth as part of a global agenda.[17] At the World Summit on Sustainable Development some countries sought to undermine the ICPD language in the negotiated text. However, the fact remains that an additional 3 billion people will require water, energy, food and shelter in the coming decades which places increasing demands on natural resources. In developed countries, consumption issues play as great a role as population growth in the degradation of ecosystems. As part of their sustainable development agendas, countries need to: (1) follow through on their commitments made at ICPD, including funding of programs aimed at addressing population growth and (2) support sustainable production and consumption approaches and remove subsidies that degrade ecosystems.

Progress in achieving the Millennium Development Goals (MDGs) is slight; thus, policy-makers must face the likely scenario that ecosystems will continue to be degraded to meet human needs (for more on the MDGs, see Schmidt-Traub and Cho's chapter in this volume). The fact that the Millennium Assessment's sub-global assessments failed to mention the indirect driver of population growth only gives more reason for concern that population growth will not be adequately factored into policies on sustainable development.

The US environmental community went through a challenging period in the 1980s and 1990s over whether and how to work on population issues. While that debate should not be revisited, groups concerned about ecosystem health need to understand the role of population growth better. The Millennium Assessment's social and behavioral responses all but ignore population policies, focusing rather on consumption. It is important to face the fact that an additional 3 billion people will mean impacts on ecosystems regardless of policies and technologies (MA, 2005a).[18]

Population growth in developing countries is stretching countries' capacities to provide services and meet basic human needs. In these countries, the impact of population on the environment is linked as both cause and effect to other problems such as poverty, inadequate health care and education, a declining natural resource base, and a highly degraded environment. Countries with predominantly arid climates face the greatest challenge, particularly as climate change may cause even greater scarcity of natural resources such as freshwater.

Experience and analysis show that rising levels of health and education, especially for women, tend to correlate with stabilizing and even declining levels of population. Addressing the problem of population growth, therefore, can be helped by solutions to the problems of the pervasive lack of decent educational and job opportunities and resistance to improvement in the status of women. There is strong evidence that improvements in women's and girls' health, education and reproductive services under international family planning programs help reduce fertility rates and slow population growth, thereby improving both human well-being and environmental quality.

To address population growth issues, it is important that programs be supported, such as the United Nations Population Fund (UNFPA)[19] as well as in-country programs aimed at women's empowerment. National level decision-makers should take into account population issues in all strategies and development plans and consider how providing comprehensive family planning and educational services can improve human well-being.

3. Develop Integrated Governance Systems for Ecosystem Security

The Millennium Assessment provides a wealth of information to governments, as mentioned above, on not only the status of ecosystem services, but also the importance of working across scales and actors. As described above, it is not useful for governments to restate past commitments. To move toward ecosystem security – preserving the natural functions of ecosystem goods and services for all future generations – governments must put into place the political, social, economic and administrative systems for conservation of ecosystem services. Good governance is often used as a concept, but seldom exists in practice. It is a long-term process, but some foundation blocks that governments should undertake to improve include (1) establishing inter-ministerial committees on ecosystem security; (2) building governmental networks around ecosystem issues; (3) reviewing and modifying enabling frameworks to facilitate integrated resource management; and (4) strengthening monitoring and enforcement systems.

Many countries have established inter-ministerial committees that failed to achieve results. This has happened for many reasons, but is often due to a lack of political will. However, this does not mean that the process should be abandoned, particularly since guaranteeing ecosystem security requires working across institutions (MA, 2005a). For real outcomes to be achieved, political champions for ecosystem security will be needed in each country (MA, 2005e). Follow-up to the Millennium Assessment by civil society should be aimed at finding these champions in non-environmental ministries and institutions such as ministries of energy, infrastructure, agriculture, finance or development. Civil society should find ways to speak to non-environmental ministries such that ecosystem security is seen as a fundamental component of the functioning of the ministries (MA, 2005a).[20]

An inter-ministerial committee on ecosystem security would offer governments a means to debate at the national level the trade-offs described in the Millennium Assessment and to move toward integrated decision-making on natural resource management. A sample agenda topic for such an inter-ministerial committee could include, for example, national energy and water development options. Discussion among the ministries would include the current mix of energy supply options, the role of water in meeting current and future energy needs, options for optimization of the national energy grid, protected and other areas where infrastructure development would have devastating impacts on people and the environment, multi-purpose renewable energy alternatives, and so on. In this way, the combination of conservation, finance, infrastructure and development issues could be considered, and movement would be made away from sectoral decision-making.

Another element of forming new governance structures is the building of formal and informal networks/associations among decision-makers that will help them carry out their responsibilities. For example, the African Ministerial Conference on Water (AMCOW) was established to strengthen intergovernmental cooperation on water and sanitation in Africa (Abuja, 2002). Similar alliances could be formed to help build trust[21] and allow for the sharing of experiences among decision-makers (MA, 2005e). A Women's Association of Water Ministers, for example, could be one such entity that could allow for information-sharing and capacity building across regions of the globe.

In addition to creating new institutional frameworks, it is also important that countries develop the legal and regulatory systems to enable integrated decision-making.[22] The Millennium Assessment provides an excellent platform from which to start. 'Natural resource management always involves conflicting interests and the main lesson learnt from this assessment of responses is that innovative ways of collaboration and conflict resolution, together with reasonable legal and economic frameworks, often appear to be crucial for effective responses' (MA, 2005e).

The process of this transformation must take into account the reality of limited political will as well as minimal financial resources. The call to civil society to find political champions was noted. The financial challenges can be overcome if civil society organizations can find ways to gain funding to promote sustainable development actions – improve human health, raise education levels, involve women in decision-making processes, and so on – as opposed to strictly environmental ones.

One challenge at the international policy level is the differentiation between hard and soft law. Some international meetings, such as the International Conference on Freshwater in Bonn, Germany (2001), are not considered UN conferences, and thus their outcomes are not binding. Civil society organizations should spend time and resources lobbying their own governments to adopt certain progressive recommendations from these international processes.

In addition to institutions and laws, it is important for monitoring and enforcement to be part of governance systems aimed at ecosystem security. International meetings and agreements often emphasize the need for planning processes (National Biodiversity Strategy and Action Plans, National Sustainable Development Plans, Integrated Water Resource Management Plans, Poverty Reduction Strategy Papers, and so on). Sound planning is key to assess trade-offs among policy actions. Iterative planning processes will be needed if countries are to manage rapid changes to ecosystem services. One additional important aspect of any sustainable development scheme is effective enforcement of environmental laws and regulations.

CSD and other institutions need to evaluate how countries are doing when it comes to monitoring and enforcement. The International Network for Environmental Compliance and Enforcement (INECE) promotes information sharing and capacity building. This network, among others, should feed into CSD and other processes. For example, the network could monitor the Convention on Biological Diversity (CBD) and how countries are enforcing biodiversity conservation legislation. Sharing of experiences through the network should take into account basic issues, including (1) how enforcement action is taken in order to accomplish deterrence; (2) costs of compliance; (3) institutional credibility; (4) relationships between enforcement personnel and regulated entities; and (5) technical feasibility. For example, rather than CSD and other international processes restating the need for environmental impact assessments (EIAs), meetings could be used to understand the challenges related to EIAs. Moreover, questions on how to move toward integrated monitoring and enforcement systems would be key. Some countries' experiences with EIAs have shown poor compliance in part due to a lack of communication. For this reason, a point of discussion among governments could be program designs needed to improve relations between governmental officials and regulated entities.

III. IMPROVING STAKEHOLDER ENGAGEMENT

As ecosystems come under increased pressure for development, the potential for over-allocation of natural resources, such as freshwater, increases. Moreover, the range of stakeholders to be included in decision-making processes becomes increasingly broad. Although this adds to the complexity of ecosystem management, the consequences of not doing so can be devastating. The absence of mechanisms for participation and redress can result in decisions that adversely impact, exclude, and consequently are opposed by, affected communities. Such decisions are rarely effective and frequently are illegitimate and unjust. They undermine the ability to integrate environmental concerns into development processes. This ultimately has impacts on long-term growth for communities across the globe. Some policy options for increasing the role of non-governmental actors, ranging from environmental groups to business, include among others (1) adopting the major group model by international institutions and processes and (2) making gender equality a reality. The first option is aimed at generating more civil society involvement at the global level within and among institutions ranging from ones that are pro-environment, such as the United Nations Environment Program, to ones less so, like the World Trade Organization. The second possibility listed above involves both bottom-up and top-down approaches to mainstreaming gender, broadening the base of participation and responding to the specific needs of women.

1. Adoption of the Major Group Model by International Institutions and Processes

Unlike many other UN and inter-governmental forums, the Commission on Sustainable Development allows for the involvement of civil society through clusters of interested groups. Agenda 21 established nine Major Groups – Women, Children and Youth, Indigenous People, Nongovernmental Organizations (NGOs), Local Authorities, Workers and Trade Unions, Business and Industry, Scientific and Technological Communities, and Farmers (Agenda 21, 1992). Throughout the World Summit on Sustainable Development process, the Major Groups were central in lobbying governments and were able to make remarks formally presenting their positions at Multi-Stakeholder Dialogue Sessions. The Major Groups also made use of side events and other outlets.[23] Through these various channels, the Major Groups were able to influence text, share their experiences with sustainable development, and make the voices heard of people and communities unable to participate in such international events. The Major Group model should be considered by other international institutions and processes as an approach to generate more dialogue and exchange by the increasing number and range of actors involved in sustainable development – development organizations, conservation groups, faith-based organizations, private companies, and so on. The UN CSD experience should offer lessons learned on how to structure platforms for civil society engagement (UN, 2003).[24]

In addition to the Major Groups, there is a broad range of issue caucuses, ranging from Production & Consumption to Biodiversity to Peace, that work through the Major Groups as well as on their own to influence governments involved in the CSD. The Sustainable Development Issues Network (SDIN) was established to help organize input from NGOs to the CSD. SDIN involves not only the NGO Major Group, but others as well, such as the Women and Indigenous Major Groups. In particular, SDIN helps give a voice to civil society organizations that are not large or international.

The Millennium Assessment rightly points out that

> NGOs can help to monitor implementation and compliance with environmental policies from local scale, to global multilateral environmental agreements. NGOs tend to support public participation (or at least transparency to increase the awareness of civil society) in government and private sector decision-making, which can help to enforce existing environmental policies and develop support for improved policies (MA, 2005c).

The Major Group model is one that should be adopted by other international institutions and processes (for example, UNEP, CBD and WTO). It is not a perfect model, but it is one that has been applied for over a decade and offers significant lessons learned through experience.

It is important for the Major Group model to be improved upon and advanced, through for example, establishment of transparent and open procedures for selection of civil society representatives to travel to international events and to speak for major groups, financial support for major groups to have discussions well in advance of international events, and procedures for the development of common position statements. As with CSD, the Major Group model has experienced some steps backward. For example, the Major Groups have not joined forces to question the retrenchment in access of civil society to CSD.[25] Further, many of the Major Groups, such as the NGO Major Group, need to adopt their own internal policies for transparency and information sharing – procedures adopted by each collaborative that contribute to an overall understanding among the Major Groups.[26] This could enhance the credibility of civil society as they lobby governments to improve their practices with regard to information disclosure, transparency and accountability.

Civil society can not only help the monitoring of progress but can also assist governments in implementation of international commitments. Voluntary partnerships are one such vehicle and should be encouraged by international institutions and processes. These cannot be a substitute for responsible government action, however.

2. Make Gender Equality a Reality

In many ways, gender issues receive significant attention among policy-makers. Most recent international processes recognize gender actions. The Chair's Summary of the Intergovernmental Preparatory Meeting of the 13th Commission on Sustainable Development, for example, notes the need to utilize gender-sensitive monitoring and evaluation systems, collect gender-disaggregated data as appropriate, develop gender-sensitive indicators, and conduct gender analyses (CSD, 2005). However, it is often a paper exercise as opposed to one in which gender mainstreaming occurs in practice. The Millennium Project Task Force on Education and Gender Equality issued a report with a number of practical steps to help reduce inequalities based on gender. However, the report also pointed out key elements needed – leadership and political will (Millennium Project). The Millennium Assessment makes some key points that relate to gender, but omits others: in a number of developing nations females produce 60 to 80 per cent of food, yet in developing countries where data is available, only 2 per cent of women own land, 5 per cent receive technical assistance, and less than 10 per cent receive credit.

Rather than debate the need to include 'gender' in policy documents, the focus should be on how to make gender equity a reality. In the water and sanitation sectors, for example, a broad range of organizations and institutions have identified clear policy actions. These include:

- Facilitating access to credit for expanding access to safe drinking water, adequate sanitation, and water for productive uses, in particular for women;
- Providing micro-credit and creative alternative financing mechanisms to gender-sensitive organizations for improving or building community-based water and sanitation services;
- Changing discriminatory laws that prohibit women from getting loans or opening bank accounts;
- Improving women's productivity in using water in agriculture and small business through training, market linkages, and access to information; and
- Supporting and promoting secure and equitable access to land and tenure arrangements that enable female producers to become decision-makers and owners (MA, 2005e; UN Task Force, 2005).[27]

IV. ENCOURAGING NEW WAYS OF THINKING

In addition to the institutions and enabling frameworks for integrated resource management, the global community needs to discuss and share new ways of thinking on how to conserve and even reverse the rapid loss of ecosystem services. Potential policy responses to encourage improved natural resource management include (1) recognizing the need for adaptive management in policy frameworks and (2) establishing payment for ecosystem service (PES) fee systems. These are just two possibilities; much work remains to be done to find new and creative ways to slow and reverse the rapid loss of ecosystems, particularly since in the next half century an additional 3 billion people will need use of limited natural resources.

1. Recognize Adaptive Management in Policy Frameworks

Science-based decision-making has been advocated since the 1972 Stockholm Conference on the Human Environment.[28] The Rio Declaration on Environment and Development established the Precautionary Principle as a means for maintaining the science-based approach while recognizing that humans cannot always wait for science to ensure environmental protection (UN Resolution, August 1992).[29] If movement is to be made toward ecosystem security, then recognition and adoption of adaptive management by all institutions involved is needed (MA, 2005e).[30] As the Millennium Assessment points out, ecosystems are vast and complex (MA, 2005a).[31] While there are numerous definitions of adaptive management, most emphasize that adaptive management is the process of adjusting management actions as new and better information emerges about the ecosystem.[32] So in some sense, this is not a policy recommendation, but rather a recommendation that all policies, projects and interventions related to ecosystem management take into account adaptive management and establish monitoring and feedback loops to continue to gather information to improve decisions.

The Millennium Assessment clearly points out the advantages of adaptive management (MA, 2005a; 2005e).[33] In the water management sector, for example, adaptive management can help scientists determine the environmental flow needs of rivers. Based on current knowledge, some actions may be taken to restore environmental flows. In other cases, additional analysis may be needed over periods of months and years. Re-allocation of water uses (for example among agricultural, industrial, municipal and environmental uses) may be required which can take substantial time due to the changes needed in laws, incentive structures and behaviors. Throughout this period, adaptive management is needed as improved monitoring and data collection systems are put into place, more information is obtained, and scientists work toward consensus through facilitated dialogues and learning network approaches. Adaptive management should take advantage of all available tools – new and innovative desk-top approaches, quick diagnostic tools, and communication tools for debate and discussion among scientists and other experts.

As part of resource management strategies, countries should set up mechanisms for adaptive management. Civil society, particularly research and scientific institutes, can offer lessons learned on how to conduct adaptive management. Implementation may take considerable capacity building, particularly in data-poor countries, but will be critical if ecosystem service management is to be improved in the coming decades when more information is gained, not only on the status of ecosystem health, but also on the threats to these services.

2. Develop Payment for Ecosystem Services (PES) Systems

One of the challenges in moving toward ecosystem security is convincing policy-makers that protecting ecosystem services will promote social and economic development. The Millennium Assessment gives an in-depth look at the range of ecosystem services and their current state, but it does not provide the economic data necessary for national leaders to commit to 'investing' in ecosystem conservation. For example, there is no data on the cost of clean water provided by watersheds. A promising scheme to promote ecosystem conservation through the financial market is an ecosystem service fee system. National governments, international financial institutions, and the private sector should mainstream ecosystem services management into all productive sectors, urban and regional planning, business planning, and insurance assessments. Civil society organizations can be partners in this work by creating schemes to support payment for ecosystem services.

Environmental service fee systems have been established in a number of places across the globe. These systems offer valuable lessons on how to create new mechanisms that encourage individuals and businesses to recognize the *value* of ecosystem services (MA, 2005e).[34] For example, in Ecuador, a Water Conservation Fund (FONAG) was created that collects a fee from users of water from the Condor Bioreserves. These funds are used to support watershed management projects.[35] EMAAP-Q

(Quito Municipal Water and Sewage Company) and EEQ (Quito Electric Company) direct a percentage of their profits to the Fund. The Andean Beer Company and the Swiss Agency for Development and Cooperation have also joined FONAG to direct resources toward watershed conservation. Although the effort can be criticized,[36] FONAG serves as a model that should be improved upon as a means to integrated watershed management – changing individual, private sector, and governmental practices to protect watersheds and meet human needs. Other types of fee systems could be established for many purposes such as to support school access to water, sanitation, and educational programs in ecologically sustainable water management.

The experiences with PES to date have not adequately prioritized economic valuation research. This should be considered a priority for future systems; this data is needed to inform policy decisions to best manage and protect ecosystems.

The structures for PES need to be established, but farmers, forest owners and other private landowners also need to play a crucial role in freshwater ecosystem service protection and management. The Millennium Assessment does note that 'a study of more than 100 of the world's largest cities, for example, found that more than 40 per cent rely on runoff producing areas that are fully or partially protected' (MEA, 2005b). When it comes to watershed management, it is critical that management of protected areas include improving watershed conservation practices and approaches. Building upon lessons learned in the water sector, in general, it is important that payment for ecosystem service schemes take a holistic approach.

V. CONCLUSION

The Millennium Ecosystem Assessment provides a comprehensive assessment of the conditions and trends in ecosystem services. The Assessment's scenarios and response options offer the sustainable development community insight into how the future may play out in ecosystem health, but they lack the depth to direct policy-makers who are facing some stark choices – how to provide food, water, energy and jobs to a growing number of people. Political decisions often are based on short-sighted goals of re-election, career advancement, party positioning, and so on. The Millennium Assessment raises the alarm that ecosystems are being degraded at a rapid rate and that this may undermine future well-being. The challenge will be to link the passion for ecosystem conservation to compassion for human well-being. This will require using what is already known about policy responses, testing new approaches, strengthening governance systems, letting go of the tendency for 'territoriality',[37] and finding ways to bring about multiple benefits with policy actions – ones that recognize the vital links among growing populations, escalating resource demands, shrinking supplies and basic human goals of health and economic security.

NOTES

1 Senior International Water Policy Advisor, The Nature Conservancy (TNC). This chapter is prepared in the author's personal capacity and does not represent the views of TNC. This chapter does not present the full range of policy responses possible, but rather offers some suggestions that could prove useful in slowing ecosystem degradation. Contact: karin.krchnak@gmail.com.

2 'Actions to increase one ecosystem service often cause the degradation of other services' (MA, 2005a).

3 'Degradation of freshwater and other ecosystem services generally have a disproportionate impact on those who are, in various ways, excluded from participation in the decision-making process.'

4 'Laws, policies, institutions, and markets that have been shaped through public participation in decision-making are more likely to be effective and perceived as just.'

5 For example, the fact that 'a child born in sub-Saharan Africa is 20 times more likely to die before age 5 than a child born in a developed country' is cause for action, but one must know the cause of this – contaminated water, lack of health services, famine conditions?

6 'In many cases either the data needed to answer the questions were unavailable or the knowledge of the ecological or social system was inadequate.'

7 As a language issue, the fact that it is described as 'environmental resources' implies that the needs of the environment are somehow separate from those of humans.

8 Ecosystem security, at the most basic level, may be understood as preserving the natural functions of ecosystem services for future generations.

9 UN General Assembly Resolution 47.191 delineated the functions of the CSD to include:

- Monitoring the progress in the implementation of Agenda 21 related to the integration of environmental and developmental goals throughout the UN system based upon information provided by agencies.
- Considering information provided by governments on their activities, challenges and concerns, including progress in the implementation of environmental conventions.
- Reviewing progress in meeting commitments set forth in Agenda 21, specifically technology transfer and the provision of financial resources.
- Reviewing and monitoring progress towards the target of 0.7 per cent of the gross national product of developed countries for official development assistance and the adequacy of funding and mechanisms, including efforts to reach the objectives agreed in chapter 33 of Agenda 21.
- Receiving and analyzing relevant input from competent non-governmental organizations, and enhancing the dialogue with non-governmental organizations and other entities outside the United Nations system.

10 'Partnerships for sustainable development – voluntary, multi-stakeholder initiatives aimed at implementing sustainable development – were an important complementary

outcome of the World Summit on Sustainable Development. At its 11th Session in May 2003, the Commission on Sustainable Development reaffirmed that these partnerships contribute to the implementation of intergovernmental commitments, recognizing that partnerships are a complement to, not a substitute for, intergovernmental commitments.'

11 Anonymous.

12 'Hence, it is crucial for decision-makers to develop institutions at the global, regional and national levels that strengthen the adaptive capacity of actors at the sub-national and local levels to develop context-specific responses that do address the full range of relevant drives.' Although perhaps there need to be new institutions at the regional and national levels, it is not clear that there need to be new international institutions. Perhaps a research institute could undertake the task of tabulating existing institutions as a means to consider which should be strengthened and which may no longer serve a functional purpose. In the water sector, for example, a comprehensive list of institutions working on global water issues is lacking.

13 This is provided merely as an example and could be modified or adapted to future CSD sessions.

14 See section III(1) below for more information on the Major Groups.

15 'Increasing consumption per person, multiplied by a growing human population, are the root causes of the increasing demand for ecosystem services.' 'The structure of the world's ecosystems changed more rapidly in the second half of the twentieth century than at any time in recorded human history, and virtually all of Earth's ecosystems have now been significantly transformed through human actions.'

16 'The MA scenarios show that this target [WSSD target on biodiversity] will be difficult to achieve by 2010 and that the pressures on biodiversity will continue to grow during the first half of the twenty-first century, particularly through population and economic growth….'

17 The review sessions of ICPD continue to contribute to the global debate. The comment is directed at non-ICPD international negotiations – particularly those relating to sustainable development.

18 'Both economic growth and population growth lead to increased consumption of ecosystem services, although the harmful environmental impacts of any particular level of consumption depend on the efficiency of the technologies used in the production of the service.'

19 UNFPA provides such critical services as educational materials for families, training for clinical and community health care providers, girls' education and supplies such as sterile birthing kits.

20 In the long term, ecosystem security should be a component of all planning. Points made by the MA that sectoralize ecosystems (such as in the comment: 'Integration of ecosystem management goals within other sectors and within broader development planning frameworks') should be avoided when convincing decision-makers that ecosystem security is basic to human health, development, and so on.

21 'Recent research on ecosystem management suggests that investment in trust-building is crucial for mobilizing institutional and organizational capacity to make responses.'

22 This should not preclude countries, however, from encouraging communities to develop best management practices as a means toward ecosystem security.

23 At the Fourth Preparatory Meeting to the World Summit on Sustainable Development, 16 issue caucuses joined together in a press conference to comment on the lack of progress in reaching agreement on goals, timetables and funding for achieving sustainable development.

24 'UN–civil society relationship in essence concerns participation. Handled well, it enhances the quality of decision-making, increases ownership of the decisions, improves accountability and transparency of the process and enriches outcomes through a variety of views and experiences. But – handled badly – it can confuse choices, hamper the inter-governmental search for common ground, erode the privacy needed for

sensitive discussions, over-crowd agendas and present distractions at important meetings.'

25 At the World Summit on Sustainable Development, the Major Groups did raise alarm over the lack of access to the Sandton Conference Center where negotiations were being held. Access issues were resolved to enable the Major Groups to access the building and thus to be able to lobby delegations. At the 13th Session of CSD, civil society was denied access to the floors of plenary rooms, being directed only to the gallery area. This new policy was justified on the grounds of crowd control. However, in past CSD sessions, it was never a problem to accommodate all interested and accredited parties to access the plenary floor. Individual civil society organizations questioned this new policy, but as a whole the Major Groups did not. Civil society may be excluded from the UN Millennium Summit (September 2005) and so far there has not been a coordinated effort by civil society to ensure open and transparent processes for inclusion of civil society's voice, despite the fact that, as the MA notes, civil society is part of the sustainable development solution.

26 The NGO Major Group has operated for example without open and transparent criteria for decision-making. As a result, some key decisions are made based on personalities.

27 Land tenure is also a broader issue that needs to be addressed for sustainable development to be achieved. As the Millennium Assessment notes: 'rural communities in the formal tribal "homelands" in South Africa had no rights of permanent residence outside those areas.' Land tenure starts a process of building wealth, one that allows for inheritance and taxation systems that provide funding to the state, but that allow families to build wealth as well. When the fall of communism occurred in 1989 in Central and Eastern Europe and the former Soviet Union, one of the first steps in many countries was implementation of land tenure programs. The rapid economic growth occurring, for example, in Central and Eastern Europe is due to a number of factors, but property ownership as a wealth-building process should not be discounted.

28 An action point, for example, was to disseminate knowledge within the scientific and technological communities and to ensure that decision-makers at all levels shall have the benefit of the best knowledge that can be made available in the forms and at the times in which it can be useful.

29 Principle 15: 'In order to protect the environment, the precautionary approach shall be widely applied by States according to their capabilities. Where there are threats of serious or irreversible damage, lack of full scientific certainty shall not be used as a reason for postponing cost-effective measures to prevent environmental degradation.'

30 'People who depend on ecosystem services for their livelihoods often accumulate ecological knowledge that external researchers lack. By adaptive management including close monitoring of a diverse set of ecological variables they are able to observe and understand early signals of ecosystem change and distinguish from natural variability.'

31 'Ecological surprises are inevitable because of the complexity of the interactions and because of limitations in current understanding of the dynamic properties of ecosystems.'

32 http://www.fws.gov/midwest/EcosystemConservation/glossary.html.

33 'Adaptive co-management systems are flexible, community-based systems of resource management tailored to specific places and situations, supported by, and working with, various organizations at different levels.' 'Active adaptive management is a tool that can be particularly valuable given the high levels of uncertainty surrounding coupled socioecological systems.'

34 'Social, behavioral and cognitive change appears to be important if we want to understand to what extent people respond to economic incentives. The changes in values and attitudes evolve through public awareness and education.'

35 In 2000, a new Ecuadorian finance law allowed companies that handle public funds to put money into a trust fund designated for a specific purpose. FONAG was created under this construct. Since the challenge for the conservation and environmental

community is often to penetrate finance ministries, efforts around encouraging similar laws and regulations could be a vehicle for furthering ecosystem valuation approaches.

36 The Fund has been criticized for not being able to finance projects quickly enough while governance issues were resolved. Also, some criticism has focused on the fund supporting small-scale projects as opposed to large-scale ones.

37 This is understood as the need for organizations to claim their 'mark.' For example, the CSD NGO Major Group process could have been stronger to date if organizations were able to put aside their need to insert their rhetoric into the debate.

REFERENCES

The Abuju Ministerial Declaration on Water: A Key to Sustainable Development in Africa, Launch of the African Ministerial Conference on Water (AMCOW), Abuju, Nigeria, 29–30 April 2002.

Agenda 21, Section III, Strengthening the Role of Major Groups (1992), online at: www.un.org/esa/ sustdev/mgroups/mgroups.htm and www.un.org/esa/sustdev/documents/agenda21/english/ agenda21chapter23.htm.

Clark, J. and Z. Aydin (2003) 'UN system and civil society – An inventory and analysis of practices'. Background paper for the Secretary-General's panel of eminent persons on United Nations relations with civil society. Retrieved 28 July 2006 at www.un.org/reform/civil-society.html.

Commission on Sustainable Development, 11th Session (2003), online at: www.un.org/esa/sustdev/partnerships/about_partnerships.htm.

Commission on Sustainable Development, 13th Session (2005), online at: www.un.org/esa/sustdev/csd/csd13/matrix.pdf.

Food and Agriculture Organization (FAO), 'Women and Food Security', online at: www.fao.org/FOCUS/E/Women/Sustin-e.htm.

Millennium Ecosystem Assessment (2005a), *Ecosystems and Human Well-being: Synthesis*, Washington, DC: Island Press.

Millennium Ecosystem Assessment (2005b), *Ecosystems and Human Well-being: Volume 1*, Washington, DC: Island Press.

Millennium Ecosystem Assessment (2005c), *Ecosystems and Human Well-being: Volume 2*, Washington, DC: Island Press.

Millennium Ecosystem Assessment (2005d), *Ecosystems and Human Well-being: Volume 3*, Washington, DC: Island Press.

Millennium Ecosystem Assessment (2005e), *Ecosystems and Human Well-being: Volume 4*, Washington, DC: Island Press.

Millennium Project (2002), online at: www.unmillenniumproject.org/who/tf3docs.htm.

UNFPA (2002), 'State of the world population 2002: People, poverty and possibilities', New York: United Nations Population Fund.

UN General Assembly Resolution (1992), 'Report of the United Nations Conference on Environment and Development, The Rio Declaration on Environment and Development', A/CONF.151/26 (Vol. 1) 12 August online at: www.un.org/documents/ga/conf151/aconf15126-1annex1.htm.

UN General Assembly Resolution A/RES/47/191 (1993), online at: www.un.org/documents/ga/res/47/ares47-191.htm, 29 January.

UN Interagency Task Force on Gender and Water (2005), 'A gender perspective on, water and sanitation', (April 2005), online at: www.un.org/esa/sustdev/csd/csd13/documents/bground_2.pdf.

11. Framing the Millennium Ecosystem Assessment messages for political resonance

Frances Seymour

The Millennium Ecosystem Assessment poses two distinct challenges for those who would seek to use the findings as a platform for political change. The first challenge is that the findings are massive and complex. They span 24 ecosystem services across multiple ecosystems, scales and time horizons. Conditions, trends and their drivers, scenarios, and responses are woven together in a many-textured fabric. By comparison, even the complex sub-theme of climate change coalesces into a simple linear story line. While the Board Statement does a heroic job of distilling down some key messages, the Millennium Assessment findings do not lend themselves to summary headlines, sound bites and clear vehicles for political mobilization.

The second challenge posed by the Assessment findings is that the main messages that emerge are not really new. At least those of us within the professional environmental community have understood this set of issues for years. For example, we have known that many key natural resources are being exploited at rates beyond their capacity to regenerate; we have known that many important ecosystem services are undervalued, and we have known that the costs and benefits of ecosystem change are inequitably distributed. Moreover, for the most part, we have known the 'right' public policy answers for addressing many of these issues (as a quick perusal of World Resource Institute's own publication history and current portfolio of objectives would demonstrate):

- We know that policies that govern the exploitation of natural resources – and especially pricing and subsidy policies – need to be reformed to internalize environmental costs. For example, we need to make national energy policy as if climate change were a real threat; we need to price water as if it were a scarce resource.
- We know that our current institutions will have to change, and new

ones will have to be created, to meet the challenges of ecosystem stewardship. The mandates and accountabilities of government agencies must be changed to reflect the sustainability imperative, and institutions that map to ecosystem boundaries must be created or strengthened. For example, the performance of Ministries of Agriculture should be assessed based on how well they help farmers shift to sustainable practices rather than simply on how much they boost commodity production; transboundary river basin authorities must be given the mandates and resources to meaningfully affect decisions across national jurisdictions.

- We know that environmental governance regimes need to be reformed to encompass the broader range of stakeholders affected by ecosystem management decisions, and to ensure that their participation is informed and meaningful. Achieving these objectives will require dramatic improvements in the openness and integrity of decision-making processes. For example, processes that result in permits to pollute or extract publicly-owned natural resources must be made much more transparent and accountable to affected communities.

The problem is not that we haven't known the basic nature of current ecosystem mismanagement or what should be done about it; instead, the problem is that we haven't been successful in harnessing existing political processes to advance our agenda. A thorough discussion of the reasons for this failure is beyond the scope of this chapter;[1] surely the long-term nature of environmental problems, the complexity of attributing cause-and-effect in ecosystem management, vested interests in current management regimes, and public complacency all play a role. Critically, the myth that environmental protection and economic growth involve zero-sum trade-offs persists; the new global discourse on poverty reduction is being conducted as if sound ecosystem management were at best irrelevant, or at worst, a competing objective.

The so-called 'response options' contained in the Millennium Assessment raise the question of what it will take to change the *politics* of ecosystem management on local, national and global scales in such a way that good *policies* can be put into place. With that observation as a starting point, this essay attempts to identify some of the challenges raised or informed by the Assessment findings that have at least the potential for political resonance, and suggest how messages might be shaped and alliances formed to build political constituencies for change.

Millennium Ecosystem Assessment findings that might have political resonance include:

- that the trade-offs between provisioning and other ecosystem services (especially regulating) have far surpassed optimum levels, and are therefore economically inefficient and wasteful;
- that ecosystem change is likely to be highly discontinuous, characterized by sudden, non-linear events and even catastrophic collapse;
- that ecosystem degradation is exposing human communities to an increasing risk of an impoverished future due to decreasing resilience and capacity to provide needed services; and
- that winners and losers from the use of ecosystem services tend to separate out into spatially and temporally distinct groups, with a tendency for more concentrated groups that are already well-off to be the short-term winners, and more dispersed, less well-off groups to be the long-term losers.

The remainder of this chapter will propose a messaging strategy relating to each of the four findings mentioned above, and conclude with two proposals for communication strategies. The basic messaging and communication strategies sketched below for raising the political profile of particular issues are not themselves new, and have been employed with mixed success by advocates at various times on various issues. Nevertheless, the Assessment provides a new language and reinforced scientific credibility to render such strategies worth trying again with renewed vigor.

FOUR MESSAGING STRATEGIES TO PROMOTE ECOSYSTEM PROTECTION

Frame Inappropriate Trade-Offs as Wasteful and Inefficient

The Millennium Assessment documents that we are systematically destroying the ability of ecosystems to give us valuable regulating and cultural services 'for free' in exchange for a short-term increase in the quantity of lower-value provisioning services, even though the cost of replacing degraded or lost regulating and cultural services with 'engineered' solutions is high or even impossible. Examples include:

- removal of natural vegetation and structural hardening along river courses and coastlines, undermining natural erosion control and hydrological regulation;
- decimation of natural pollinators and enemies of agricultural pests through habitat disruption and overuse of pesticides; and
- loss of climate regulation, such as alteration of local rainfall patterns due to land cover change.

These findings suggest that one strategy for changing the politics of ecosystem management is to frame the inappropriate trade-offs between provisioning and regulating/cultural services as examples of government waste and inefficiency. For example, why should tax payers accept government expenditures on bricks-and-mortar solutions to flood control – which have both high capital and high maintenance costs – when protecting natural vegetation along river courses and coastlines can provide those same services for free? Similarly, why should we accept agriculture sector subsidies for production practices that degrade ecosystems, when those same subsidies could be invested in ecosystem protection and restoration that enhance long-term food security?

In the United States, for example, if we could be successful in characterizing perverse ecosystem management trade-offs as equivalent to other notorious instances of 'waste, fraud, and abuse' periodically exposed in Federal agencies, perhaps the public's and legislators' tolerance for such practices would decline. The environmental community has attempted to make this argument in the past – for example, in the context of the US Army Corps of Engineers' approaches to river channelization, and coastal communities' demand for hardened structures to stabilize beaches – but the approach certainly has more resonance in the aftermath of Hurricane Katrina.

This strategy is probably most relevant at the local and national levels, in response to ecosystem trade-offs that are contained within a single political jurisdiction. At the grassroots, advocacy targets would be taxpayers, as well as communities specifically threatened by ecosystem mismanagement. At the 'grass tops', advocacy targets would be officials of budgetary and finance agencies, and relevant legislative committees and processes. The strategy would require, first, establishing the economic value of ecosystem services, and second, investment in data collection and analysis for different ecosystem management strategies to make the case against waste and inefficiency.

While more difficult to pursue when trade-offs occur across multiple jurisdictions, this strategy might also have some traction at global and regional levels. Internationally, the strategy is probably best pursued in the context of development finance and specialized UN agencies, and in particular, looking for implicit perverse trade-offs across agency-specific and sector-specific approaches. With Jeffrey Sachs assigning price tags to achievement of various development objectives (Sachs, 2005), demonstrating the efficacy of lower-cost, ecosystem-based approaches or costing out the negative trade offs on ecosystems for different development strategies is one way to insert the (otherwise missing) environment into the Millennium Development Goals discussion.

For example, a significant Assessment finding is the increased prevalence of disease associated with the loss of certain ecosystem services, such as the links between forest degradation and malaria. One could imagine an analysis of the costs of responding to such health problems using a curative approach, and compare it to the costs of investing in the maintenance of ecosystem services as a preventive approach. Such an analysis could conceivably catch the attention of a Bill Gates – whose philanthropy has thus far focused on direct health interventions such as vaccinations – and prompt more attention to and investment in 'upstream' approaches (that is, maintaining ecosystem services) as well.

In the same way that the 'Pollution Prevention Pays' campaign succeeded in establishing the principle that prevention of pollution is almost always cheaper than treatment,[2] the broader goal here would be to establish an analogous principle that 'Ecosystem Preservation Pays', for example, compared to the cost of treating the adverse health effects of ecosystem degradation. (Of course, the cost–benefit analysis will be sensitive to the time horizon for incurring those costs and the discount rate applied.)

Regionally, economic analysis of the implicit trade-offs being made through infrastructure development might provide another target of opportunity. In the Latin America region, for example, governments and financiers (including the Inter-American Development Bank) could be pressed to demonstrate that the improved transport services and regional integration promised by the Initiative for the Integration of Regional Infrastructure in South America (IIRSA) are not counterbalanced by the losses in ecosystem services likely to accompany significant investments in road and dams. Techniques already exist for such purposes; for example, the non-profit group IMAZON (http://www.imazon.org.br/home/index.asp?lg=en) has used GIS mapping to assess the potential impact of new roads on deforestation in the Amazon. In Southeast Asia, governments and financiers (including the Asian Development Bank) could similarly be pressured to calculate the losses in ecosystem services likely to result from regional infrastructure development in the Greater Mekong Sub-region.

Turn Ecological 'Trigger Points' into Political Trigger Points

A second clear message coming out of the Assessment is that the over-exploitation of ecosystem services is increasing at risk of abrupt and potentially catastrophic ecosystem failures. The best current examples come from the fisheries sector – for example, the notorious 1992 collapse of the Newfoundland cod fishery – but the Assessment suggests that we should not be surprised by discontinuities in ecosystem services at all scales resulting from such trends as nutrient loading and climate change.

We should think more about how to transform ecological 'trigger points' into *political* trigger points for reforming ecosystem management. In the same way that ecosystems can respond to stress in a non-linear fashion, political systems can do so as well. Indeed, one of the most important opportunities to change the politics of ecosystem management will be in the immediate aftermath of extreme events. While it may be too late for the ecosystem that has crashed, the opportunity is not wasted if policy-makers responsible for similar ecosystems elsewhere are spurred into action. Already we have seen examples of governments responding to ecosystem failures by imposing swift policy changes: in the 1990s, the Philippines and other countries in Asia imposed logging bans in response to major floods, landslides and resulting loss of life attributed to deforested hillsides.

Non-linear responses of political systems to ecosystem failures can present both threats and opportunities. While certainly not hoping for tragedies, nor adopting a 'we told you so' attitude, we should be ready with solutions to propose should such events occur. If we are not, either opportunities will be missed, or governments will embark on misguided responses. An example of the latter would be the initial domestic and international responses to the catastrophic forest fires in Indonesia in 1998. Not only were the fires initially blamed on shifting cultivators (rather than land-clearing for commercial plantations, as was later demonstrated to be the primary cause); the bulk of government attention and donor response focused on building fire-fighting capacity, rather than addressing the root causes related to perverse forest policies and management practices.

The US response to 9/11 demonstrates the potential for non-linear institutional change in response to catastrophic events: previously disparate and uncoordinated federal agencies have been joined together in a mammoth Department of Homeland Security. Leaving aside a judgment on the efficiency and effectiveness of that particular institutional solution, we simply note that opportunities to reform the disparate and uncoordinated agencies charged with ecosystem management will come with trigger events, and we should be prepared to offer specific proposals for institutional redesign.

The trigger point strategy can apply at all levels, depending on the nature of the ecological trigger point being surpassed. On local scales, disruptions to hydrological systems – experienced as flooding and drought – are probably the most likely to mobilize public attention and political response to protect and/or restore natural water regulation systems. At national scales, major natural disasters – including severe storms – could be used as moments to focus attention on the need to maintain and invest in ecosystem resilience.

On the global scale, major disruptions to marine and atmospheric ecosystems are most relevant. The international community demonstrated its capacity to mobilize an effective response to the discovery of the hole in the ozone layer, although it is important to note that the costs of failing to respond would have been borne disproportionately by citizens of industrialized countries (whose temperate latitudes and light pigmentation would render them vulnerable to increases in skin cancers). Similar responses have not been forthcoming when the costs of ecosystem failure are concentrated in developing countries, as is the case with desertification.[3]

One could easily imagine jingoistic, us-versus-them responses should ecosystem failures with costs experienced in one country be attributed to the actions of governments or citizens of another country. For example, the collapse of an international fishery could easily prompt violent conflict among nations, and international disputes over access to fisheries have come to the brink of doing so in the recent past. As a result, proponents of improved ecosystem stewardship need to be prepared not only with technical solutions, but also with political solutions that promote collaborative problem-solving rather than confrontation.[4]

As mentioned in the introduction, such political solutions may involve the strengthening or creation of new regional and/or global institutions. It is striking the degree to which the international community has invested in institutions to mediate disputes over international finance and trade, while relatively little institutional infrastructure has been developed to mediate disputes over ecosystem mismanagement with transboundary implications.

Frame Ecosystem Maintenance as Insurance Against Risk

Another key message emerging from the Millennium Assessment – related to the trigger point issues mentioned above – is that by degrading an increasingly broad range of ecosystems on an increasingly global scale, the human community is dramatically increasing the risk of both acute and chronic shortages of ecosystem services. The resilience of individual ecosystems is compromised through loss of biodiversity. With few buffers left – in the form of relatively unexploited ecosystems – there are fewer 'Plan Bs' available to provide alternative services to human populations when ecosystems fail.

An alternative strategy for capitalizing on extreme events and one also applicable to more chronic ecosystem degradation is to frame more environmentally sound responses as a social insurance scheme. By 'taxing' ourselves, that is, by forgoing or using provisioning services more wisely now (which, if over-exploited, might push regulatory services over the edge in the near term, while undermining the future capacity of ecosystems to continue to provide provisioning services), we are purchasing insurance against an ecologically impoverished future.

The ongoing debate in the United States over the future of social security provides an interesting, if imperfect analogy. The continuing existence of the program demonstrates that the public places a value on managing the risk of impoverishment in old age, and is willing to pay into an insurance system now in exchange for greater economic security in the future. If we believe that ecosystem protection is essential to long-term economic security, perhaps the issue can be in similar terms.

Elements of the analogy are as follows:

- current Social Security payroll taxes are analogous to a reduction in current ecosystem provisioning services that the public would be willing to forgo;
- the Social Security 'trust fund' (if it actually were in a 'lock box') is analogous to healthy ecosystems managed to maintain their capacity to provide services sustainably decades into the future; and
- entitlement to future Social Security benefits is analogous to securing the future stream of ecosystem services that would be preserved by forgoing some provisioning services now.

A key attribute of Social Security that may explain its political resilience is that everyone who pays into the system can (in theory) expect to receive benefits; thus it is not, strictly speaking, an insurance scheme.

A flashpoint of the Social Security debate has been the concept of 'personal accounts', which effectively shift risk from the common pool to the individual, but are attractive due to the perceived likelihood of enjoying the benefit even if the larger system goes bankrupt. Even though many ecosystems services are by their nature not amenable to privatization, it would nevertheless be important to frame the message in terms that convey assurance that the benefits of ecosystem stewardship would be broadly shared by everyone, even if not all individuals face the same degree of risk from their degradation.

Ecosystem services, like Social Security, also have equity dimensions: increases in payroll taxes, and decreases in the flow of provisioning services, could both be regressive, unless taxes and/or benefits are structured to be progressive. Thus, any use of this strategy would need to ensure that reductions in current consumption fall disproportionately on those elements of society best able to absorb such reductions. (For example, large-scale commercial and sport fishermen would absorb greater proportional reductions in catch than small-scale, subsistence fishermen; low 'lifeline' tariffs would apply to low-income users of water, gas and electricity, while higher charges would be applied at higher levels of use.) In addition, implementation of the strategy should not result in ecosystem degradation simply being shifted to other countries.

One obvious ally in this strategy is the insurance industry itself. Indeed, insurance companies – seeing soaring payouts for extreme events – were among the first to take the issue of climate change seriously.

Selling improved ecosystem stewardship as an insurance policy – and especially achieving reductions in and more equitable use of provisioning services – will be most feasible in situations in which the community of stakeholders is clear, the trade-offs between near-term ecosystem exploitation and longer-term ecosystem integrity are understood, and the required reduction in current consumption does not cause severe hardship (either because compensation mechanisms can be marshaled to reduce such hardship, or prices of the ecosystem service in question rise to reflect increased scarcity). For example, a community of fishermen holding licenses to exploit a particular fishery could agree to reduce the annual catch in order to allow ecosystem recovery and gradually adjust off-take to sustainable levels.

Where the community of stakeholders is more diffuse and diverse, the trade-offs uncertain, and the reduction in consumption represents higher costs, such agreements will be more difficult to reach. Unfortunately, such conditions are the rule rather than the exception for most ecosystem management challenges at all but the most local scale. The establishment of the Social Security system in the United States reflected a social compact arising after the Depression, a consensus that Americans shared a collective interest in funding a basic safety net for everyone. Could one imagine such a sense of community arising at regional or even global levels, sufficient to support collective action to reduce pressures on ecosystems to sustainable levels? The challenge is sobering. Nevertheless, this is exactly the kind of visionary thinking that needs to be developed in advance of the kind of political 'trigger point' described above, so as to take advantage of political will in the moments that it emerges.

Cast Ecosystem Stewardship as a Fairness Issue

The Assessment makes clear that the costs and benefits of current patterns of ecosystem management are not distributed equitably. Winners tend to be more concentrated groups of stakeholders that are already well-off and realize their benefits in the short term, while losers tend to be more dispersed groups of stakeholders that are less well-off, and who bear costs over the long term. The fundamental unfairness of this arrangement is a message to build on, and more importantly, an injustice with which to awaken latent constituencies for reform.

Unfortunately, environmentalism has been successfully caricatured as the province of latte-sipping elites, and environmental protection as a zero-sum trade-off with economic growth, employment and poverty reduction. In the many cases in which the facts will support turning this argument on its head, it will be a powerful one for aligning the economic interests of non-elites with the interest of improving ecosystem stewardship.

Other than making sacrifices for their own children, human beings do not have a very good track record of altering behavior now to serve the interests of generations yet unborn. Thus I would recommend focusing on the three dimensions of fairness outlined below – local, national, and global – rather than inter-generational equity.

At relatively local scales, there would appear to be significant headroom for mobilizing constituencies around the unfairness of current and proposed ecosystem management regimes. For example, typical approaches to providing urban transport services are profoundly anti-poor and anti-environment: road-building primarily serves the needs of those wealthy enough to afford a car, while paving the way for an emissions-intensive future and urban sprawl. Alternative approaches focused on urban planning around transport corridors, mass transit, and pedestrian-

and bicycle-friendly options, would provide more equitable access to transport while minimizing ecological footprints.

The fairness argument is also likely to have traction at the national level, at least with respect to some ecosystem management challenges. In developing countries, the inequitable distribution of revenues from the exploitation of natural resources has prompted calls for decentralization (not to mention numerous separatist movements). In Indonesia, for example, resentment of Jakarta-based elites – who reaped fortunes from timber under the New Order regime while insulated from the environmental damage caused by logging in remote provinces – gave energy to decentralization legislation in the immediate post-Suharto period. As subsequent experience in Indonesia has demonstrated, however, a broader suite of reforms is necessary for such energy to translate into improved ecosystem management.

Perhaps the most significant challenge posed by the results of the Millennium Assessment is the need to develop systems of distributive justice on a global scale. Analysis of the scenario results implies, among others, a need for a significant investment in global cooperation, but any such cooperation will need to be based on a negotiated sense of 'what's fair?' in allocating the costs and benefits of any ecosystem management regime. We already know how difficult this is: witness the ten-year negotiation over the law of the sea treaty, and the halting progress of the Kyoto process.

It may be that the public and politicians find the fairness argument compelling when applied to equity across countries only when the very poorest countries are getting a very bad deal. For example, the Jubilee Debt Campaign (http://www.jubileedebtcampaign.org.uk) – an international network formed in the 1990s to advocate debt cancellation – has been successful in mobilizing support for debt relief by focusing on the plight of countries in sub-Saharan Africa that have been victimized by kleptocratic leaders and the vagaries of global commodity prices. The aspects of perverse agricultural and fisheries subsidy regimes in developed countries most likely to provoke outrage are those that disadvantage the poorest countries – for example, cotton producers in Mali, or coastal fishermen in sub-Saharan Africa. Otherwise, there seems to be a tendency to see other countries as competitors without a moral claim to fairness: a campaign to ensure that China gets its fair share of the world's ecosystem services would not be likely to get very far, even though a significant number of the world's poor live there.

By contrast, there does seem to be some public appetite for changing consumption patterns and public policies when such changes are associated with both improved environmental management and a better deal for poor individuals and communities (as opposed to nations). For example, a number of consumers in developed countries have demonstrated willingness to pay a price premium for 'fair trade' and/or sustainably-produced coffee and other tropical commodities. However, such voluntary, labeling-based approaches may be limited to those cases in which a clear linkage exists between a product and a specific vulnerable ecosystem or community.

There may be some scope for casting equity of access to ecosystem services on a global scale as a moral issue when applied to individuals and communities rather than countries. That idea is explored in the context of climate change in the section below.

COMMUNICATING THE MESSAGES: TWO STRATEGIES

Catalyze a Norm Shift

One possible response to ecosystem degradation not covered explicitly by the Assessment is to frame it as a moral and ethical issue, and promote changes in individual behavior and social policy on that basis.

Over the last 200 years, the global community has seen a number of shifts in norms; behaviors and policies that were once socially acceptable have gradually come to be classified as unthinkable (Finnemore and Sikkink, 1998). In the nineteenth century, slavery was acceptable; by the twentieth, it was not. Torture and execution have similarly gone from being tacitly acceptable to outrage-provoking. Women and minorities were once not allowed to vote; now they hold public office. Societies that tolerate slavery, condone torture and the death penalty, and/or deny political participation based on gender or race are increasingly seen and treated as pariahs in the international community.

Within the United States, we have experienced such norm shifts in our lifetimes. Well into the 1950s and early 1960s, schools and public accommodations were segregated by race. While that legacy is still with us, overt, publicly-sanctioned discrimination is no longer tolerated. Some of these shifts have even been driven by environmental consciousness: it used to be acceptable to throw rubbish out of a car along the highway; it used to be considered glamorous to wear exotic furs. Now most people would think twice before doing either.

The public and politicians have also shown a willingness to project such changing values abroad, even when no narrow national self-interest is served; the anti-Apartheid campaign provides an example.

Perhaps, then, we should think about how to promote the changes in individual behavior and social policy needed to protect ecosystems as moral and ethical imperatives, and plot systematically to catalyze a global norm shift on that basis. Climate change would appear to be the issue most ripe for this treatment: the richest people in the world generate the most emissions, and the richest countries bear the most responsibility for cumulative emissions that are causing climate change, while the poorest countries and poorest individuals within countries are most vulnerable to the effects of climate change, and least able to adapt to those effects.

If we can do a better job of drawing a dotted line between incremental miles driven by Americans in Hummers and SUVs and incremental deaths due to flooding in Bangladesh and drought in Africa, we might have a basis for an historic norm shift. Graphic photographs of wildlife in leg traps and clubbed baby seals have rendered it politically incorrect to wear fur for fashion in many sub-cultures of the United States. If we can make the link to increased suffering and death of the world's poorest people, it will no longer be socially acceptable to indulge in conspicuous consumption of emissions-intensive fuels. The objective would be to make driving when public transport is available, or riding in an SUV – much less a Hummer limousine – the social equivalent of wearing a leopard-skin coat made with child labor.

Making such links will require a step change in our community's ability to communicate simple messages to the public and politicians. The final section of this chapter is devoted to that challenge.

Convey Clear Messages and Graphic Images

In order to just get the attention of policy-makers and the general public, much less convince them to make the radical changes in policies and behaviors necessary to reverse current trends of ecosystem degradation, the environmental community will have to do a much better job of communicating our messages. Simple analogies, graphic images, reference points, and other tools are needed to convey the nature of the problems and be convincing about the feasibility of proposed solutions.

To talk about the abrupt changes resulting from climate change, we could, for example, craft an image of a canoe with two paddlers heading towards a major waterfall. The paddler in the back of the canoe – representing industrialized countries – is bigger and stronger, and has been paddling furiously toward the waterfall for many years. The paddler in the front of the canoe – representing developing countries – is smaller and weaker, and has mostly been a passenger up to now, but is beginning to help accelerate the speed of the canoe toward the waterfall. This analogy achieves the following objectives:

- it conveys the truth that both industrialized and developing countries are 'in the same boat' when it comes to climate change;
- it conveys our progress toward abrupt and catastrophic ecosystem failure (represented by the waterfall);
- it suggests that the industrialized counties would be foolish not to start reversing the direction of their paddling (meaning emissions growth) to slow down forward progress toward the waterfall, regardless of what developing countries do; and
- it suggests that developing countries would be foolish not to moderate the vigor of their paddling (meaning emissions growth), regardless of what industrialized countries do.

To convey a sense of the increasing risk of ecosystem failure, we could develop an ecosystem version of the 'Doomsday Clock' popularized by the *Bulletin of the Atomic Scientists* (http://www.thebulletin.org/doomsday_clock/timeline.htm). In the Doomsday Clock, the proximity of the minute-hand of the clock to midnight represents the level of danger of nuclear catastrophe – progress on disarmament treaties turns the clock back, while proliferation and testing moves the clock forward. One could imagine a whole suite of 'ecosystem failure' timepieces – one each for climate, fisheries, nutrient loading, and so on – updated periodically to reflect the latest scientific information about conditions and trends in ecosystems, as well as progress in making and implementing policies to stave off disaster. Indeed, the 'ticking clock' analogy may be more effective for conveying progress toward ecosystem collapse than for nuclear catastrophe, as one could actually estimate (however imperfectly) the time remaining at current rates of exploitation before a specific threshold is crossed.

It would be important for the ecosystem clocks to convey the need for specific action, rather than just raise the ambient level of anxiety – a criticism leveled at the color-coded terror alert system implemented in the United States. Accordingly, actual and prospective shifts in the clocks should be keyed to specific actual and prospective shifts in policy. The degree of certainty could somehow be conveyed using different colors or sizes in the image. Updating the clocks, and providing the accompanying interpretation, could be a continuing role for contributors to the MA.

To help people connect the dots between excessive consumption in industrialized countries and impacts on people and ecosystems in developing countries, we could develop a web-based calculator that would convert units of specific policies and behaviors into specific units of impacts. For example, using a long string of transparent assumptions, one could convert 'miles driven by Americans in SUVs' into 'excess deaths from storm events' due to climate change.

CONCLUSION

In summary, the Assessment findings reinforce what we already know is wrong with current ecosystem management paradigms, and what we already know about how to fix them. The challenge is to develop the political strategies and galvanize the public support necessary to put more sound policies and practices in place.

This chapter suggests framing political approaches to ecosystem issues around efficiency, trigger points, social insurance, fairness and ethical norms, and closes with a call for improved messaging and imaging of the problems and proposed solutions. Clearly, these ideas need to be further fleshed out, and contextualized into diverse political contexts on different scales, North and South. In the absence of any genuinely new ideas, the best we can do is run faster and smarter with what we have got.

NOTES

1 Shellenberger and Nordhaus (2004) in their essay, 'The death of environmentalism: Global warming politics in a post-environmental world', provide one analysis of the failure of the environmental movement in the United States.

2 More information on the 'Pollution Prevention Pays' campaign can be found at the United States Environmental Protection Agency's website: http://www.epa.gov/opptintr/p2home/.

3 The Global Environmental Governance (GEG) unit of the Center for Science and Environment in Delhi provides a useful ongoing resource regarding the disparity of response to global environmental problems depending on which societies are affected. The GEG website can be found at http://www.cseindia.org/programme/geg/geg-index.htm.

4 Conca and Dabelko (2002) argue in their book, *Environmental Peacemaking*, that cooperation on environmental matters can also promote broader peacemaking.

REFERENCES

Conca, K. and G. Dabelko (2002), *Environmental Peacemaking*, Washington, DC: Woodrow Wilson Press.

Finnemore, M. and K. Sikkink (1998), 'International norm dynamics and political change', *International Organization*, **52** (4), 887–917, Autumn.

Sachs, J. (2005), *The End of Poverty*, New York: The Penguin Press.

Shellenberger, M. and T. Nordhaus (2004), 'The death of environmentalism: Global warming politics in a post-environmental world', El Cerrito, CA: Breakthrough Institute.

PART III

Building institutions to sustain ecosystem services

12. Meeting the Millennium Ecosystem Assessment challenge: The case for biome stewardship councils

Sudhir Chella Rajan

BACKGROUND

The distressing news from the Millennium Ecosystem Assessment is that human impacts have been detrimental in practically every biome, from Mediterranean forests to the tundra. Even in the Assessment's two relatively optimistic scenarios, TechnoGarden and Adapting Mosaic, these impacts are projected to worsen over the next half century. Nevertheless, what is also clear is that at the local level there are many opportunities to achieve positive synergies and enhance ecosystem services to improve human welfare. The most promising pathways open to us are to create the appropriate incentives and institutional arrangements at the local and regional levels to tap into these synergies, and to find ways to coordinate activities across ever larger scales up to the global level.

Drawing parallels between the Assessment's findings and the more specific concern about climate change is instructive. The Intergovernmental Panel on Climate Change (IPCC) was also a large-scale collaborative scientific endeavor to gauge the global scope and impacts of significant human-induced change on the natural environment. The similarities in effort, rigor and objectivity are expressed in the legitimacy of both enterprises as scientific assessments. The IPCC, however, was initiated and to a large degree supported by national governments, whereas the Millennium Assessment was developed and orchestrated by civil socicty organizations with the endorsements of intergovernmental agencies. The differences in how each was initiated have, in turn, led to dissimilar responses on the part of political actors.

The response to climate change has predictably relied on the commitments and actions of individual countries that are parties to intergovernmental agreements. This is possible because greenhouse gas emissions can, for the most part, be related to national accounts, and responsibilities can be assigned based on quantifiable targets. The Millennium Ecosystem Assessment, on the other hand, reveals a much more complex and web-like structure in the links between ecosystems and human activity at multiple scales. It is difficult to treat ecosystem integrity as a fungible quantity that can be regulated and given a tradable value in a manner commensurate to, say, greenhouse gas emissions. Similarly, while the monotonic relationship between economic growth and energy use, and therefore greenhouse gas emissions, largely holds true through the so-called 'IPAT' equation,[1] it is much harder to make this connection between wealth creation and ecosystem damage. While population, affluence and technology do collectively influence ecosystems, other more qualitative factors such as cultural practices and social relations can be equally important in determining the kinds of ecosystem change that result. Poverty itself can be a cause of the degradation of natural resources, but persistent environmental degradation in both rich and poor countries can contribute to poverty, particularly for those whose livelihoods depend directly and proximately on the security of natural capital.

Equally importantly, natural resource and ecosystem boundaries often have little or no necessary relation to national borders. A given country's policies do of course matter very significantly in determining the health of its ecosystems, but often less so than the practices of groups living within an ecological region. By comparison, while climate impacts are obviously transnational in scope, national economic and energy policies are profoundly important for determining levels of impact (see Table 12.1).

All this suggests that it is much more difficult to formulate a response to the findings of the Millennium Ecosystem Assessment in a way that is exactly parallel to the response to the threat of climate change. Perhaps as a reflection of the enormous difficulty in defining an appropriate political platform for initiating decision-making around ecosystem management, therefore, the Assessment's Responses Working Group has provided a generic, if not facile, set of recommendations. These include the call for improved coordination among the different levels of decision-making processes; integration; transparent participatory approaches; explicit consideration of trade-off and synergies; and the creation of the proper enabling conditions such as good governance and appropriate institutions.

While a successful response to the global ecological crisis will undoubtedly require these elements in some broad sense, a far more specific set of policy options could be developed if clearer connections were drawn between the widespread threat to ecosystems (and thereby to the welfare of humanity at large) and the need for a correspondingly expansive and globally articulated strategy. The Assessment's findings are, after all, only the latest in a long series of reports describing global threats to human welfare with strong local impacts, such as infectious diseases, speculative finance, sectarian violence and terrorism. None of these can be easily circumscribed by national boundaries and therefore point in the direction of a novel mixture of local and global policies.

Table 12.1 Main differences between climate change and ecosystem damage

Climate change	Ecosystem damage
Human-induced drivers are simple to understand and relatively well-known: greenhouse gas emissions, which are in turn tied to energy use and land-use change	Human-induced drivers are complex: shifting populations and livestock; industrial, agricultural practices; land-use change; climate change; etc.
Easily quantifiable sources of emissions, making it relatively easy to develop cap-and-trade policies	Ecosystem 'bads' hard to quantify in commensurate terms across contexts.
National policies can make a direct and significant impact; straightforward ways to separate out large polluters from small polluters.	National policies need to be extremely well coordinated across different types of ecosystems. May not be easy to make a quantifiable distinction among countries on the basis of whether or not their actions will have significant impact on global ecosystem damage.

At the very least, this class of problems requires regional if not international collaboration to characterize the dangers and develop strategies to address them. More often than not, the most effective solutions entail concerted action that transcends official country-level collaboration. Often working in a crisis management mode, networks of governmental, inter-governmental and non-governmental actors with specific expertise are typically called upon to attend to problem clusters in various locations around the world. For instance, medical teams from private hospitals, Médecins Sans Frontières and the International Red Cross with logistical support from United Nations relief agencies work independently of national governments to provide medical services for people who are rendered stateless by war or severe natural calamity. Similarly, intelligence gathering by local police agencies across countries working with the United Nations Office of Drugs and Crime and the US State Department is necessary to apprehend international drug traffickers.

Although such networks generally seem to take shape only for makeshift emergency operations, they often gain sufficient importance to become more permanent institutional arrangements with systematic relationships built across agencies and organizations for long-term strategic cooperation.

An intriguing question that needs further exploration is whether analogous trans-boundary solutions in the form of quasi-governmental networks are appropriate and feasible for addressing the clear and present danger to the world's ecosystems. The fact that regional and global institutional arrangements are essential for sustainability is no longer controversial, but the formation and character of a legitimate and effective global administration are far from obvious. This chapter proposes an adaptive governance framework for trans-border ecosystem management.

BIOME STEWARDSHIP COUNCILS

A biome is a natural ecological unit that is derived from the understanding that forms of life associated together in a similar physical natural environment, latitude, altitude and terrain share certain common elements by virtue of belonging to a single habitat. Any group of living and non-living things interacting with each other is considered an ecosystem, and biomes can be understood simply as large ecosystems comprising habitats of similar climate and vegetation. The close linkages and interdependencies among the soil, flora and fauna within a biome can produce homeostasis, an open system whose structure and functions are maintained through multiple equilibria.

Historically, humans have adapted best to their biomes whenever they have engaged in mutually supportive interactions. However, to say this is not to reduce the complexity of social organization into biological terms involving the mere extraction of economic value from natural capital. It is, rather, to acknowledge that the prospects to enhance human well-being over the long term depend crucially on the ability of societies to shift their practices towards sustainable forms after recognizing both the constraints that their natural environment places on them as well as the opportunities it affords. Indeed, this is the single most consistent message running through the myriad pages of the Millennium Assessment, which shows that although current patterns of human activity are causing over 60 per cent of the earth's economically valuable ecosystem services to be degraded, there are many opportunities to reverse this trend while enhancing human welfare.

The MA identifies 14 types of biomes around the world (Figure 12.1). To the extent that each biome shares roughly similar types of life forms, soil and climate, it is fair to assume that in their geographical extent they would provide comparable forms of ecosystem services. Undoubtedly, the forms and rates of extraction of social and economic value will vary widely within the same biome depending on the way the societies are organized, their technological capabilities and wealth, policies with regard to environmental protection, and lifestyles. In other words, while it is true for instance that montane grasslands and shrublands in the Andes and East and Central Africa share similar ecological features and climate, there are significant differences in social and economic conditions, not to mention political systems and cultural practices. Given the similarities in natural capital, however, it is quite likely that residents in both parts of the world will share many living and working conditions related to farming and crop type, livestock, housing construction, access to water, and so on, although the differences in the ecological efficiency of the use of natural capital will largely be a function of the existing human, institutional and technological capital. In short, there will be many significant opportunities for social formations within a given biome to learn usefully from each other on best practices to obtain ecosystem services.

The governance framework to carry this out is to develop a Biome Stewardship Council (BSC) for each biome defined in the Millennium Assessment with an additional one for coastal and marine fisheries.[2] In other words, there will be 15 BSCs, each in charge of managing a single global biome. To begin with, each BSC would consist of a group of individuals elected or nominated by local community organizations that reside in the regions making up a biome, including a few government officials who are deemed to be particularly sensitive to ecosystem protection. To ensure public confidence, the rules for developing BSCs would need to come from input from extensive consultation from stakeholders who may, for instance, decide that a 'biomic constitution' is warranted. Whatever the final outcome of these consultations, it is important that certain key principles be adhered to, namely that the council members a) be properly representative of the stakeholder groups resident to the biome; b) function through internal rules of legitimacy; and c) uphold basic values of ecological stewardship.

Each of these principles can be satisfied in multiple ways, but the most effective solution would be one that has simple rules developed through stakeholder input that are then consistently applied. For instance, within each biome the numbers of council members could be in proportion to the

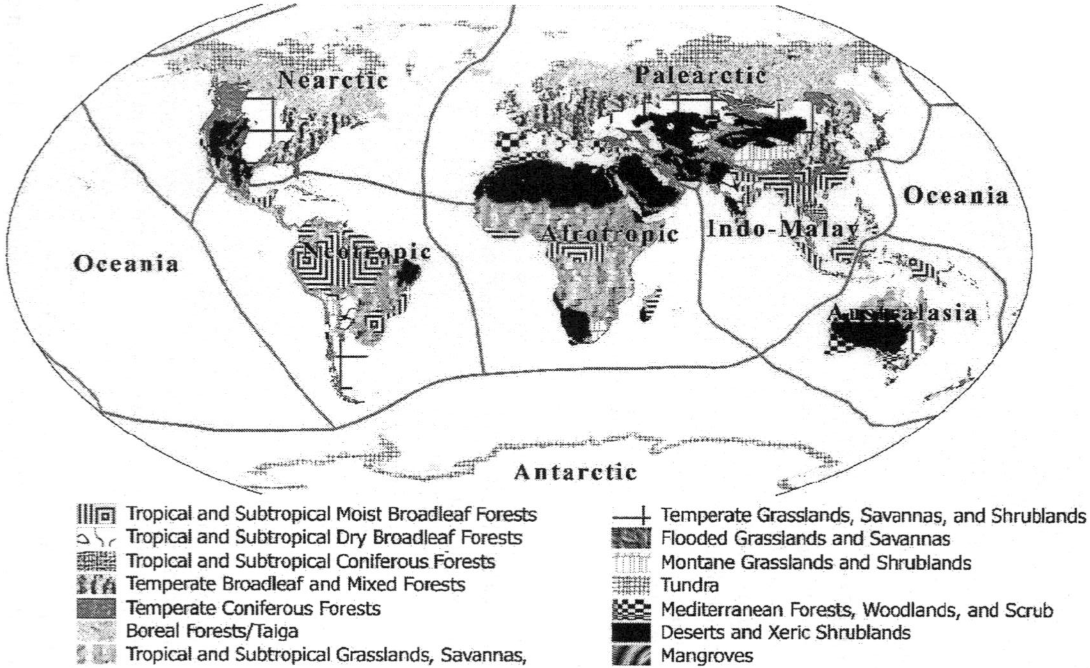

Source: Olson et al. (2001)

Figure 12.1 Major biomes of the world

corresponding populations, with additional representatives from the member governments. Thus, if there are 200 million people resident in biome A and there are 20 countries intersecting the biome, one model would have a single popular (for example, elected) representative for every 2 million residents plus one government (for example, nominated) representative for each country, resulting in 120 council members.[3] Initially, at least, it is conceivable that the popular candidates are proposed by local community organizations within the sub-regions on the basis of their service to the community and involvement with promoting sustainability.

Council rules may require open (for example, internet-based) meetings with time made available for public comment and involvement. BSCs should have explicit mandates to receive (and be responsive to) inputs from local communities. A clear chain of two-way communication between local stakeholder groups and the BSC should be developed from early on, with training programs built into the process to enhance local capacity for participation. The council members could determine their own rules of procedure for internal discussion, formation of committees and office-bearers, voting on proposals, and so on, as developed in other legislative organizations around the world. Council members should also have staggered terms and term-limits, with provisions for rotating terms for committee membership, and generally adopt rules to reduce the opportunity for the consolidation of power while remaining an effective and informed governing body.

FUNCTIONS OF THE BSC

The principal mandate of the BSC should be to maximize both ecosystem protection and human welfare within the biome. Its starting framework should therefore be organized around basic principles of sustainable development (as defined by the Brundtland Commission). To the extent possible, the nomination or election of council members should be based on their stated and demonstrated commitment to these principles at local, regional and global scales. Since biomes typically stretch across many countries, and given the strong representation of local communities, it is unlikely that any single government's recalcitrance will disrupt the BSC's overall commitment towards sustainability.

Each BSC should be provided with well-trained support staff in the form of a small technical secretariat whose functions would include data collection and analysis as well as local capacity building. Large international environmental and development organizations should lend their support and technical resources to these secretariats, as should academic and professional ecologists and the staff of well-organized environment and forestry departments of national governments. Such support could include training as well as the provision of information and models and consulting services, as necessary. To retain the legitimacy of the council, however, the involvement of outside organizations should be restricted to the sharing of resources and capacity building. Special efforts also need to be made to ensure gender parity for both the BSC and secretariat staff with special attention given to the traditional role of women as custodians of natural resources.

The idea of developing formal transnational organizations to take on the responsibility for monitoring, evaluating and eventually regulating ecosystem health and services is not entirely unprecedented. The Arctic Council can be viewed as an early example of a BSC, in which governmental and non-governmental actors are focused on the social, economic and environmental conditions of the Arctic tundra.[4] However, it does not include popular representation or regulatory function. The BSCs would also differ from the forestry and environment departments of the countries they represent. In the early years, they would complement each other, but over time, it is meaningful to consider the BSCs relegating the latter to a relatively minor role. In the next section, the logic of this apparent concession of sovereignty will be explained, but first a scenario for how the BSC's functions will evolve will be laid out.

One of the first major responsibilities of the BSCs would be to review the Millennium Ecosystem Assessment's sub-global assessments that are relevant to the respective biomes and to enhance them with local knowledge to understand threats, opportunities and best practices. Subsequently, the councils should develop guidelines and capacity-building exercises that seek to tap into the self-interest of local residents to become custodians of their local surroundings. These early tasks will be useful in building trust in the organization and will help the BSC itself develop its own institutional capacity. While the BSCs will not have political authority in the sense of being able to enforce changes in practices, their quasi-public role as elected bodies will provide it with the legitimacy needed to advise national and local governments as well as local residents on how to improve the environmental efficiency of extracting ecosystem services. Its muted though discernible presence as a responsible and elected body will in turn provide multilateral organizations and national governments with the confidence to cede new functions to them, much the same way that public–private partnerships are developing between governments and non-governmental organizations (NGOs) and profit-driven entities.

Over a period of time, the BSCs themselves would authorize performance-based incentives to local community groups in order to maximize ecosystem protection without diminishing services. For instance, much like utility commissions that are mandated to set rates for water and electricity services, the committees set up by the BSCs would be able to set up a system of rewards and penalties depending on how ecosystem services are managed for, say, a given watershed or coastal zone. The authority for such rule-making would be contingent on two crucial, though not unrealistic, legal and political steps taking place. The first would be the negotiation of an international treaty based on the Convention on Biological Diversity to provide BSCs with the authority to form trans-border rules relating to ecosystem protection. The second is the ratification of the treaty, which would entail the loan of member government resources and personnel for enforcement of the BSC's rules.

To the extent that the rule-making power of the BSCs will be limited to adaptive and flexible forms of performance-based pricing and regulation of services tied to such issues as the prevention or management of floods, drought and land degradation without causing direct interference in more politically sensitive concerns such as land-tenure or water rights, it seems likely that they will have effective authority to implement their policies. But also, because they would have gained legitimacy as elected bodies and through their performance and stakeholder involvement during their formative years, they would face less resistance from individual governments than would, say, civil society organizations that operate on a global scale. Similarly, by learning from existing participatory forms of ecological stewardship such as joint forest management, the BSCs will also gain legitimacy and political power 'from below', especially if they continue to ensure room for participatory mechanisms to filter up into their rule development.

BSCs may occasionally find themselves being at cross-purposes with a country's environmental policies and even with one other. For instance, a given country's national wildlife protection policy may be inconsistent with a BSC's global concern to allow sustainable hunting of certain species to provide food and income for local populations. Similarly, species encroachment or resource depletion across neighboring biomes may result from a BSC's policies, but have adverse implications for its neighboring BSC. It is obviously impossible to anticipate the full range of such conflicts, but setting in place adequate mechanisms for arbitration and negotiation may alleviate their intensity. It may, for instance, be good practice to set up an ombudsman's office at the international level early in the process of setting up BSCs. Similarly, biennial summits of all the BSCs may be useful for trading ideas and building trust among all the BSC members.

Funding for the BSCs will be likely to come from two different sources. In the early years, multilateral donor agencies and private foundations will provide a pool of funds for creating the BSCs and their secretariats. Given their largely advisory rather than regulatory role during this period, the resources required for the BSCs will be modest relative to the substantial sums of money expended by the donor community on disparate and often poorly coordinated conservation efforts. Together, the 15 BSCs will probably support fewer than 1000 full-time staff and, with the addition of projects and infrastructure, will probably require no more than about $150 million per year during the first five years or so of their operation. During this period, efforts could be made to design and collect a small excise on revenue streams generated by the ensuing improvements in ecosystem services with no net losses for the stakeholders. In addition, revenucs from ccotourism as well as a reasonable fraction of national governments' forest protection budgets pooled together could contribute to the BSCs. Alternatively, if a Tobin tax[5] were to be levied on all international transactions, a small portion could be used to fund the BSCs on the grounds that much of global commerce is responsible for increasing the ecological footprint of modern lifestyles.

HOW WILL THE BSCS ACQUIRE SOVEREIGNTY?

The single most common criticism that a proposal like the one outlined above will face is that it is naive to expect countries to cede their sovereignty to a regional or international authority, even if it only covers a few relatively minor sectors from the standpoint of national security. Under what possible circumstance, one might ask, will the United States Forest Service and the Environmental Protection Agency allow a handful of international agencies to take over their key functions? Similarly, why would Sweden abide by the ruling of the Tundra BSC?

One answer to this criticism is quite straightforward. Since around the time of the Bretton Woods agreement, the world has seen several international regimes to which countries cede their sovereignty. Thus, the Nuclear Non-Proliferation Treaty has required all but eight countries (three that did not sign it plus five that were 'grandfathered' as nuclear powers) around the world to be subject to what one might ordinarily call intrusive regulations. More recently, the World Trade Organization has routinely developed rulings with which even the most powerful countries in the world have been forced to comply. The European Union is yet another example of rule-making at the supra-national level, a concession that was agreed to by the member states on the understanding that it was advantageous to all to have a continent-wide regime to govern certain issues like standards for education, health and the environment.

This is not to suggest that these institutions are not resisted; there are of course bitter conflicts related to all the examples mentioned above, but the commitment of national leaders to maintain them is sufficiently strong that it seems unlikely that they will unravel easily. Moreover, in each case, their legitimacy has already been sealed by the countries themselves, whose legislatures ratified them in the past. In other words, the political question of whether or not to allow outside institutions to play a role in domestic affairs has already been addressed historically, arguably in a reasonable way, inasmuch as the parties that took the decision were legitimate representatives. Barring the rare case of a later government backtracking on an international agreement formed by a previous one,[6] many supra-national regimes seem surprisingly durable.

A second issue has to do with the relatively unobtrusive framework of the BSC proposal. Because of its phased introduction, it should be unthreatening in the early years precisely because no government function would be replaced. Rather, it would operate through a series of learning steps and confidence-building measures, developing trust both from its constituency and from government agencies and the executives in the countries in which it operates. To the extent that its subsequent development would itself be contingent on an international agreement being drawn, the BSC would not be threatening to national governments. Furthermore, since the precise boundaries of biomes are hard to delineate and because they are non-contiguous across continents, the BSC need not ever be viewed threateningly as a political unit having territorial borders, but rather as a functional governance entity concerned with particular types of ecosystems. Conversely, a single country may have many biomes, so that the national government may justifiably think of itself as literally 'outsourcing' its ecosystem management functions to multiple BSCs.

Third, national governments are themselves forming a variety of networks while recognizing the need for broader strategic cooperation on critical issues. Examples can be found in the Alliance of Small Island States, which is focused on addressing the impacts of climate change, and the G20, which is developing a concerted developing country trade strategy. Sometimes these networks are less formally defined and comprise government officials and legislators. For instance, the Global Legislators for a Balanced Environment (GLOBE) was founded in 1989 primarily in the form of an environmental NGO with parliamentarians who seek to share information and potentially develop coordinated policy on the environment. There are several other networks of legislators, judges and bureaucrats through which the participants formulate new ideas together and pursue common goals, resulting in what has been termed 'disaggregated sovereignty', a situation in which the state is not disappearing, but where its component institutions are actively seeking collaborative relationships across borders (Slaughter, 2004).

Whereas traditional concepts of sovereignty emphasized separation into territorially independent groups, new ideas about sovereignty focus on the positive capacity to participate collectively in international institutions to address global and regional problems collectively by accepting mutual obligation (Chayes and Chayes, 1995). In this context, the BSCs can be seen as constituting yet another network that has the unique feature of strengthening existing ad hoc transnational arrangements among governmental and non-governmental agencies, but with the additional legitimacy of democratic input. In some ways, nevertheless, the BSC involves a more radical shift in sovereignty than, say, climate treaties, whose enforcement would take place primarily at the national level. For, even if national governments were to provide support for enforcement capability, the BSCs would have the ultimate authority to penalize and reward parties under its jurisdiction, albeit on a narrower basis.

In general terms, the tendency is for these types of institutional arrangements to confer rights and solicit responsibilities at different levels based on the principle of subsidiarity where governing bodies perform tasks that are best suited to their scale of operation. Thus, local authorities may be associated primarily with the management of the local economy, policing functions, and service provision of education, housing, transportation and health care; eco-regional governments would be responsible for the planning aspects of many of these same concerns within the context of ecosystem demands and constraints; and continental scale administrations and global politics would be relevant for resolving inter-regional conflicts and managing commerce, and relate generally to matters of worldwide concern.

Lastly, it is important to point out the ethical basis for developing global institutions that may end up restricting a few privileges, especially for the wealthy countries that are typically most resistant to such limits, in order to enhance ecosystem services and human welfare for all. Thomas Pogge thus explains why it is no longer tenable to pretend that one's supposed private actions are immune from global consequence.

> We are not bystanders who find ourselves confronted with foreign deprivations whose origins are wholly unconnected to ourselves. First, their social starting positions and ours have emerged from a single historical process that was pervaded by massive grievous wrongs. Second, they and we depend on a single natural resource base, from the benefits of which they are largely, and without compensation, excluded. Third, they and we coexist within a single global economic order that has a strong tendency to perpetuate and even aggravate global inequality (Pogge, 2001: 11).

CONCLUSIONS

At the dawn of the twenty-first century, there is little doubt that the future of humanity rests on our ability to retool our institutions to maintain the vitality of the fragile ecosystems that support us. It is also increasingly clear that national governments alone cannot be entrusted to the task, given their competing priorities and the poor correlation between political and natural boundaries. Furthermore, the non-linearities of ecosystem change and the complex relationships between human activity and ecosystem services call for political responses that stretch the capacity of any given nation-state, but also entail local knowledge that is often too fine-grained to be reliably addressed by any global inter-governmental entity. Finally, the conventional picture of distinct societies, cultures and publics having (relatively) self-contained obligations towards each other and the environment is increasingly hard to defend. An incremental adaptive institutional framework that is supra-national, but sub-global offers, therefore, a practical way to govern the global commons.

NOTES

1 This is really an identity which relates Income (I) in a given society with the product of Population (P) and measures of Affluence (A) and the level of Technology (T).

2 Note that the biomes described in the MA correspond roughly to those defined by the World Wildlife Fund, which are themselves agglomerates of more than 825 terrestrial eco-regions (http://www.worldwildlife.org/science/ecoregions.cfm). Although the eco-regions are even more uniform biological units than biomes, many are sub-national and it would be potentially self-defeating to treat them as independent governance entities.

Also, the borders of biomes should not be considered static boundaries, but as corresponding roughly to crossover regions across areas with different bioclimatic characteristics.

3 A similar institutional form for a Global Parliamentary Assembly has been suggested by Falk and Strauss (2001), except that they exclude the involvement of government representatives. In the case of the BSC, state representation is a double-edged sword: in some situations it could help strengthen its power by lending official support for recommendations developed by the Council; in others it could weaken it through obstruction. Nevertheless, the dominance of popular representation in this particular example is expected to mitigate the negative effects of governmental presence in the BSC.

4 The Arctic Council is an inter-governmental forum for cooperation among national governments and indigenous peoples spanning the Arctic region. In addition to the Foreign Ministers of eight countries, there are Permanent Representatives from six organizations for indigenous peoples in the Council. The Council currently functions as a high-level forum for developing joint policy guidelines for national governments to implement, based on research findings it sponsors. More information on the Arctic Council and its activities and governance structure can be found at http://www.arctic-council.org.

5 Tobin taxes refer to the proposal made by the Nobel laureate, James Tobin, to assess a small tax (less than 0.5 per cent) on all foreign exchange transactions to deter speculation on currency fluctuations. The rate would be low enough to be insignificant for longer term investors, but still serve to thwart large-scale short-term currency manipulations.

6 This did happen with the Kyoto protocol by the Bush and Howard governments in the US and Australia, respectively. However, it is important to note that neither country's legislature had yet ratified it.

REFERENCES

Chayes, A. and A.H. Chayes (1995), *The New Sovereignty: Compliance with International Regulatory Agreements*, Cambridge, MA: Harvard University Press.

Falk, R. and A. Strauss (2001), 'Toward global parliament', *Foreign Affairs*, **80** (1), 212–20.

Olson, D.M., E. Dinerstein, E.D. Wikramanayake, N. Burgess, G.V.N. Powell, E. Underwood, J.A. D'Amico, H.E. Strand, J.C. Morrison, C.J. Loucks, T.F. Allnutt, J.F. Lamoreux, T.H. Ricketts, I. Itoua, W.W. Wettengel and Y. Kura (2001), 'Terrestrial ecoregions of the world: a new map of life on earth', *Bioscience*, **51** (11), 933–8.

Pogge, T.W.M. (2001), *Global Justice*, Oxford: Blackwell.

Slaughter, A.M. (2004), *A New World Order*, Princeton, NJ: Princeton University Press.

13. Smart globalization: Opportunities for business, scientific and political leadership

David Jhirad

INTRODUCTION: ECOSYSTEM THREATS AND OPPORTUNITIES

The Millennium Ecosystem Assessment (MA) is the most comprehensive scientific audit of the effect of human activities on the world's ecosystems. The audit reveals that the way society has obtained its food, fresh water, timber, fiber and fuel over the past 50 years has seriously threatened the earth's capacity to deliver sustainable growth. The MA report states that 'any progress achieved in addressing the goals of poverty and hunger eradication, improved health, and environmental protection is unlikely to be sustained if most of the ecosystem "services" on which humanity relies continue to be degraded' (MA, 2005: 2). A somber conclusion is that the world's nations are embarked on a path that runs counter to the Millennium Development Goals agreed to by world leaders at the United Nations in 2000.

The twentieth century has witnessed a period of unprecedented economic growth that lifted masses of people out of abject poverty. It has brought technological innovations and societal advances in public health, agriculture, energy and transport systems. The current wave of economic globalization has led to an extraordinary enhancement of human well-being, and longer, better lives for many of the world's people. But the benefits of globalization have not yet reached far enough: over three billion people live in poverty, deprivation and disease, and their natural trust fund – the fields, fisheries, forests and waterways on which they depend – is increasingly at risk.

The MA makes a compelling and urgent case for greatly improved management of our vast natural capital asset to provide income, wealth and benefits for future generations. Building on this premise, this chapter argues that recent breakthroughs in the business, scientific and political arenas provide hopeful signs that such an outcome, though ambitious, is achievable. Increasingly, there is a realization that preserving and enhancing our planetary trust fund is vital for sustained economic growth and global security.

Major transformations are occurring in private and public decision-making, incorporating incentives that embrace longer time frames and geographic scales. In addition, science and technology innovations are providing powerful decision tools for analyzing and predicting ecosystem behavior. Such developments suggest new norms for action and new forms of global governance that could be effective and inclusive.

This chapter provides examples of major multinational companies and banks that are incorporating ecosystem risks and opportunities in their core business strategy. Increasingly, corporate leaders have recognized the business risks posed by global climate change, loss of biodiversity and water scarcity, and have seized opportunities for competitive advantage through effective management of ecosystem risk. Furthermore, large corporations as well as small and medium enterprises are beginning to prosper in new markets that create livelihoods for the poor, while delivering goods and services with social and environmental benefits. While stressing the importance of steering private investment in pro-poor directions that enhance ecosystem management, the chapter points out that good governance is the key to poverty reduction and effective management of ecosystems. Collectively, such approaches constitute 'smart globalization'.

An essential but often ignored component of 'smart globalization' is the need for the best available data on ecosystems' goods and services, much of which is incomplete or missing. The possibilities of 'tipping points' caused by unpredictable, abrupt change provide a powerful imperative to conduct careful and continuous measurement and monitoring of the health of planetary ecosystems. The chapter briefly surveys some encouraging approaches to ecosystem measurement, which is also central for efficient market functioning and sound natural resource management. But there is an urgent need to institutionalize this process and make it on-going.

The chapter argues that lasting solutions to ecosystems' degradation demand effective and innovative global governance mechanisms, such as the creation of a Leaders Forum of 20 nations (L20). Such a mechanism would allow major emerging economies to share global challenges and contribute to the solution. It would also permit an approach that transcends national, ministerial and sectoral boundaries, and that fosters 'forward engagement' with the long-term future. The L20 already exists at the level of finance ministers, where it is known as the G20, and could be created 'by simple invitation without complex legal action and without opening up the Pandora's box of deciding who is in and who is out of the group' (Linn and Bradford, 2006). Once formed, the L20 could be an influential vehicle for the reform of the international financial institutions and UN agencies.

Ensuring sustainable global economic growth requires a combination of smart globalization, continuous monitoring and management of ecosystems, and the formation of a Leaders Forum. The pivotal role of ecosystems in generating rural employment worldwide is illustrated below.

GENERATING EMPLOYMENT THROUGH ECOSYSTEMS

As the MA points out, the transformation of ecosystems over the past five decades dwarfs the cumulative impact over the preceding centuries. This degradation is undercutting rural livelihoods (MA, 2005: 2). The World Resources Report (WRI, UNDP, UNEP and World Bank, 2005) documents the universal importance of income from ecosystems' goods and services – environmental income – to poor households in Asia and Sub-Saharan Africa. Income from crops, home gardens, livestock rearing and wild products from woodlands can constitute up to 70 per cent of family incomes in rural areas. Also, about half of all jobs worldwide depend on agriculture, forestry and fishing. Yet agricultural subsidies and other import restrictions in developed countries make it difficult for developing country farmers to compete on the world market (WTO, 2003: 10, 22). Figures from the UN Food and Agriculture Organization (FAO) show that in 2001, the percentage of the active work force employed in agriculture, fisheries and forestry varied from 7 per cent in the developed countries to 54 per cent in the developing countries, for a global average of 44 per cent. Table 13.1 shows the dependence of employment on ecosystems in the major regions of the world, and in selected countries.

The challenge for the future is to accelerate poverty reduction and enhance human well-being, while repairing and enhancing the ecosystems that create wealth. Addressing this challenge will require better and smarter globalization which does not undermine the capacity of ecosystems to produce services, and equitably distributes the costs and benefits.

Table 13.1 Ecosystems bring jobs

Percent of global workforce employed in agriculture, fisheries and forestry, 2001	
Region/Country	Percent of active workforce
World	44
Developed Countries	7
Developing Countries	54
Asia and Pacific	60
Cambodia	70
China	67
India	59
Nepal	93
Latin America and the Caribbean	19
Bolivia	44
Guatemala	45
Haiti	62
Near East and North Africa	33
Afghanistan	67
Turkey	45
Yemen	50
Sub-Saharan Africa	62
Burkina Faso	92
Ethiopia	82
Niger	88
Tanzania	80
Countries in transition	15
Albania	48
Azerbaijan	26
Tajikistan	33

Source: FAO (2004) pp. 169–174, Table A4

The challenge for the future is to accelerate poverty reduction and enhance human well-being, while repairing and enhancing the ecosystems that create wealth. Addressing this challenge will require better and smarter globalization which does not undermine the capacity of ecosystems to produce services, and equitably distributes the costs and benefits. A sophisticated market economy, based on sound and equitable governance, and driven by science and technology innovation, has proved to be a powerful combination for generating lasting prosperity.

The MA explored two development paths, one in which the world becomes increasingly globalized and the other in which it becomes increasingly regionalized. It also examined two different approaches to ecosystems management, one in which actions are reactive and most problems are addressed only after they become obvious, and the other in which ecosystem management is proactive and policies deliberately seek to maintain ecosystem services for the long-term (MA, 2005: 71). This chapter draws on the best features of global interconnectedness and trade, driven by the innovative and inspired leadership of business, scientific and political leaders. Examples of businesses that are incorporating ecosystems risk and opportunity into their global corporate strategy are discussed in the section that follows.

MANAGING ENVIRONMENTAL RISKS: INCENTIVES FOR COMPETITIVE ADVANTAGE

The MA outlines actions that business can take to improve their bottom line, reduce degradation of ecosystems, and benefit human well-being. The MA states that business can be a 'very positive force' in pursuing new business opportunities and markets, reducing its environmental footprint, developing and commercializing new technology, and establishing effective partnerships.

Major companies such as General Electric, Walmart and Coca-Cola are implementing serious environmental initiatives as a central part of their core business strategy. This 'green revolution' among corporate leaders flows from a recognition of the business risks posed by global climate change, loss of biodiversity and water scarcity, and the opportunities to seize competitive advantage through strategic management of these risks.

Shareholders – and many mainstream environmental organizations – were buoyed by General Electric's 'Ecomagination' announcement in May 2005 that it would reduce its carbon dioxide emissions and double its research and development budget in energy efficient, clean technology to $1.5 billion by 2010. Jeff Immelt, the CEO, committed GE to a set of green goals going far beyond current US government policy. He pledged that by 2012, GE would boost its energy efficiency by 30 per cent, and cut its greenhouse gas emissions from over 40 per cent above today's levels, representing 'business-as-usual' to 1 per cent below.

As Immelt put it, 'We are living in a carbon-constrained world where the amount of CO_2 must be reduced.' He made the point that investors abhor policy uncertainty, and that innovation had suffered as a result. 'America is the leading consumer of energy. However, we are not the technical leader. Europe is the major force for environmental innovation.' This illustrates that major companies can be trailblazers even in the face of government inaction and paralysis. As Immelt remarked, 'We are investing in environmentally cleaner technology because we believe it will increase our revenue, our value and our profits, not because it is trendy or moral, but because it will accelerate our growth and make us more competitive.' Lash and Wellington (2007) have provided a review of leading industrial responses to climate change in their recent paper on competitive advantage on a warming planet.

The results of this initiative have been spectacularly successful. GE has reported over $10 billion in revenues from ecomagination products and services in 2005 – offering a clear path to a planned $20 billion in annual sales in 2010. Orders have doubled to nearly $17 billion for a range of 30 eco-technologies, such as compact fluorescent light-bulbs, photovoltaics, biofuels, wind turbines, desalination platforms, fuel-efficient jet engines, and high-efficiency gas turbines. (GE's ecomagination report is available at www.ge.com/ecoreport.)

Walmart has also made major commitments to improve environmental performance. It plans to invest $500 million annually in energy programs that include cutting its energy use by 30 per cent, obtaining renewable electricity from solar photovoltaics and wind farms, and doubling the fuel efficiency of its shipping fleet. Walmart also seeks to green its supply chain by requiring environmentally friendly and sustainable products. As their CEO Lee Scott puts it, 'We believe that these initiatives will make us a more competitive and innovative company' (Esty and Winston, 2006).

MANAGING ENVIRONMENTAL RISKS: INCENTIVES FOR BANKS AND CAPITAL MARKETS:

Many banks have realized that environmental and social risks, while often difficult to quantify, pose threats to their reputation and profitability. One of the early leaders to embed social and environmental criteria deeply in its investment portfolios was ABN AMRO, a Dutch Bank with over half a trillion euros in assets. In recognition of its leadership, the Bank earned the World Environment Center's 2006 Gold Medal for International Corporate Achievement in Sustainable Development. Citigroup executives have also made commitments to screen projects for adverse impacts on ecosystems, and to deny loans to companies with illegal logging violations. Citigroup is greatly expanding its investments in renewable energy companies and projects, and has started to report on greenhouse gas emissions from all power projects in its investment portfolio (Esty and Winston, 2006; Citigroup, 2006).

These two banks were among ten banks that announced an agreement in 2003 known as the 'Equator Principles', which set up new standards for bank decision-making on project finance. Developers of infrastructure projects such as power stations and pipelines are required to demonstrate that the environmental and social impacts of such projects have been carefully assessed and mitigated. There are now over 40 banks that are signatories to the Equator Principles, representing an overwhelming portion of international project finance.

In the energy sector alone, the International Energy Agency (IEA) estimates that 20.2 trillion dollars will be required for global infrastructure investment over the next 25 years (IEA, 2006). The IEA's position is that the current global energy path is unsustainable from a financial, security and climate stabilization perspective, and that energy efficiency is the centerpiece of a transition to sustainability. The major challenge faced by all nations is to re-direct this massive capital flow to clean, efficient energy and transport systems. An energy revolution of this magnitude could profoundly reduce the stress on ecosystems, increase security, and stabilize greenhouse gas emissions.

Many of these powerful initiatives by business leaders and international organizations could be replicated by all multinational companies, as well as thousands of small and medium enterprises. But for such initiatives to be truly legitimate and successful on a global scale, they need to see a business opportunity in the neglected market of 4 billion people at the bottom of the global economic pyramid (BOP).

CRAFTING ENTERPRISE SOLUTIONS TO POVERTY REDUCTION

This BOP market, with a per capita income of less than \$1 500/year, represents an aggregate market of over 9 billion dollars. This market represents a great opportunity for the private sector to increase revenue, drive economic growth, create jobs, and stimulate technological innovation. The myth is that the poor have no money. The reality is that the collective buying power of low-income households in developing countries is highly significant. By working in concert with governments and International Financial Institutions (IFIs), the private sector is beginning to prosper in new markets that create livelihoods for the poor, and deliver goods and services with social and environmental benefits. Both large corporations and small and medium enterprises participate actively in this process (Prahalad, 2005; Prahalad and Hammond, 2002).

To be pro-poor, investors and borrowers need to incorporate environmental sustainability in their activities. The developers of power, oil, gas and mining projects will need to do a better job of managing risks to human health, as well as damage to rivers, fisheries and other ecosystems. Borrowers from the Equator banks may have to drop or change their plans to conform to environmental standards, as occurred in many of ABN AMRO's projects last year. However, while steering private investment in pro-poor directions that enhance ecosystems management is critical, it cannot achieve the desired outcome where bad governance is pervasive.

Private investment in hydrocarbons and other extractive industries has sometimes been associated with corruption, environmental degradation, social dislocation and impoverishment. Changing this will require more transparency, public participation and accountability. The Extractive Industries Transparency Initiative (EITI), launched by the British Government, is already proving successful. Royal Dutch Shell and BP have agreed to disclose detailed payment information on their oil operations in Nigeria and Azerbaijan, respectively. Investors representing over $7 trillion have endorsed EITI, and civil society organizations are using EITI as an instrument for government accountability. Endorsement of EITI by G-8 nations and oil producing countries would make a decisive difference to the lives of the poor who live in the 60 countries that depend on oil, gas, and mining revenues (Soros, 2005: 43).

For private investment to promote poverty reduction and environmental sustainability, significant changes are required in the current norms and institutions that govern development finance. Public international financial institutions – such as the multilateral development banks (MDBs) and bilateral export credit and guarantee agencies (ECAs) – could work with client governments and the private sector to integrate environmental, social and equity objectives into investment decision-making. The international community must also mobilize the concessional finance necessary to enable societies to choose environmentally and socially superior alternatives. Finally, bilateral donor agencies can act aggressively to strengthen the capacity of developing countries to support a just and fair private sector.

This smarter approach to economic globalization can work when the poor are empowered through access to information, participation and justice, and when they have legally recognized resource rights that allow them to manage, sell, rent and invest in ecosystem services. By partnering with the private sector to make credit available, and by improving the marketing and transport of goods, the poor can gain income and benefit from the wider marketplace that globalization affords.

The success of private investment approaches to poverty reduction and environmental sustainability will hinge on assigning financial and economic value to ecosystem services. What follows are some examples of effective economic cost–benefit analysis of ecosystem services that have emerged in recent years.

FORGING PRAGMATIC, MARKET-BASED SOLUTIONS

Market-based approaches, where informed by socially and environmentally responsible public policy, have been effective in forging solutions to some environmental problems. Many things in nature are clearly priceless. But economic cost–benefit analysis also has a crucial role to play in sound public policy. Emissions trading has been successful in reducing sulfur dioxide and nitrogen oxides, and tradable fishing quotas have reduced over-fishing (Aulisi et al., 2005: 11; Kura et al., 2004: 92; Ellerman et al., 2000: 315; NRC, 1999: 192). Innovative approaches are being used to assign value, and therefore to protect, 'ecosystem services', including those provided by coral reefs, mangrove swamps, water filtration, flood prevention and wildlife preservation. Rural people need to participate in and benefit from these approaches, and this will only happen if they have a degree of control over the process and the ecosystem 'assets' (WRI et al., 2005).

'Mainstreaming' sustainability means putting ecosystems goods and services on the financial balance sheet, and including it in risk assessment. Since the disruption of such services can be a significant component of project risk, insurance companies have a financial incentive to protect and enhance ecosystems' integrity. Economic valuation is only part of the solution, because not everything is for sale. But the growing consensus about where and when to apply such analysis is an important step forward in operationalizing the MA findings. But for markets to operate efficiently, there is a need to stop navigating blindly, and to establish better measurement and monitoring systems for ecosystems. Several innovative developments in measurement and monitoring are currently underway.

ESTABLISHING BETTER MEASUREMENT AND MONITORING SYSTEMS

Approaches built on good data and analytic rigor offer the best hope of changing policy as well as voters' minds. The MA surveyed the best available environmental data. But much of the information needed is incomplete or missing. As the report put it, 'Our knowledge of ecosystems has increased dramatically, but it simply has not kept pace with our ability to alter them.'

The possibility of unpredictable, abrupt change provides a powerful imperative to conduct careful environmental measurement. Scientists, engineers and management pundits always say, 'what matters gets measured'. Social scientists and even political scientists have turned to quantitative measurement and hard numbers. In contrast, environmental data is often primitive.

The MA was a pioneering effort to establish systematic data sets on ecosystems and the services they provide across the world. But there is a need to institutionalize this process and make it on-going. What is remarkable is that governments have been making decisions on environmental policy in the absence of valid baselines. Even deforestation estimates are only accurate to within plus or minus 50 per cent.

A host of technologies that can aid efficient market functioning, promote sound governance of natural resources, and protect the interests of the poor are emerging. Low-cost environmental data collection using high-resolution satellite mapping and remote sensing is one example. Tracking and monitoring devices are helping to reduce over-exploitation of fisheries. In Malaysia, conservationists use satellite tracking to keep count of elephants (WWF, 2005). Rural Indian farmers with high-speed access to the Internet receive on-line updates about market prices and the weather, making them more competitive (Annamalai and Rao, 2003: 1). Increasingly low-cost and accessible technologies are beginning to measure trends in deforestation, soil erosion and climate change. India, China and Brazil have launched their own satellites, and are sharing data with other developing countries. A possible model for international efforts to monitor and track ecosystem dynamics is the early warning system set up in the wake of the Asian Tsunami of 2004.

IMPLEMENTING EARLY WARNING SYSTEMS

The 2004 tsunami has spawned international, cooperative efforts to set up an early warning and detection system in the Indian Ocean, to mirror existing systems in the Pacific. There are also monitoring stations in the Pacific to track El Niño, an ocean current. In the early 1980s, many countries on the eastern shores of the Pacific suffered from severe weather associated with El Niño. But early warnings from the new monitoring system in 1997 and 1998 helped farmers and citizens avoid damage.

According to the US National Oceanic and Atmospheric Administration (NOAA), California incurred over a billion dollars less in damage in 1997–98, thanks to the early warning system for El Niño. NOAA believes that there is enormous potential to link national and regional monitoring systems, and to integrate different kinds of data. NOAA and many private academic and governmental partners in the US are constructing ocean observation systems that integrate biology, oceanography, chemistry and socioeconomic data. This data will allow systematic reporting on the status of marine and coastal ecosystems through Integrated Ecosystems Assessments (Crane, 2006).

Perhaps the biggest technological project in the history of ecosystems is nearing fruition. Sometimes referred to as the Hubble telescope of ecology, the US National Ecological Observatory Network (NEON) would cover the United States with a network of ecosystem observation systems. NEO will set up measurement stations in each of 20 'domains', representing every climate and ecosystem zone in the United States. By measuring a host of variables, including carbon dioxide flux, nitrogen oxides, water vapor, solar irradiation, ozone concentrations, temperature, wind speed, levels of bacteria and pollen, the goal is to understand how ecosystems respond to human-induced changes in climate, land use and invasive species (Hopkin, 2006). The funding requested for NEON is $12 million in the 2007 budget for the National Science Foundation. This figure rises to well over $100 million in 2011. If funded, this project could offer opportunities for international collaboration in ecosystem monitoring and measurement.

Bilateral and multilateral agencies indicate they are willing to invest resources in this area, reversing years of neglect. NOAA, for example, has indicated that it will provide environmental satellite data free to poor countries. All governments, businesses and international institutions need to make major investments in acquiring, managing, sharing and using ecosystem data. Such data needs to be coordinated and crystallized into 'operational intelligence', so that all these institutions will no longer navigate blindly. Institutions such as the National Academies of Science in each country, or the Intergovernmental Panel on Climate Change (IPCC), could be charged with the responsibility for painting a complete picture by 'connecting the dots'.

Hopefully, it will not be long before existing, fragmented databases, including poverty maps and ecosystem services, are overlaid routinely on the sites of timber harvests, mining operations, or industrial plants to identify how these developments might affect poor families in the region. An excellent example is the Kenya Atlas (WRI, ILRI, DRSRS and CBS, in press).

Science and technology innovation provide an essential platform for a smart globalization strategy. And the dynamic and unpredictable nature of critical ecosystems is a powerful imperative for effective communication between scientists and the global policy community.

BUILDING PARTNERSHIPS WITH THE SCIENTIFIC COMMUNITY: UNDERSTANDING AND ANTICIPATING TIPPING POINTS

The MA demonstrates the shocks and surprises that can occur when ecosystems are subjected to prolonged stress and perturbation. As the report puts it, 'the living machinery of Earth has a tendency to move from gradual to catastrophic change with little warning. Such is the complexity of the relationships between plants, animals and microorganisms that these "tipping points" cannot be forecast by existing science'. The MA cites the example of the sudden collapse of cod stocks off the east coast of Newfoundland. There are few signs of the fish returning even 13 years after the closing of the main fishery.

The earth's climate is also a complex, non-linear system capable of experiencing abrupt change (Hansen et al., 2006). Small risks of catastrophic consequences – even if they cannot be precisely quantified – should not be ignored. Governments have a responsibility to take action to eliminate even the low probability of catastrophe. Such responsibility was recognized during the cold war when formulating policy governing the use of nuclear weapons.

Climatologists and the international policy community have focused mainly on gradual climate change scenarios. These scenarios rest on mathematical models that fail to capture the discontinuous and extreme events that characterize the earth's climate history. By investigating planetary history recorded in tree rings, ice cores and sediments, scientists have discovered several instances of significant abrupt climate changes over the last 100 000 years. Temperature changes in some regions of up to 16° degrees Celsius (28° Fahrenheit) have been observed, sometimes in the span of a decade. Altered climate patterns could last for many decades, as they did when the heat conveyor of the ocean, such as the Gulf Stream, collapsed about 8200 years ago.

A recent report to the Pentagon provides an alternative to the commonly accepted scenarios of gradual climate change. In one scenario, patterned after an event that occurred 8200 years ago, polar glaciers continue to melt, the waters of the North Atlantic lose their salinity, and the flow of the Gulf Stream ceases. Average annual temperatures drop by up to 5° F over Asia and North America, and 6° F in Northern Europe. The Southern Hemisphere experiences an increase of up to 4° F, all in a single decade. The report explores the destabilization of the geo-political environment brought about in such a scenario. It points out that wars could break out among old and new adversaries over access to water, food sources and energy supplies, and millions of refugees would cross national borders. *Fortune* magazine describes this scenario as 'The mother of all national security problems'.

The National Academy of Sciences has examined the earth's climate archives to provide the scientific evidence for abrupt climate change. Many scientists see the need for a new generation of mathematical models that accurately represent the behavior of complex, non-linear systems like the earth's climate. With such systems, smooth, predictable behavior is the exception rather than the norm. Even if current models cannot reproduce past discontinuities or predict the future, climate science points to an uncertain but non-zero risk of sudden and cataclysmic change. All ecosystems share such characteristics, whether local fisheries or the climate of the planet.

But the missing ingredient is often political leadership and vision. World leaders, the United Nations and the G-8 countries have recognized the imperatives of climate stabilization and the achievement of the Millennium Development Goals. Former President Clinton has launched the 'Clinton Global Initiative', which aims to form a global consensus to eradicate poverty, spread good governance, stabilize greenhouse gas emissions, and achieve religious reconciliation. The findings of the MA are critical elements in implementing these objectives. In addition, there is a compelling case for a new leaders' forum, to find global answers to global problems.

ESTABLISHING A NEW GLOBAL LEADERS FORUM: THE L-20

Paul Martin, the Prime Minister of Canada, has made a strong case (Martin, 2005) for a Global Leaders Forum, called the L20. This is based on the success of the G20,[1] a group of finance ministers and central bank governors from the G-8 and from emerging centers of economic power.

Over the last six years, as Paul Martin puts it,

> the G-20 countries have agreed on a set of common principles for sound domestic economic management: broadened the so-called Washington consensus on economic development to include emphasis on investment in social sectors such as health and education as necessary components of sustainable policies (this expanded set of priorities came to be known as the Montreal consensus); agreed on strong measures to fight the financing of terrorism; and adopted the Accord for Sustained Growth, which outlines policies to promote monetary and financial stability, enhance domestic and international competition, and empower people to participate successfully in markets.

The G20 countries represent nearly 90 per cent of the world's economic output, 75 per cent of all trade, and two-thirds of the world's population. The G20 illustrates the importance of international political leadership, because some decisions can only be made at this level.

Johannes Linn and Colin Bradford (Linn and Bradford, 2006) have argued that the many global challenges facing political leaders cannot be solved by 'yesterday's institutions'. They also suggest that the 'Group of Eight' be expanded to 20 key leaders as a new forum for global negotiation and decision-making, known as the L20. They state,

> the G-8 summit, in particular, is a forum of the eight industrialized countries that were the dominant powers of the mid-twentieth century. By excluding the emerging powers of the twenty-first century, it has become increasingly ineffective, unrepresentative and illegitimate. US global interests are best served by increasing the inclusiveness and effectiveness of the global steering process rather than sticking with the obsolete, overly formalized and unrepresentative G-8.

There are other good reasons to set up a Leaders Forum for mainstreaming the findings of the MA. First, major emerging economies could share this global challenge and contribute to the solution. Second, the challenges posed by the MA require an integrated approach that transcends national, ministerial and sectoral boundaries, and that fosters engagement with the long-term future. Third, the G20 already exists at the level of finance ministers, and could be created 'by simple invitation without complex legal action and without opening up the Pandora's box of deciding who is in and who is out of the group' (Linn and Bradford, 2006). Once formed, the L20 could be an influential vehicle for the reform of the international financial institutions and UN Agencies.

Such forms of effective and inclusive global governance will be required, as well as new norms for action. Profound changes are also on the horizon in private and public sector decision-making, incorporating incentives and values that span the globe and embrace longer time frames. Science and technology innovations are providing powerful decision tools for analyzing and predicting ecosystems behavior, and such tools can reinforce core values of morality, accountability and ethical stewardship.

The transition from an old paradigm in crisis to a new paradigm is not a simple incremental process. Writing about the structure of scientific revolutions, Thomas Kuhn (1970) states

> it is a reconstruction of the field from new fundamentals, a reconstruction that changes some of the field's most elementary theoretical generalizations as well as many of its paradigm methods and applications... When the transition is complete, the profession will have changed its view of the field, its methods, and its goals.

A similar process has been generated by the scientific findings of the Millennium Ecosystem Assessment.

Expanding the initiatives outlined in this chapter will require the building of powerful new global alliances around an emerging set of transformational ideas. The massive challenges of the twenty-first century require no less. These daunting challenges include stabilizing the global climate, enhancing the earth's ecosystems, catalyzing sustainable energy and transport systems, and eradicating destitution, deprivation and disease. These problems are too important for partisanship, or anti-corporate ideology. The new alliance for the Earth needs big ideas, harnessed to a vision of a better future, and built on positive and practical solutions, not dire warnings of gloom and doom. Such solutions will require unique global partnerships that will harness the engines of markets, science, technology, entrepreneurship and empowered civil societies.

NOTE

1 The G-20 includes Argentina, Australia, Canada, China, France, Germany, India, Indonesia, Italy, Japan, Mexico, Russia, Saudi Arabia, South Africa, South Korea, Turkey, the United Kingdom, and the United States. The EU is also a member, as is the managing director of the IMF, the President of the World Bank, the chairperson of he IMF's International Monetary and Financial Committee, and the chairperson of the IMF's and World Bank's Development Committee.

REFERENCES

Annamalai, K. and S. Roa (2003), *What Works: ITC's e-Choupal and Profitable Rural Transformation*, Washington, DC: WRI.

Aulisi, A., A. Farrell, J. Pershing and S. VanDeveer (2005), *Greenhouse Gas Emissions Trading in the US States: Observations and Lessons from the OTC NOx Budget Program*, WRI White Paper, Washington, DC, WRI, online at http://pdf.wri.org/nox_ghg.pdf.

Citigroup (2006), *Investing in Solutions to Climate Change*, Citigroup, online at http://www.wri.org/climate/pubs_description.cfm?pid=4201.

Crane, M. (ed.) (2006), *Earth Systems Monitor*, **15** (2).

Ellerman, A., P. Joskow, R. Schmalensee, J. Montero and E. Bailey (2000), *Markets for Clean Air: the US Acid Rain Program*, Cambridge: Cambridge University Press.

Esty, D. and A. Winston (2006), *Green to Gold*, New Haven: Yale University Press.

Food and Agriculture Organization of the United Nations (FAO) (2004), *State of Food and Agriculture 2003–2004: Agricultural Biotechnology – Meeting the Needs of the Poor?*, Rome, FAO, online at http://www.fao.org/waicent/faoinfo/economic/esa/en/pubs_sofa.htm.

General Electric (GE) (2005), *Taking on Big Challenges*, GE, online at http://www.ge.com/ecoreport.

Hansen, J., M. Sato, R. Ruedy, K. Lo, D. Lea and M. Medina-Elizade (2006), 'Global climate change', *Proceedings of the National Academy of Sciences*, **103**(39), 26 September, online at http://www.pnas.org/cgi/content/abstract/103/39/14288.

Hopkin, M. (2006), 'Spying on nature', *Nature*, **444**, pp. 420–21, November 23.

International Energy Agency (IEA) (2006), *World Energy Outlook 2006*, Paris: IEA.

Kuhn, T. (1970), *The Structure of Scientific Revolutions*, 2nd edn, New York: New American Library.

Kura, Y., C. Revenga, E. Hirsoshino and G. Mock (2004), *Fishing for Answers: Making Sense of the Global Fish Crisis*, Washington, DC: WRI, online at http://pubs.wri.org/ pubs_description.cfm?PubID=3866.

Lash, J. and F. Wellington (2007), 'Competitive advantage on a warming planet', *Harvard Business Review*, **85**(3).

Linn, J. and C. Bradford (2006), *Pragmatic Reform of Global Governance: Creating an L20 Summit Forum: Policy Brief #152*, Washington, DC:

Brookings Institute. Online at http://www.brookings.edu/comm/policybriefs/pb152.htm.

Martin, P. (2005), 'A global answer to global problems', *Foreign Affairs*, May/June.

Millennium Ecosystem Assessment (MA) (2005), *Ecosystems and Human Well-Being: Synthesis*, Washington, DC: Island Press, online at http://www.milleniumassessment.org/products.aspx.

National Research Council (NRC) (1999), *Sharing the Fish: Toward a National Policy on Individual Fishing Quotas*, Washington, DC: National Academy Press.

Prahalad, C.K. (2005), *The Fortune at the Bottom of the Pyramid*, Upper Saddle River: Wharton School Publishing.

Prahalad, C.K. and A. Hammond (2002), 'Serving the world's poor, profitably', *Harvard Business Review*, **80**(9), 48–58.

Soros, G. (2005), 'Transparency can alleviate poverty', *Financial Times*, **43**, 17 March.

World Resources Institute (WRI), International Livestock Research Institute (ILRI), Department of Resource Surveys and Remote Sensing, Ministry of Environment and Natural Resources of Kenya (DRSRS) and Central Bureau of Statistics, Ministry of Planning and National Development of Kenya (CBS) (in press), *Nature's Benefits: An Atlas of Ecosystems and Human Well-Being in Kenya*, Washington, DC and Nairobi: WRI.

World Trade Organization (WTO) (2003), *Annual Report 2003*, Geneva: WTO, online at http://www.wto.org/english/res_e/reser_e/annual_report_e.htm.

World Wildlife Fund (WWF) (2005), *AREAS Project: Technology*, WWF, online at http://www.worldwildlifefund.org/action/areasproject/technology.cfm.

WRI, United Nations Development Programme, United Nations Environmental Programme and the World Bank (2005), *World Resources 2005: The Wealth of the Poor – Managing Ecosystems to Fight Poverty*, Washington, DC: WRI, online at http://multimedia.wri.org/worldresources2005.cfm.

14. A commission to spur investments in ecosystem services for the poor

Frances Irwin

INTRODUCTION

Reducing poverty requires turning our growing understanding of the relationship between healthy ecosystems and human well-being into investments in ecosystem services. The Millennium Ecosystem Assessment tells us that regions facing the greatest challenges in overcoming ecosystem degradation and maintaining their remaining natural assets are the same regions facing the greatest development challenges such as improving income, health and education (MA, 2005a: 61). It is South Asia and sub-Saharan Africa that have the largest number of undernourished people and the regions where food production lags, for example. The degradation of ecosystem services – food, fuel, drinking water, spiritual and religious values, and other benefits nature provides – is harming many of the world's poorest people and is sometimes the principal factor in causing poverty (MA, 2005a: 13).

The Assessment finds that changes in policies, institutions and practices show promise of reversing ecosystem degradation and meeting development needs. Not surprisingly, the changes required would be significant (MA, 2005a: 20, 92). Most poor countries now lack fundamental governance measures such as providing the public with access to information (McIntosh, 2006). Local communities may not have the right to use resources such as fisheries and forests. Decisions to invest seldom factor in longer-term environmental costs such as the loss of flood protection when a mangrove is turned into a shrimp farm or the loss of water purification services from wetlands destroyed by the construction of dams. Without improvements in governance, the Assessment warns that degradation of ecosystem services could grow significantly worse and the ambitious Millennium Development Goals will not be met (MA, 2005a: 61).

Adopted in 2000 by world leaders as part of the United Nations Millennium Declaration, the eight Millennium Development Goals set an agenda for cutting world poverty by half by 2015. Through a global partnership, they aim to achieve universal primary education, promote gender equality, reduce child mortality and improve maternal health, combat disease, and ensure environmental sustainability in the next decade.[1] The Goals refocus the development agenda and, by providing specific time-bound targets, begin to provide a means of public accountability. However, the links between reducing poverty and investing in ecosystem services are not yet widely understood and put into practice. As a result, few countries are on track for meeting the Millennium Development Goal for environmental sustainability.

This chapter recommends using momentum from the Millennium Ecosystem Assessment and other initiatives to establish a Commission on Macroeconomics and Ecosystem Services for Poverty Reduction. Such a Commission would work with national finance and planning agencies to stimulate broad national processes to propose, fund and track use of environmental investments that reduce poverty. This chapter first briefly discusses the need to increase investments in ecosystem services. It then details three barriers to making these investments and outlines how the Millennium Ecosystem Assessment and work by groups such as the Poverty–Environment Partnership[2] provide a growing base for overcoming these barriers. It concludes by drawing on the experience of the earlier Commission on Macroeconomics and Health to propose how a Commission on Macroeconomics and Ecosystem Services for Poverty Reduction might stimulate national investment in ecosystem services clearly linked to reducing poverty.

INVESTMENTS IN ECOSYSTEM SERVICES ARE INADEQUATE

Countries are not investing in ecosystem services at a level that will increase the wealth of the poor. The Millennium Ecosystem Assessment finds that 'few responses to poverty reduction consider the importance of ecosystems to improve basic needs of the poorest' (MA, 2005b: 9). Low-income countries prepare Poverty Reduction Strategy Papers as the basis for development assistance. With the important exception of water and sanitation, these chapters have not included practical strategies for achieving the Millennium Development Goal on environmental sustainability.[3] Increasingly, however, researchers are documenting that an important way of reducing rural poverty is to enable the poor to use ecosystem services to improve their livelihoods by, for example, providing clear rights to use a fishery or a forest or to farm a plot of land. In addition, researchers point out that soil fertility, water quality and other ecosystem services play a fundamental role in meeting Millennium Development Goals such as reducing hunger (WRI, UNDP, UNEP and World Bank, 2005: 154–61).

Overall, civil society analysts report that the portion of official development assistance for the environment is falling – from nearly 20 per cent in 1995 to 9 per cent a decade later. At the same time, many poorer countries are becoming more dependent on resources gained from extraction and other uses of land and natural resources (Toulmin and Bigg, 2004: 8,10). It is also these regions that have the majority of the remaining unconverted natural lands and their assets. The Millennium Ecosystem Assessment shows that the projected loss of these assets in the first half of the twenty-first century is highest in tropical forests and grasslands (MA, 2005a: 32). The Brazilian Amazon, for example, comprises about one third of the world's tropical forest. Between 2000 and 2005, Brazil accounted for 42 per cent of net global forest loss, with most of this deforestation occurring in the Brazilian Amazon (Barreto et al. 2006: 12). One important function of the Amazon is the regulation of the hydrological cycle – 40 per cent of the rainfall in the rest of Brazil, critical for the country's agriculture and hydropower sectors, comes from the Amazon (Lovejoy, 2006).

Even rough estimates of the needed investments to meet environmental goals for poverty reduction are difficult to make. However, the background report prepared for the Poverty–Environment Partnership for the 2005 World Summit on the Millennium Development Goals does suggest a range. It says that based on the best available evidence, investments in ecosystem services of $60 to $90 billion per year are needed globally to reach poverty reduction targets over the next 10–15 years. This does not include another $80 billion per year to address climate change (UNDP et al., 2005a: 11). One way of putting this number in perspective is that industrial countries spent US$283 billion to support their farmers through agricultural subsidies in 2005 alone (OECD, 2006).

BARRIERS TO INVESTING IN ECOSYSTEM SERVICES

Why are investments in ecosystem services not growing to match the needs? At least three barriers help account for the failure to invest in ecosystem services. First, the detailed investment case that these services are crucial to reducing poverty is just beginning to be made to national political, planning and finance leaders. Second, those who make decisions about the economy and development, and those who make decisions about managing the environment and natural resources lack the knowledge, forums and incentives to work together in many countries. Third, civil society groups and local communities can play an important role in insisting that investments in ecosystem services be made and in monitoring what happens on the ground. However, they have little ability to take part in making decisions in many poorer countries which frequently lack governance processes such as public access to information, participation and justice.

The Absence of a Compelling Investment Case

A convincing case for investment in ecosystem services to reduce poverty is in the early stages of being made and has not yet reached political and economic leaders. Most of these leaders are unaware of the relationships between healthy ecosystem services and a healthy economy with fewer people living in poverty. Too often environment is still seen as a luxury – something to worry about after the economy has developed. Those who do grasp the relationship face challenges in deciding what investments to make and in quantifying the returns on investments in concrete economic terms. As a result, economic and development initiatives such as structural adjustment and poverty reduction have not incorporated investments in ecosystem services. This is partly because environmental and natural resource management leaders have not presented the needs as an integrated set of investments in human well-being. Environmental issues emerged as a wide range of different impacts – air pollution, water pollution, endangered species, disappearing wetlands, drying land, a hole in the ozone layer, a changing climate and many others. In response, countries have adopted innumerable laws and international conventions focused on dealing with individual types of resources and of environmental effects, some immediate but others appearing only over decades, even centuries.

In an effort to provide a more unified approach, some researchers have framed the challenge as maintaining and restoring natural capital – living systems that provide ecosystem services. They have begun to develop a more coherent view of nature's services (Daily, 1997). They have also begun to value these assets (Costanza et al., 1997) and suggest strategies for creating an industrial revolution based on natural capital (Hawken et al., 1999; Daly, 1996) Others have introduced tools such as national environmental accounts that can be used to make the connection between the environment and the economy.[4]

The Millennium Ecosystem Assessment builds on this work in framing ecosystem services as benefits to human well-being. This framework presents ecosystems as natural assets essential to economic growth and poverty reduction. However, more integrated approaches have not yet reached the corridors of most national planning and finance ministries. Recognizing the need to present the economic rationale for investment in a way that convinces political leaders and macroeconomists, the Poverty–Environment Partnership addressed this challenge at the 2005 World Summit. It provided an analysis of economic evidence that investments in sound and equitable environmental management produce high economic and social returns for developing countries (UNDP et al., 2005a). The Partnership also convened a High-Level Policy Dialogue to discuss the case for investing in environment to reduce poverty and next steps after the Summit (Bass et al., 2006: 52–5).

Failure of Environmental Ministries to Speak the Language of Finance and Economics

To implement the Millennium Development Goals, national governments need to agree on both longer-term and immediate goals, determine the policy reforms and environmental investments needed, and obtain and allocate the funds. However, financial and development experts usually lack an understanding of the relevance of ecosystem issues, the difficulty of connecting cause and effect in ecosystem change, or the long distances and time frames often involved. They do not have the capability to translate general environmental plans into specific items for a national development plan budget or program. Environmental and natural resource agencies lack staff trained to estimate the costs and benefits of environmental interventions to support development. In fact, national plans for environmental sustainability or for implementing international conventions on wetlands, biodiversity or desertification seldom even include budgets with proposals for specific investments, researchers found (see Schmidt-Traub and Cho's chapter in this volume). Few environmental leaders are familiar with the worlds of macroeconomists and development planners. Costa Rica's environment minister noted at the 2005 World Summit that 'we need to learn the language of finance and economics, and demonstrate the economic benefits of the environment' (*The Economist*, 2005).

The handbook for preparing strategies to achieve the Millennium

Development Goals notes that there is no clear operational guidance to countries on what they should aim to achieve and how they can differentiate strategies that are environmentally sustainable. It lists six areas for investment including agricultural production systems, forests, freshwater resources and ecosystems, fisheries and marine ecosystems, air and water pollution, and global climate change. It then suggests that each country identify its own goals and quantitative targets (Bahadur et al., 2005: 98). The UN Millennium Project's plan to achieve the Millennium Development Goals calls for scaling up environmental investments. However, it stresses that 'the lack of good data and indicators on the environment hides the extent to which most developing regions have suffered extensive environmental degradation in the last decade and are not on track to achieving environmental sustainability' (UN Millennium Project, 2005).

In the health arena, in contrast, a series of reports coming out of a university-led process provided detailed data on specific interventions in the health of children and how they can be delivered. Funded by foundations and the World Bank, they were published in the medical journal *The Lancet* in 2003 (Black et al. 2003). Similar detailed cost estimates for specific environmental interventions to achieve environmental sustainability have not been available (Sachs, 2005: 293).[5]

Limited Public Role in Setting and Achieving National Development Goals

The Millennium Ecosystem Assessment stresses the need for increased transparency and accountability of governmental and private sector management of ecosystem services. Broader participation in making decisions increases understanding of impacts and distribution of costs and benefits of differing trade-offs in ecosystem services. It is also likely to result in a broader range of options in ways to achieve goals by drawing on the knowledge of local people as well as technical experts (MA, 2005a: 93–4). Access to information as well as an opportunity to take part in the planning, implementation and monitoring are crucial to allowing other parts of government and civil society to help identify environmental investments and influence the budget process. With information on implementation, people can then track progress, lessening the likelihood of corruption and improving the likelihood that the Millennium Development Goals will be met. The Assessment's Southern African team highlighted the importance of information. 'Nothing may be more crucial to the sustainable management of ecosystem services than the free flow of information and the enabling of individual as well as institutional flexibility, creativity, and innovation' (SA/MA, 2004).

The limited public role in setting and tracking the implementation of Millennium Development Goals in many countries hinders the ability to identify and successfully make environmental investments. Of the 40 countries classified by the World Bank as Heavily Indebted Poor Countries (HIPC), only one has a freedom of information law (World Bank, 2006a; Banisar, 2006). More are beginning to focus on the role such laws play in effective governance. Uganda is the one HIPC which has just adopted a freedom of information law; Indonesia, as noted below, is strengthening its laws. Several countries, including Ghana and Honduras, are considering freedom of information laws as part of developing poverty reduction strategies. In both countries the bills have become stuck in the legislative process, however (McIntosh, 2006).

The next section discusses whether these barriers – early stage of making a compelling case for increased investments, inability of environmental ministries to speak the language of development and finance, and gaps in governance policies and practices – might be effectively addressed by a Commission on Macroeconomics and Ecosystem Services for Poverty Reduction.

IS THE TIMING RIGHT?

Is the timing right for a major initiative to stimulate investments in the environment to reduce poverty? The importance of linking the environment and development has been recognized for decades, and so has the need for open decision-making. The Commission on Environment and Development, chaired by Gro Brundtland and known as the Brundtland Commission, pointed out in its 1987 report *Our Common Future* that the institutions working on environment and development tend to be 'fragmented, working to relatively narrow mandates with closed decision processes'. The report underlined the need to give agencies responsible for increasing investment, employment, food, energy and other economic and social goods a mandate to become responsible and accountable for pursuing their goals in ways that enhance natural resources. The Commissioners concluded: 'The real world of interlocked economic and ecological systems will not change; the policies and institutions concerned must' (World Commission on Environment and Development, 1987: 9–10, 310–12).

Five years later in 1992, more than a hundred nations participating in the Conference on Environment and Development adopted the Rio Principles (UNEP, 1992), which include both integration of decision making on environment and development and public access to information, participation and justice. Since then many countries have prepared national strategies on sustainable development or environmental action plans. Some have established environmental accounts, have increased their use of economic measures to achieve environmental goals and have taken steps to increase the public's role in making decisions. As noted above, however, few of the early efforts to implement the Millennium Development Goals, often through Poverty Reduction Strategies which are the key development policy documents introduced in 1999, included environmental investments. Nearly 20 years after the Brundtland Report was issued, its lead author noted the failure '"to merge environment with economics in our process of decision-making", whether in the cabinet chambers of government, the board rooms of industry, or the kitchens of our homes. This is in some ways the most important imperative of all,' he said. 'If we change the way we make decisions, we will change the decisions we make: if we don't, we won't' (MacNeill, 2006). There clearly remains a need to increase the priority placed on ecosystem services among national political and economic leaders if we are to reduce poverty.

Are there reasons to think that this is a promising time to make another significant effort to address the disconnect between the environment and development? Might the Millennium Ecosystem Assessment and the Poverty–Environment Partnership provide a critical mass of activity on which a focused initiative such as the Commission could draw to make the case for environmental investments and stimulate national planning, budget and implementation processes with the public playing a significant role?

Millennium Ecosystem Assessment

The Millennium Ecosystem Assessment introduces a useful conceptual framework, presents a multi-scale analysis of ecosystem service status and trends, and provides a group of scientists experienced in working across disciplinary and national boundaries. By connecting ecosystem services to human well-being, the Assessment makes the environment more congruent with the economy. It makes it possible to talk in terms of investments in ecosystem services thus facilitating communication with development planners and finance officials. By placing four types of ecosystem services – provisioning, regulatory, cultural and supporting –

within the same framework, it becomes possible to identify trade-offs and synergies among services for specific decisions. Although an accounting system to support such a framework is in its infancy (Boyd, 2006) analyses of the economic value of services are increasingly being made at lower levels. An analysis of investments to restore watersheds in India provides an example. In the village of Darewadi in the state of Maharashtra, villager income from crops totaled about 2000 rupees. Five years later, after its watershed was regenerated, villagers' income grew more than five times to 11000 rupees as they increased production of cereals and vegetables and added milk production as well as cash crops (WRI et al., 2005: 129).

The Millennium Ecosystem Assessment also sets out the current global condition of ecosystem services. Through its scenarios it shows the level of the challenge society faces in reversing degraded services while meeting the needs of growing populations and economies. Under all four scenarios, significant growth in consumption of ecosystem services is expected, along with continued loss of biodiversity and further deterioration of services due to habitat loss (MA, 2005a: 17).

In addition, as Norgaard points out in this volume, a significant result of the Assessment is the ability many participating scientists have acquired to work together to collectively develop deeper understandings of the ways in which human and ecological systems interact across ecosystem and geographic scales. In an unusual approach for a global scientific assessment, the Assessment included civil society and the private sector along with governments as stakeholders at the global level. It required sub-global assessments to 'centrally involve the intended users as stakeholders and partners throughout the process, from methodological design through the review process' (MA, 2003).

Poverty–Environment Partnership

As noted earlier, aid agencies, multilateral banks, UN and international civil society organizations formed a partnership early in the millennium to address poverty–environment issues within efforts to achieve the Millennium Development Goals. Recognizing the confusion about and neglect of the environmental sustainability goal, the partners aim to present best evidence on the economic importance of environment to poverty reduction and pro-poor growth, and identify priority areas for improved investment to achieve both the sustainability goal and the other goals. Their 2005 report documents that poor households and developing countries rely heavily on stressed environmental assets and that investments in increasing access to water supply and sanitation, soil conservation, and energy yield, or are likely to yield, high rates of return. It calls for improvements in governance including integration of environmental assets into national planning and mobilization of more resources (UNDP et al., 2005b).

The groups that belong to the Poverty–Environment Partnership are increasingly coordinating their work and engaging in joint projects. Two members – UNDP and UNEP – have merged their poverty–environment programs into a single Poverty and Environment Initiative that will support country-led efforts to integrate environmental sustainability into policy and planning processes for poverty reduction and pro-poor growth.[6] This decision builds on earlier projects that clarified the connections between the environment and poverty. Overgrazing, for example, reduces grass growth, degrading the soil and adding to erosion. As a result, the land can support fewer livestock, meaning reduced income for the herders. Similarly, overuse of pesticides and fertilizer leads to pollution, depleting water resources and interfering with irrigation. In both cases, levels of poverty are increased (Governing Council of UNEP, 2005: 4, 7).

The World Bank, another member of the Partnership, has also performed much relevant research. It issued what it called a millennium capital assessment in 2006. This report highlights the importance of natural resources as a source of wealth in low-income countries. It found that for many poorer countries, natural resource wealth can, with the right institutional structures and policies, be used to generate incomes, sustain wealth, and provide savings and investment opportunities (World Bank, 2006b). Another World Bank analysis examined environmental fiscal reform. It looked at how to develop revenue from taxes or user fees on natural resources or the environment that can be invested in ecosystem services as well as in as education and health (World Bank, 2005).

Bilateral donors have supported relevant projects. The UK Department for International Development published *Linking Poverty Reduction and Environmental Management* in 2002 (UK Department for International Development et al., 2002). The same year, the Agency for International Development African Bureau issued *Nature, Wealth and Power* to encourage dialogue on natural resources in rural development (USAID et al., 2002).

Some international non-governmental groups started work making the link more than a decade ago. Worldwide Fund for Nature initiated a Macroeconomics Program for Sustainable Development 15 years ago. It has worked to incorporate social equity and environmental sustainability into national and international macroeconomic policies and practices. In *Escaping Poverty's Grasp*, program director David Reed lays out the environmental foundations of poverty reduction, drawing on experience in China, Indonesia, El Salvador, South Africa and Zambia. The report proposes using a common set of analytical standards for projects aimed at poverty reduction and environmental management in rural areas. This would include addressing issues at the micro (local), meso (sub-national), and macro (national) level (Reed, 2006). The World Resources Institute also has a long history of working on environment and development issues. Much of this work is reflected in the World Resources Reports (WRI et al., 2004; WRI et al., 2005; Henninger and Snel, 2002; Wood et al., 2000; Revenga et al., 2000; White et al., 2000; Matthews et al., 2000; Burke et al., 2000; WRI et al., forthcoming). WRI has also done extensive work in mapping ecosystems and also their relationship to human well-being.

Civil society groups at the local and national level also play a crucial role. They are particularly important in improving governance. For example, national coalitions of these groups participate in The Access Initiative,[7] which seeks to improve the policies and practices of national governments in providing access to information, participation and justice. The coalitions evaluate the performance of their government using a standard methodology. An overlapping group of civil society groups is working with national governments and donors through a related effort called the Partnership for Principle 10. All members make commitments to improve their performance in providing access to information, participation and justice. Bolivia, Chile, Indonesia, Mexico, Uganda, the United Kingdom and Ukraine are formal state members. Uganda met its commitment by implementing a new Freedom of Information Act. Indonesia, which just joined in 2006, has committed to increasing the public's involvement in the environmental impact assessment process, incorporating public participation guidelines into new local environmental regulations, and responding to public grievances in environmental cases (WRI, 2006).

Much analysis is available. In some areas, the level of activity is growing. A high-level effort to make the link with national political leaders and planning and finance ministries has the potential to put this work into practice to reduce poverty and improve the health of ecosystem services.

ESTABLISH A COMMISSION ON MACROECONOMICS AND ENVIRONMENT FOR POVERTY REDUCTION

If investments in ecosystem services are to be increased, a focused effort is needed to build a process for identifying investments, quantifying their costs and their economic benefits, incorporating them in national budgets, and tracking implementation. This will require strengthening the links among decision-makers in the worlds of macroeconomics, planning, environment and natural resources. A Commission on Macroeconomics and Ecosystem Services for Poverty Reduction could provide a forum to accomplish this. Over perhaps two years, a Commission could raise the understanding and political visibility of the fundamental role of healthy ecosystem services in reducing poverty. It could provide the impetus for countries to introduce processes to make environmental investments. The experience of the Commission on Macroeconomics and Health provides an encouraging precedent.

Convening a Commission

The Commission on Macroeconomics and Health offers insight into key questions about how a Commission on Macroeconomics and Ecosystem Services for Poverty Reduction might be convened, funded, and carry out its work. Gro Brundtland, then Director-General of the World Health Organization, convened the Commission on Macroeconomics and Health. The 18 members were leaders in health and economics from academia, governments and international agencies. Chaired by macroeconomist Jeffrey D. Sachs, the Commission also had a Secretariat appointed by the WHO Director-General. Funding came from foundations and national governments.

A Commission on Macroeconomics and Ecosystem Services for Poverty Reduction might be convened by the UN Development and Environment Programmes. It might work closely with countries who have taken the lead in this area – Costa Rica, the Netherlands and Sweden, for example, and with a few others who are particularly interested in moving forward. Members of the Poverty–Environment Partnership are well-positioned to play key roles. This would include international financial institutions such as the World Bank and non-governmental groups. Funds to support a Commission might come from national donors funding research on environment and poverty reduction as well as foundations.

Setting Goals for a Commission

A Commission on Macroeconomics and the Environment for Poverty Reduction could also learn from the Commission on Macroeconomics and Health in adopting its basic goal.

> Develop and communicate the case for investing in ecosystem services as fundamental to reducing poverty and achieving economic development.

The Commission on Macroeconomics and Health made the case that health is fundamental to development. It issued a report at the end of 2001 documenting that improving health is not only a fundamental goal of development; it is also a means to achieving development goals including poverty reduction. It showed that improvements in health mean higher incomes, higher economic growth and reduced population growth. A basic goal of a Commission on Macroeconomics and Ecosystem Services would be to make the parallel case that the health of ecosystem services is essential to reducing poverty in a way that communicates to financial and development planners. While the data are not as advanced as in the human health field, recent work such as analyzing the benefits of watershed regeneration discussed earlier demonstrates that a compelling case can be made.

The Commission on Macroeconomics and Health called for low- and middle-income countries to partner with high-income countries to 'scale up access of the world's poor to essential health services, including a focus on specific interventions'. It estimated that donor grants would need to increase from $6 billion per year to $27 billion per year for a sustained period of perhaps 20 years (Sachs, 2001: 4, 12, 16). It recommended that countries establish their own National Commissions on Macroeconomics and Health or similar bodies, chaired by ministers of health and finance that included civil society to provide leadership in scaling up.

Unlike the health arena, the environmental field does not have either quantified data on the costs and benefits of specific environmental investments or a well-developed process of scaling up these investments. The Commission on Macroeconomics and Health drew on the work of *The Lancet* Child Survival series[8] for health investments. Environmental investments are likely to be more site-specific than those for health. Prevention and treatment of malaria is largely the same regardless of country; approaches can be standardized. This is typically not the case with environmental problems. In addition, environmental investments are more political than those for health in that they usually involve trade-offs among development goals. How will use of water be divided among users for irrigation and in urban areas? How will use of forests be shared between timber and tourism? These decisions require both expertise from a number of sectors and an open governance processes. Thus, a Commission on Macroeconomics and Ecosystem Services for Poverty Reduction might frame additional goals in ways such as the following:

> Identify specific environmental interventions such as watershed or wetland restoration and quantify their costs and benefits;
>
> Develop processes for scaling up these interventions; and
>
> Establish national processes (including environment, natural resource, planning and finance ministries) to develop and carry out the planning, budgeting and implementation, including broad public participation at all stages.

Carrying out the Work

The Commission could work through task forces organized largely around the goals.

- *Make and communicate the case that healthy ecosystem services are essential to development.* This task force would work with financial leaders and development planners to demonstrate that investments in ecosystem services are fundamental to poverty reduction and need to be increased.
- *Identify specific environmental interventions.* This task force would collect best practices in environmental management, including a rigorous assessment of the links between inputs and the results for ecosystem services and reducing poverty. In their review in this volume, Schmidt-Traub and Cho propose that a joint task force of UN groups, the World Bank and secretariats of the environmental conventions might handle this task. Another option is to work closely with a few countries. One part of this work would be making available the work on valuation of ecosystem services in a way useful to national planners and finance people.
- *Sources of funding.* This task force would identify the total funds needed and potential sources including national funding, for example through changes to tax systems, payment for ecosystem services, as well as donor and private sector funding.
- *Improvements in governance.* This task force would propose ways to

form coordinating bodies such as national councils or regional resource centers on economic growth and poverty reduction. It would also focus on measures to improve local rights of access to resources, *public* access to information, participation and justice, ways to set and track indicators (including improving Millennium Development Goals and targets on environmental sustainability), and uses for tools such as mapping and environmental accounts (see Box 14.1).

BOX 14.1 SHARPENING TARGETS FOR ENVIRONMENTAL SUSTAINABILITY

Working with individual countries and their citizens, the Commission might come up with ways to sharpen environmental sustainability targets and indicators. For example, the *World Resources 2005 Report* proposes the rewording of Target 9 under the Millennium Development Goal for environmental sustainability to focus specifically on ecosystem capacity, cover all levels of governance, and include the poor. The target now calls for integrating 'the principles of sustainable development into country policies and programs and reversing the loss of environmental resources'. If changed, it would instead read:

> Maintain or restore the capacity of ecosystems to provide critical ecosystem services, and integrate the principles of sustainable development into local, national and international policies and programs.

Specific indicators under the revised target could include:

- extent and condition of communal fisheries (coasts and inland);
- extent and condition of forested areas held in common;
- watershed conditions on communally held lands (vegetation cover, water availability, groundwater trends);
- soil fertility on private farmlands; and
- land degradation.

The *WRR* authors also suggest an additional target:

> Ensure the poor have access to environmental resources and decision-making.

Specific indicators for this goal might include:

- proportion of rural households with access to secure land tenure;
- proportion of rural households with access to environmental information (for example extension services pollutant or environmental health alerts, environmental impact studies on proposed concessions or developments'; and
- participation in local decision-making (WRI, UNDP, UNEP and the World Bank, 2005, 158).

The process of preparing reports would give participants an opportunity to develop a common language and increase the corps of people skilled in using it that could in turn assist national ministries of planning, finance and environment and natural resources to work more effectively together.

Implementation of the Commission Results

From the beginning, a Commission on Macroeconomics and Ecosystem Services for Poverty Reduction would consider how its work could be most effectively implemented. The UNDP and UNEP might, for example, convene meetings of interested countries or facilitate peer-to-peer learning through individual countries.

Again the experience of the Commission on Macroeconomics and Health demonstrates the possibilities. Based on the Commission's 2001 report, WHO held a consultation in 2003 that brought together high-level health, finance and planning officials from 40 low- and middle-income countries. They discussed investing in the health of the poor by ensuring more resources for health overall, by making health central to national macroeconomic frameworks, and by making donor funding more predictable. Among the results by 2005 were increased pledges for health spending in Sri Lanka and Indonesia; a health investment plan in Ghana; a Commission on Macroeconomics and Health in Mexico that led to 'a more inclusive and long term vision of social needs and the better design of policy objectives'. In the Caribbean, a Commission on Health and Development was developing evidence for placing health centrally in development (WHO, 2005). To inform policy-making more broadly, a study tracked financing sources, agents, and providers of funds in three countries and made suggestions for systematic National Health Accounts (Michaud, 2005).

An indication of the effectiveness of the commission is the World Health Organisation's decision to convene a second commission. Launched in 2005, the Commission on Social Determinants of Health will recommend interventions and policies to improve health and narrow health inequalities. This Commission, which is looking at communities such as informal workers, indigenous peoples and slum dwellers (WHO, 2006), offers a commission focused on ecosystem services experience with a networked approach.

CONCLUSION

An effort to establish a Commission on Macroeconomics and Ecosystem Services for Poverty Reduction would, of course, face significant hurdles. As Janetos points out in his chapter included in this volume, national governments in most cases took part in the Millennium Ecosystem Assessment through their representatives in meetings of parties to international conventions. Thus there are relatively few high-level national political champions to encourage participation in activities building on the Assessment. Including participation from local communities and civil society will require time and persistence.

Nevertheless, the Assessment clearly shows that the same areas that face the most significant environmental degradation also face major development challenges. The growth of the Poverty–Environment Partnership indicates that an increasing number of international and national donors, UN agencies and non-governmental groups see the value of expanding the focus of their work on the environment beyond addressing environmental impacts to making investments to reduce poverty. Doing this means working more closely with development planners and macroeconomists. A Commission could turn these links into living connections by helping countries facing both ecosystem degradation and development challenges to devise and carry out strategies that invest in ecosystem services for the poor.

NOTES

1. See http://www.un.org/millenniumgoals/.
2. See http://www.undp.org/aboutpep.
3. See discussion in 'Special Section on Global Development Policies: Making the MDGs and PRSPs work for the Poor and the Environment' (WRI, UNDP, UNEP and World Bank, 2005) on experience with Poverty Reduction Strategies and ways to improve them.
4. See http://unstats.un.org/unsd/envAccounting/seea.htm.
5. Sachs notes that investments for environmental sustainability have not been quantified.
6. For more see http://www.undp.org/pei.
7. For more information see http://www.accessinitiative.org.
8. See, for example, Black et al. (2003).

REFERENCES

Bahadur, C., M. Kruk and G. Schmidt-Traub (2005), *Preparing National Strategies to Achieve the Millennium Development Goals: A Handbook*, Millennium Project.

Banisar, D. (2006), 'Freedom of information around the world 2006: a global survey of access to government records laws', *Freedominfo.org*, online at: http://www.freedominfo.org/documents/global_survey2006.doc.

Barreto, P., C. Souza, R. Nogueron, A. Anderson and R. Salomao (2006), *Human Pressure on the Brazilian Amazon Forests*, Washington, DC: WRI.

Bass, S., T. Bigg, J. Bishop and D. Tunstall (2006), 'Sustaining the environment to fight poverty and achieve the Millennium Development Goals', *Reciel*, **15** (1), 39–55.

Black, R.E., S.S. Morris and J. Bryce (2003), 'Child survival I: Where and why are 10 million children dying every year?', *The Lancet*, **261**, 28 June.

Boyd, J. (2006), 'The nonmarket benefits of nature: what should be counted in green GDP?', Discussion Paper, Washington, DC: Resources for the Future.

Burke, L., Y. Kura, K. Kassem, C. Revenga, M. Spalding and D. McAllister (2000), *Pilot Aanalysis of Global Ecosystems: Coastal Ecosystems*, Washington, DC: WRI.

Costanza, R., R. d'Arge, R. de Groot, S. Farber, M. Grasso, B. Hannon, K. Limburg, S. Naeem, R.V. O'Neill, J. Paruelo, R.G. Raskin, P. Sutton and M. van den Belt (1997), 'The value of the world's ecosystem services and natural capital', *Nature*, **387**, 253–60, 15 May.

Daily, G.C. (ed.) (1997), *Nature's Services: Societal Dependence on Natural Ecosystems*, Washington, DC: Island Press.

Daly, H.E. (1996), *Beyond Growth: the Economics of Sustainable Development*, Boston, MA: Beacon Press.

The Economist (2005), 'Greening the books: Ecosystem services', *The Economist*, 17 September, p. 82.

Governing Council of UNEP (2005), 'Background chapters for the ministerial-level consultations on energy and environment for development, chemicals management as well as tourism and the environment', Nairobi, UNEP.

Hawken, P., A. Lovins and L.H. Lovins (1999), *Natural Capitalism: Creating the Next Industrial Revolution*, Boston, MA: Little, Brown.

Henninger, N. and M. Snel (2002), *Research Report: Where are the Poor? Experiences with the Development and Use of Poverty Maps*, Washington, DC: WRI.

Lovejoy, T. (2006), 'The Amazon forest: Let it rain', *International Herald Tribune*, 27 July.

MacNeill, J. (2006), 'The forgotten imperative of sustainable development', speech presented on 20 August on receiving the Elizabeth Haub Award for Environmental Diplomacy, White Plains, NY: Pace University School of Law.

McIntosh, T. (2006), 'Freedom of information laws added to the development agenda', *Freedominfo.org*, 22 March, online at http://www.freedominfo.org/features/20060322.htm.

Matthews, E., R. Payne, M. Rohweder and S. Murray (2000), *Pilot Analysis of Global Ecosystems: Forest Ecosystems*, Washington, DC: WRI.

Michaud, C. (2005), *Flow of Donor Funds in Cambodia, Indonesia and Sri Lanka: Synthesis of Key Findings*, Geneva: WHO.

Millennium Ecosystem Assessment (MA) (2003), *Sub-Global Assessment Selection Process*, Washington, DC: Island Press.

Millennium Ecosystem Assessment (MA) (2005a), *Ecosystems and Human Well-Being: Synthesis*, Washington, DC: Island Press.

Millennium Ecosystem Assessment (MA) (2005b), *Ecosystems and Human Well-Being: Volume 3: Policy Responses*, Washington, DC: Island Press.

Organisation for Economic Co-operation and Development (OECD) (2006), *Agricultural Policies in OECD Countries: at a Glance – 2006 Edition*, Paris: OECD.

Reed, D. (2006), *Escaping Poverty's Grasp: The Environmental Foundations of Poverty Reduction*, London and Sterling, VA: Earthscan.

Revenga, C., J. Brunner, N. Henninger, R. Payne and K. Kassem (2000), *Pilot Analysis of Global Ecosystems: Freshwater Systems*, Washington, DC: WRI.

Sachs, J.D. (2001), *Macroeconomics and Health: Investing in Health for Economic Development*, Geneva: WHO.

Sachs, J.D. (2005), *The End of Poverty: Economic Possibilities for Our Time*, New York: Penguin Press.

The South Africa Millennium Ecosystem Assessment (SA/MA) (2004), *Ecosystem Services in the Gariep Basin*, Stellenbosch University: Sun Press.

Toulmin, C. and T. Bigg (2004), *Financing for Environment and Development*, draft.

UK Department for International Development, Directorate General for Development of the European Commission, UNDP and the World Bank (2002), 'Linking poverty reduction and environmental management: Policy challenges and opportunities', Washington, DC: World Bank.

UNDP, UNEP, IIED, IUCN and WRI (2005a), *Investing in Environmental Wealth for Poverty Reduction: Environment for the MDGs, Poverty–Environment Partnership*, New York: UNDP.

UNDP, UNEP, IIED, IUCN and WRI (2005b), *Sustaining the Environment to Fight Poverty and Achieve the MDGs: The Economic Case and Priorities for Action. A message to the 2005 World Summit*, New York: UNDP.

UNEP (1992), *Rio Declaration on Environment and Development*, online at: http://www.unep.org/Documents.multilingual/Default.asp?DocumentID=78&ArticleID=1163.

UN Millennium Project (2005), *Investing in Development: A Practical Plan to Achieve the Millennium Development Goals, Overview*, New York: UN Millennium Project.

USAID, Center for International Forestry Research, Winrock International, WRI and International Resources Group (2002), *Nature, Wealth, and Power: Emerging Best Practices for Revitalizing Rural Africa*, Washington, DC: IRG.

White, R., S. Murray and M. Rohweder (2000), *Pilot Analysis of Global Ecosystems: Grassland ecosystems*, Washington, DC: WRI.

WHO (2005), *MacroHealth Newsletter*, **12**, January.

WHO (2006), *Commission on Social Determinants of Health: Questions and Answers*, online at http://www.who.int/social_determinants/strategy/QandAs/en/print.html.

Wood, S., K. Sebastian and S.J. Scherr (2000), *Pilot Analysis of Global Ecosystems: Agroecosystems*, Washington, DC: International Food Policy Research Institute and WRI.

World Bank (2005), *Environmental Fiscal Reform: What Should be Done and How to Achieve It*, Washington, DC: World Bank.

World Bank (2006a), 'Heavily indebted poor countries (40 countries)', online at http://we.worldbank.org/WBSITE/EXTERNAL/TOPICS/EXTDEBTDEBT/0,

,contentMDK:20260049~menuPK:528655~pagePK:64166689~piPK64166646~theSitePK:469043,00.html.

World Bank (2006b), *Where is the Wealth of the Nations? Measuring Capital for the 21st Century*, Washington, DC: World Bank.

World Commission on Environment and Development (1987), *Our Common Future*, Oxford, UK: Oxford University Press.

WRI (2006), *Indonesia Joins Partnership to Improve Environmental Governance, Public Access to Information*, press release 26 April.

WRI, ILRI, Poverty Analysis and Research Unit, CBS, Ministry of Planning and National Development and Department of Resource Surveys and Remote Sensing, Ministry of Environment and Natural Resources (forthcoming), *Nature's Benefits: An Atlas of Ecosystems and Human Well-Being in Kenya.*

WRI, UNDP, UNEP and the World Bank (2004), *World Resources 2002–2004: Decisions for the Earth: Balance Voice and Power*, Washington, DC: WRI.

WRI, UNDP, UNEP and the World Bank (2005), *World Resources 2005: The Wealth of the Poor – Managing Ecosystems to Fight Poverty*, Washington, DC: WRI.

Index

ABN AMRO 334, 336
Abuja Ministerial Declaration on Water 283
Access Initiative 34, 41–2, 356
access rights 41
accountability, improving 50–57, 336
accumulation, global ecosystem change 127–33
action agenda, ecosystem services 21–65
  barriers, actions and examples 24–7
  institutional implementation 27–8
  vision 21–4
    align economic and financial incentives with ecosystem stewardship 57–65
    develop and use information about ecosystem services 28–35
    improve accountability for decisions that affect ecosystem services 50–57
    manage ecosystem services across multiple levels and timeframes 42–50
    strengthen the rights of local people to use and manage ecosystem services 35–42
Action Impact Matrix 40, 73–105
  analytical process 73–4
  conclusions and recommendations 99–100
  identifying and managing trade-offs 31
  mainstreaming and implementing the MA findings 76–80
    at global level 75–6
    at national level 74–5, 82
    policy-relevant issues 76–8
    role of MA conceptual framework 78–80
  methodology and applications 80–82
  overview and framework 73–6
  sample application
    to global and sub-global levels 92–9
    to Sri Lanka at the national level 83–92
  underlying principles 101–2
Adams, W.M. 9
Adapting Mosaic scenario 217, 218, 219, 220, 221, 315
adaptive management 288–9
Advocates Coalition for Development and Environment 52
African Ministerial Conference on Water (AMCOW) 50, 283
agency responsibility 250
Agenda 21 146, 276, 285
agriculture
  and employment 331, 332
  policies on sustainable 120
  water pollution 90
  water prices 89–90
  World Bank policy priorities 202–4
AIM *see* Action Impact Matrix
air quality 33–4
Alliance of Small Island States 325–6
Amazon 348
ambient taxes, water pollution 90
Andean Beer Company 290
Andean traditional view of water 46
Angola 56
Annamalai, K. 338
Annan, K. 257
anthropocentric viewpoint, Millennium Assessment 76–8
anti-environmentalism 192–3, 208, 209
applied knowledge 163, 164
aquaculture 4, 5–7, 57, 205
Arctic Council 322
Argentina 30, 58, 136, 223
auction economy 225

Aulisi, A. 337

Baer, P. 248
Bahadur, C. 351
Bainomugisha, A. 52
Bandara, C.M.M. 104
Banisar, D. 55
banks
  incentives for managing environmental risks 334–5
  see also World Bank
Barbier, E.B. 5
Basnayake, B.R.S.B. 105
Bass, S. 40
Bateman, M.E. 215–34
Bazilchuk, N. 56
best practices 163, 164–5, 185–6
Bigg, T. 348
bilateral donor agencies 336
biodiversity
priorities for World Bank policy 206–7
see also Convention on Biological Diversity; Global Biodiversity Assessment
biomass development, energy shortage problem 152
Biome Stewardship Councils (BSCs) 318–21
  conflicts 323–4
  functions 321–4
  funding 324
  rules 319–21
  sovereignty 324–7
biomes 129–30, 318–19, 320
biophysical transformations, globalization 126–37
Black, R.E. 351
Bojo, J. 175
Bolivia 42, 183, 356
BOP market 335
Boyd, J. 53, 354
BP 159, 336
Bradford, C. 331, 342
Brazil 61–2, 81, 133, 136, 227, 338, 348
bridging organizations 47–8
Bruce, J. 36
Brundtland Commission 9, 321, 352, 353
BSCs *see* Biome Stewardship Councils
BSR (Business for Social Responsibility) 62
Bulletin of Atomic Scientists 33, 309
Bunch, R. 35
Burke, L. 356
business
  action agenda for ecosystem services
    bring local voices to the table to influence development projects and policies 39–42
    form bridging organizations 47–8
    include ecosystem risk in financial evaluations 60–61
    incorporate ecosystem stewardship in performance objectives 63–5
    increase corporate transparency 55–7
    vision 22
  incentives for managing environmental risks 333–4
  involvement in Millennium Assessment 3
  uncertainty, reducing 228–9
Business and Industry Report 34
business school coalitions, dissemination of information 35
buying power, of poor 226, 335–6

Canada 56, 129, 133
Canaima National Park 38, 142
cap and trade schemes 220, 226–7
capacity building 185, 284
capacity constraints 179
capital markets, incentives for managing environmental risks 334–5
carbon dioxide emissions, pricing 202
Caribbean Sea sub-global assessment 47
cell phones 30, 222–3
CEPAL 134
Chayes, A.H. 326
Chigunta, F. 58
child mortality 96, 177
Child Survival Series 177, 186, 358
Chile 42, 81, 356
China

biophysical transformations 127
business schools and dissemination of information 35
environmental education 163–4, 165
environmental impact assessments 155, 157, 158
importance of applied knowledge 164
investment in biomass through government subsidies 162
need for incentives to internalize coal pollution costs 201
number of dams 133
satellite tracking 338
sub-global assessment 259, 262–3, 265
Cho, A. 169–87
Citigroup executives 334–5
citizens
empowerment to respond to ecosystem change 145–6
participation in development policy 249–50
tailoring information for 33–5
civil society
action agenda for ecosystem services
bring local voices to the table to influence development projects and policies 39–42
decentralize decisions about ecosystem services 37–9
eliminate perverse subsidies and reform taxation policies 58–60
ensure individuals and communities have secure rights to ecosystem services 36–7
establish conditions for cooperation with communities 43–7
form bridging organizations 47–8
frame messages that resonate with the public 32–3
hold elected officials accountable 51–3
identify and manage trade-offs 31–2
include ecosystem risk in financial evaluations 60–61
incorporate ecosystem stewardship in performance objectives 63–5
increase corporate transparency 55–7
perform regular monitoring and assessment 28–31
publicly decide and track investment in ecosystem services for development 53–5
raise priority of working across levels in national institutions 49–50
support markets and payments for ecosystem services 61–3, 289
tailor information for citizens, producers and purchasers 33–5
vision 22
assistance to governments in implementation of international commitments 286
and ecosystem security 282
involvement in Millennium Assessment 3
offering lessons on adaptive management 289
Partnership for Principle 10 (PP10) 42, 356
civil society organizations (CSOs)
recognition of gravity of environmental crisis 210
use of EITI as an instrument of government accountability 336
Clean Development Mechanism (CDM) 159, 195
climate change
differences between ecosystem damage and 317
need for adaptation strategies 116
need for norm shift 308
response to 315–16
scenarios 340–41
social and ecosystem interconnectivity 136
water resources 103–4
Clinton Global Initiative 341

co-generation of energy 202
co-management practices 48–9
coal burning/pollution 194, 201
coastal habitat destruction 88
coastal systems, and economic growth 86
Coca-Cola 333
Cochabamba Water War 46–7, 143
cod stocks, collapse of 12, 340
collaboration, need for 302, 317
collective learning 238, 242–3
Columbia 62, 186
Commission for Africa 179, 185
Commission on Health and Development (Caribbean) 361
Commission on Macroeconomics and Ecosystem Services for Poverty Reduction 55, 357–61
Commission on Macroeconomics and Health 178, 357, 358
Commission on Social Determinants of Health 361
Commission on Sustainable Development 276–80, 284, 285, 286, 287
communal management, natural resources 39
communication
  increasing 221–3
  of knowledge 163–6
  political reform 167–8
CONDESAN 46
Conferences of Parties (COPs) 3
conflicts *see* environmental conflicts
conservation of energy 201
conservation groups 26, 36, 38, 274–5
consultative processes, importance of 181–2
consumer sovereignty 210
consumers, corporate accountability 56
consumption issues 59, 196, 224–7, 280–82, 306–7
Controlled Substance Act (US, 1972) 224–5
Convention on Biological Diversity (CBD) 8, 110, 146, 263, 268, 284, 286, 323
Convention on Climate Change (UN) 146, 192
Convention on Migratory Species 110
Convention to Combat Desertification (UN) 110, 146
cooperation 43–7, 119–21
coordination, lack of 173–5
coral reefs, loss of 4
corporate funds, socially responsible 209
corporate transparency, increasing 55–7
corporations
  disservice to ecosystem management 228
  increasing disclosure 230–31
  involvement in Millennium Assessment 257–8
  lengthening time horizons in decision-making 229–30
  priorities for World Bank policy 205–6
Correa, H. 44, 45, 125–47
corruption 27, 336
cost-benefit analyses 154
Costanza, R. 8, 349
Country Environment Analysis papers 182
Crane, M. 339
critical ecosystems
  determining at national level 83–4
  determining the most important global goals 92
  effect of MDGs on 92–5
  effects on MDGs 95–9
Cruz, W. 81
CSD *see* Commission on Sustainable Development
cultural services 1, 9

Daily, G.C. 7, 9, 349
Daly, H.E. 349
dams 5, 132, 133
Danielsen, F. 7
Darewadi 354
data collection 221–3, 338
dead zones 4–5
debt relief campaign 306
decision makers
  informal networks/associations 283
  information on ecosystems as an aid

to 9–10
requirement for modeling effects of consumption taxes 227
targeting 190–92
decision-making
decentralization 37–9
improving accountability 50–57
integrated 273, 274, 283
involvement of local people in 36
multi-scale analysis 114–15
participatory approach 152
proactive 218
reactive 216–17
science-based 288
short-term 229–30
deforestation 86, 95–6, 348
deliberative democracy 249–50
demand side, nurturing 155
Department for International Development (DFID) 178
developing countries 281, 335–6
development
choices, implications of MA scenarios on 118
decisions, usefulness of multi-scale analysis 114–15
drivers of, and ecosystem change 117–18
and environmental conflicts 141
GDP as key indicator of 165, 167
goals *see* Millennium Development Goals; national development goals
integrating MA through a sectoral approach 118–19
projects, inclusion of local people in 39–42
development cooperation strategy 119–21
development strategies
bias towards large developers 152
national MDG-based 179–86
Dewey, J. 240
Dharmasena, G.T. 103
dialogue 44–5, 47, 147, 285, 350
Diamond, J. 192
Diaz, R. 5
direct investments, lack of, in environmental management 177–8
Directorate-General for Development (DGFD) 178
disaggregated sovereignty 326
disclosure, increasing 230–31
disease 96, 300, 358
distributive justice 306
Doomsday Clock 33, 309
drylands, vulnerability 13, 134–5
Dryzek, J.S. 238, 249
Dubash, N. 40

early warning systems, implementing 339–40
Earth Institute (Columbia University) 186
Ecoimagination report (GE) 333, 334
ecological clocks 33, 309
ecological trigger points, turning into political trigger points 301–2
economic activity, concentration of 128, 129
economic growth, moderating negative effects on coastal and marine systems 86
economic incentives, aligning with ecosystem stewardship 57–65
economic interests, and economic degradation 154
economic patterns, ecosystem change and accumulation 127–33
economic priorities, for World Bank 199–200
economics, failure of ministries to speak the language of 350–51
Economist 62, 350
ecosystem change
costs and benefits of 11–12
and drivers of development 117–18
see also global ecosystem change
ecosystem damage
differences between climate change damage and 317
and wealth creation 316
ecosystem degradation
barrier to achieving development goals 12–13
changes required to reverse 346
costs and benefits 33
disease 96, 300

economic interests of large industries 154
financial risks and opportunities 60
framing as a moral and ethical issue 307–8
human well-being and political reform 167
increasing risk to humans 32
need for major governance changes 13–14
and poverty 13, 113, 316, 346
regions confronting largest problems 21
shame-and-blame 190
using graphic images to convey message of 308–10
World Bank's contribution to 194–5
ecosystem failures
conveying a sense of the increasing risk of 309
non-linear responses of political systems 301
ecosystem health
and development, comparative analysis 117–18
global trade 116–17
human well-being 112–13, 120, 121
public indicators of 52–3
understanding and evaluating 9–10
ecosystem maintenance, as insurance against risk 32–3, 303–5
ecosystem management 215–34
cost-benefit issues 305
creating a new strategy 219–31
documenting and sharing best practice 185–6
equitable burden-sharing 79–80
free flow of information as crucial to 351–2
the immediate future of 216–19
incentives for sustainable 157–61
strategy for changing politics of 299–301
the world in 2050 231–4
see also environmental management
Ecosystem Preservation Pays 300
ecosystem risks
ecosystem maintenance as insurance against 32–3, 303–5
incentives for managing 333–5
inclusion in financial evaluations 60–61
taken seriously by insurance companies 209
ecosystem security 273–94
encouraging new ways of thinking 288–90
improving stakeholder engagement 284–7
integrating resource management 275–84
ecosystem services
agenda *see* action agenda
assessment *see* Millennium Ecosystem Assessment; scientific assessments
conflicts over use of specific 143
definitions 8–9
distinguishing among 114
effect on MDGs 95–9
effects of MDGs on 92–5
human use and condition of 1, 10, 11
human well-being and 78–9
importance of income from 331
investment in *see* investment(s)
payment for 61–3, 200, 289–90
scenarios effect on 217
valuations 53, 161, 219–20
value, disparity between actual and perceived 7
Ecosystem Services Review 29
ecosystem stewardship
aligning economic and financial incentives with 57–65
as a fairness or justice issue 33, 305–7
need for collaborative problem-solving 302
selling as an insurance policy 304–5
ecosystems
generating employment through 331–3
hierarchical framing 244–5
human use and radical alteration of 1, 10, 11
identifying impact on development goals 86–8
identifying impact of development

goals on 84–6
interconnectivity, correspondence with social interconnectivity 136–7
restoring or rehabilitating as a development goal 119
threats and opportunities 329–31
see also critical ecosystems
Ecosystems and their Services around the Year 2000 150
Ecuador 61, 289, 290
education
information and knowledge dissemination 35, 163–4, 165–6
political reform 167–8
and population levels 281
EEQ (Quito Electric Company) 290
Ehrlich, A.H. 238
Ehrlich, P.R. 238
El Niño 339
elected officials, accountability 51–3
elite groups, as barrier to sustaining ecosystem services 27
Ellerman, A. 337
EMAAP-Q (Quito Municipal Water and Sewage Company) 289–90
Emery, A.R. 249
employment
and coastal and marine habitat destruction 88
generating through ecosystems 331–3
promotion, preservation of wetlands 86
energy research 223–4
energy sector
amount needed for global infrastructure investment 335
biomass development and greenhouse gas emissions 152
development of alternative sources 224
priorities for World Bank policy 191, 196, 200–202
protocols for sustainable 159
see also coal burning; oil and gas
enterprise solutions, poverty reduction 335–7
Environment Working Group, implementation of MDGs 54
environmental conflicts 137–44
between BSCs and national policies 323–4
diversity of stakeholders 139–41
diversity of views 141–3
multi-scale 138–9
natural resource management 283
over ecosystem services 43–4
unbalanced power relations 143–4
environmental degradation *see* ecosystem degradation
Environmental Enterprise Corp (EEC) 35
environmental fiscal reform 355
environmental governance 235–53
and science
need to rethink relationship between 237
outdated understandings of 239–41
reconfiguring 247–50
scientific assessments
achieving collective understanding 242–3
conclusions 250–52
epistemological challenges 243–7
need for collective learning 238
review of ecosystem policy and management 237
shift from disciplinary to interdisciplinary perspective 235–6
environmental impact assessments 154–7, 284
environmental impacts, globalization, Latin America 44, 140
environmental knowledge 38, 144
environmental management
lack of 178–9
need for scientific research and technology development 186
neglect of direct investment in 177–8
understanding, lack of 176–7
US hostility to 209
see also ecosystem management
environmental ministries, failure to speak language of finance and

economics 350–51
environmental policy
focus on process targets 175
monitoring implementation and compliance 286
subject to international treaties/ conventions 174
World Bank priorities *see* World Bank
environmental risks *see* ecosystem risks
environmental service fee systems 289–90
environmental sustainability 169–87
factors contributing to lack of progress 173–9
capacity constraints and lack of financial resources 179
insufficient focus on national implementation and coordination 173–5
lack of operational budgets or environmental management 178–9
lack of operational objectives and prioritization 175–6
limited understanding of what works 176–7
neglect of direct investment in environmental management 177–8
getting started 186–7
as main goal of development 196
national MDG-based development strategies
aligning international support with 184–6
importance of consultative processes 181–2
steps in creating 179–84
carry out an MDG needs assessment 182–3
compile and review existing strategies 182
craft a short-term MDG-based strategy 184
develop a medium-term framework for action 183–4
launch an effective and inclusive process 181–2
need for UN harmony on 200
plan of action, lack of 172–3
sharpening targets for 360
environmentalism 305
Equator Initiative Award 39
Equator Principles 209, 335
equity issues, implementation of MA 78, 79–80
Escaping Poverty's Grasp 356
Espinoza, G. 138
Esty, D. 334, 335
ethical basis, global institutions 326–7
ethical investing 209
ethical issue, framing ecosystem degradation as 307–8
Etzioni, A. 249
European Commission (EC) 178
European Union (EU) 52, 218, 325
Eurostat 52
eutrophication 116
export credit and guarantee agencies 336
externalities, internalization, as World Bank priority 199
extractive industries 56, 336
see also oil and gas
Extractive Industries Transparency Initiative (EITI) 336
Extractive Industry Review 194, 202

facilities managers 64
fairness, ecosystem stewardship 305–7
farm policies, water pollution 91
Fiji 29–30, 36, 45
finance, failure of ministries to speak the language of 350–51
financial advisors 64
financial challenges, ecosystem security 283
financial evaluations, inclusion of ecosystem risk in 60–61
financial incentives, aligning with ecosystem stewardship 57–65
financial markets, disservice to ecosystem management 228
financial resources, lack of 179
Finnemore, M. 307

fish landings 4, 12
fish stocks 12, 58, 340
fisheries 29–30, 49, 204–5, 332
see also aquaculture
Fishkin, J.S. 249
Florini, A. 51
Food and Agriculture Organization (FAO) 4, 37, 257, 331
food security 88–90, 95–6, 103–5
forest certification 56
forest ecosystems, international trade 95
forestry
employment 332
policies on sustainable 120–21
forests
degradation and disease 300
imagining revenue potential of 63
information through satellite imagery 34
and poverty alleviation 84–6, 95
World Bank policy priorities 203
see also deforestation; reforestation; tree plantations
fragile lands, population distribution 134, 135
free trade 220–21
Freedom House 51
freshwater fisheries 205
Frier, C. 64
fuel efficiency 200
FutureShop 225

G20 countries 331, 342, 343
Gardiner, P. 35
Gates, W. 300
GDP, as key indicator of development 165, 167
Gelbspan, R. 238
gender equality 287
General Electric 333–4
geospatial mapping 31–2, 300
Ghana 55, 81
Ghanime, L. 175
Giddens, A. 125
Gillis, N. 35
Gleneagles G8 Summit 169
global accords 235–6, 241
Global Biodiversity Assessment 268, 271
global ecosystem change 125–47
as biophysical reflection of globalization 126–37
correspondence with scale and rate of globalization 137
cost and benefit accumulation correspondence with political and economic patterns 127–33
distribution of vulnerabilities, correspondence with the distribution of socioeconomic vulnerabilities 133–5
social interconnectivity, correspondence to increase in interconnectivity in ecosystems 136–7
conclusions and recommendations 144–7
local environmental conflicts 137–44
Global Environmental Facility (GEF) 157, 160, 194, 195
Global Environmental Outlooks 150
Global Forest Watch (WRI) 34
Global Leaders Forum (L20) 330–31, 342–3
Global Legislators for a Balanced Environment (GLOBE) 326
global level
ecosystem management, fairness issue 306
mainstreaming MA results, through MDGs 75–6, 92–9
Global Orchestration scenario 217, 220–21, 222
Global Reporting Initiative (GRI) 56, 230
global trade, ecosystem health 116–17
globalization
and global ecosystem change 125–47
social and environmental impacts, Latin America 44, 140
see also smart globalization
Goldman Sachs 40, 41, 60, 61, 64
Goodland, R.J.A. 190–212
Gough, M. 238
governance

developing integrated systems 282–4
ecosystem degradation and need for major change 13–14
as key to ecosystem health and human well-being 121
need for participation, transparency and accountability 155
need for World Bank to invigorate improvements in 197
see also environmental governance
Governing Council of UNEP 355
Government of Sri Lanka (GOSL) 83
grain-based diets, promotion of 203–4
graphic images, for conveying message of ecosystem degradation 308–10
green architecture 206
green revolution 333
green taxation 199
green technology 60
greenhouse gas emissions 103–4, 152, 202
Greenpeace (Argentina) 223

habitat change, impact of 3–5, 136
Hammond, A. 336
Hansen, J. 340
Hawken, P.A. 161, 162
health, and population levels 281
Heavily Indebted Poor Countries (HIPC) 55, 352
Henninger, N. 356
hierarchical framing, ecosystems 244–5
High-Level Policy Dialogue 350
Honduras 55
Hopkin, M. 339
household ecological habit 167
human development reports 150
human use, and ecosystems/ecosystem services 1, 4–7, 10, 11
human well-being
cost to natural capital 11–12
ecosystem degradation and political reform 167
ecosystem health 112–13, 120, 121
ecosystem services 78–9
energy availability 223
prominence given to, by MA 76–7
human-nature balance, implementation of MA 76–7
Hutton, J. 35

IKEA 34
IMAZON 300
Immelt, J. 333–4
incentives, environmental management 157–61, 333–5
income, from ecosystem services 331
India 133, 201, 338
indigenous peoples, consent for projects 40
Indonesia 42, 197, 356
industry-friendliness, of World Bank, need for recalibration 197
information
crucial for ecosystem management 351–2
on ecosystem services
as an aid to decision makers 9–10
developing and using 28–35
legitimacy and credibility of MA as source of 110–12
sharing 284
information and communication technology 30, 43
Initiative for the Integration of Regional Infrastructure in South America (IRSA) 300
input-based schemes, water pollution 90
Institute of Environment and Development (IED) 48, 163, 165
institutional cooperation, conditions challenging 45
insurance industry, environmental risks 209
integrated decision-making 273, 274, 283
Integrated Ecosystems Assessments 339
integrated responses, necessity of 115
inter-ministerial committees 282–3
interactions
and policy options
prioritization and determination of 88–91
perform detailed studies and analyses 91

update and refine results of studies 91–2
use of AIM approach to understand 81
Interagency Task Force on Gender and Water (UN) 37, 287
interdisciplinary scientific assessments *see* scientific assessments
Intergovernmental Panel on Climate Change (IPCC) 76, 192, 256, 258, 260, 261, 266, 267, 268, 269, 271, 315, 340
International Accounting Standards Board 57
International Conference on Population and Development (ICPD) 280
international conventions 50, 268
International Energy Agency (IEA) 335
International Financial Institutions (IFIs) 336
international institutions, adoption of major group model 285–6
international level, ecosystem management 160
International Monetary Fund (IMF) 197, 200
International Network for Environmental Compliance and Enforcement (INECE) 50, 284
international networks 50
international organizations
action agenda for ecosystem services
bring local voices to the table to influence development projects and policies 39–42
decentralize decisions about ecosystem services 37–9
establish conditions for cooperation with communities 43–7
form bridging organizations 47–8
identify and manage trade-offs 31–2
incorporate ecosystem stewardship in performance objectives 63–5
perform regular monitoring and assessment 28–31
publicly decide and track investments in ecosystem services for development 53–5
raise priority of working across levels in national institutions 49–50
vision 23–4
influence on global sustainability 160
MA need for endorsement of 268
support for BSCs 322
international support, national MDG-based development strategies 184–6
international trade, forest ecosystems 95
international treaties, limited impact on policy 174
investment(s)
development of medium-term frameworks 183–4
ecosystem management, at international level 160
in ecosystem services
barriers 348–52
for conservation of 162
inadequacy of 347–8
timing for stimulation of 352–3
tracking 53–5
environmental 358–9
in green technology 60
need for increase in, to meet MDGs 179
see also direct investments; ethical investing; private investment
IPAT equation 316
Irwin, F. 21–65, 346–62
IUCN 9

Janetos, A.C. 254–72
Japan 129, 133
Jhirad, D. 329–44
Johannesburg Plan on Implementation (JPOI) 277, 279, 280
Johannesburg Securities Exchange (JSE) 56, 230
Johannesburg World Summit on Sustainable Development (2002) 169

Jubilee Debt Campaign 306
Juma, C. 183
justice 153–4, 155, 306

Kenya 31
Kenya Atlas 340
knowing, research and education on deliberative processes of 247–8
knowledge
  communication of 163–6
  local vs. scientific, Canaima National Park 142
  see also environmental knowledge; science
Krchnak, K.M. 273–94
Kuhn, T. 343
Kura, Y. 337
Kyoto Protocol 159, 160, 194
KZN Biobase 35

La Vina, Antonio 108–24
Laguna Lake Development Authority 47
The Lancet 177, 186, 351, 358
land conversion 4, 13
land development practices
  use of water fees to help change 62
  using mobile phones to monitor 30
land registration 37
land tenure 36–7, 39
large-scale developments, environmental conflicts 141
Laris, P. 142
Lariviere, R. 143
Lash, J. 334
Latin America
  business schools and dissemination of information 35
  economic analysis of trade-offs being made through infrastructure development 300
  overview of environmental problems 138
  percentage of population living on fragile lands 134, 135
  social and environmental impacts of globalization 44, 140
law, differentiation between hard and soft 283
Leaders Forum of 20 nations (L20) 330–31, 342–3
Lee, L. 175
Lewis, H. 142
Li, L. 48, 149–68
Linking Poverty Reduction and Environmental Management 355
Linn, J. 331, 342
local communities
  action agenda for ecosystem services
    bring local voices to the table to influence development projects and policies 39–42
    decentralize decisions about ecosystem services 37–9
    eliminate perverse subsidies and reform taxation policies 58–60
    ensure individuals and communities have secure rights to ecosystem services 36–7
    establish conditions for cooperation with communities 43–7
    hold elected officials accountable 51–3
    identify and manage trade-offs 31–2
    incorporate ecosystem stewardship in performance objectives 63–5
    perform regular monitoring and assessment 28–31
    support markets and payments for ecosystem services 61–3
    tailor information for citizens, producers and purchasers 33–5
    use co-management practices 48–9
    vision 23
  ecosystem services, decision-making 36
  empowerment to respond to challenge of globalization and ecosystem change 145–6
local environmental conflicts *see* environmental conflicts

local level, ecosystem management 159, 305–6
local ownership, environmental sustainability 186
local people, strengthening rights to use and manage ecosystem services 35–42
Locally Managed Marine Areas in the Pacific 29–30, 36, 45
logging bans 197, 301
Lovins, A. 200
Lucas, N.J. 45, 125–47

MA *see* Millennium Ecosystem Assessment
McIntosh, T. 55, 352
MacNeill, J. 353
Major Group model, adoption by international organisations 285–6
malaria 96, 300, 358
Malaysia 30, 338
Mallaby, S. 193
manager's performance objectives, incorporating ecosystem stewardship goals in 63–5
mangroves, loss of 4, 5–7
mapping tools 31–2, 300
marine habitat destruction 88
marine systems, and economic growth 86
market support, for ecosystem services 61–3
market-based solutions 337–8
Martin, P. 342
Matthew Effect 160
Matthews, E. 356
May, P.F. 227
Mbow, C.T. 142
MDGs *see* Millennium Development Goals
MDMS (making development more sustainable) approach 81, 101
measuring systems, establishing better 338
medium-term investment frameworks 183–4
Meridian Institute 29
Merill Lynch 60
Mexico 42, 133, 356
migration levels, free trade and 221
millennium capital assessment 355
Millennium Challenge Corporation (MCC) 37
Millennium Development Goals (UN) 53, 83, 169, 170
  agenda for poverty reduction 346–7
  ecosystem degradation as barrier to achieving 12–13
  effect of critical ecosystems on 95–9
  effect on critical ecosystems and services 92–5
  financial assistance to meet 185
  Goal 7 *see* environmental sustainability
  implementation 54
  limited public role in tracking 352
  linked with key ecosystem services 80–81
  mainstreaming MA findings 73
  priorities for World Bank policy 198
  progress in achieving 169, 281
Millennium Ecosystem Assessment (MA) 2, 21, 273, 353–4
  ability to evaluate policy choices 264–6
  achievements 172–3
  analysis and recommendations
    communicate knowledge 163–6
    institutionalize ecosystem impact assessments 154–7
    policy actions 267–72, 316
    provide incentives 157–61
    provide justice 153–4
    remove subsidies and tax pollution and waste 161–2
    work for political reform 166–8
  authorization 110–11
  case for biome stewardship councils 315–28
  categorization of ecosystem services 9
  comparison to business audit 18–19
  conceptual framework 7, 8, 78–80
  credibility and legitimacy 110–12, 255–7
  ecosystem definition 8–9
  feedback from 3–7
  findings 10–14

challenges posed by 150–53, 296–7
communicating the messages 307–10
ecosystem threats and opportunities 329–31
integrating into national development plans 117–21
mainstreaming and implementing *see* Action Impact Matrix
marketing 266–7
political interpretation 126–44
political resonance 298
future demand for ecosystem services 63
governance and conduct 111
implications for the World Bank 194–6
lessons
priorities and imperatives 122–4
reflections on 254–72
limited mandate 173
long-term sustainability and effectiveness of 266–7
messaging strategies to promote ecosystem protection 298–307
methodology for tracking status and trends of ecosystem services 28–9
multi-scale approach 3, 137–8
political actors, requirements of 262–4
political targets and clientele 257–62
response options 297
Responses Working Group 316
reviewing assessment experience 123
scenarios 13–14, 151
approach to ecosystem management 215–34
implications on development choices 118
scientists *see* scientists
sub-global *see* Sub-Global Assessments (SGAs)
substantive issues 115–17
uniqueness of 3
usefulness for national planning 112–15
warning about condition of ecosystem services 1–2
Millennium Project 169, 179, 180, 181, 183, 184, 185, 186, 351
collaboration with decision makers 173
programmes 96–9
Task Force on Education and Gender Equality 287
Task Force on Environmental Sustainability 173, 176, 177
Millennium Promise 186
Millennium Village Initiative 186
Minister of National Development Planning (MTDP), policy priorities 108–24
Ministry of Environment and Natural Resources 84
mixed agriculture, as World Bank policy priority 204
mobile phone technology 30, 222–3
monitoring systems, establishing better 28–31, 338
Monterey Conference on Financing for Development 169
Montreal Protocol 236
moral issue, framing ecosystem degradation as 307–8
Moss, R. 265
Mozambique 37, 197
multi-scale analysis, development decisions 114–15
multi-scale assessments 3, 137–8
multi-scale environmental conflicts 138–9
Multi-Stakeholder Dialogue Sessions 285
multilateral development banks (MDBs) 336, 354
Multilateral Environmental Agreements (MEAs) 88, 96, 157, 174, 182, 186, 187
multiplier effect, consumption tax 226
Munasinghe Institute for Development (MIND) 81
Munasinghe, M. 1–19, 73–105

Namibia 39

National Academies of Science 340, 341
national budgets, World Bank scrutiny of 199
National Commissions on Macroeconomics and Health 358
national development goals
determining the most important 83
identifying effect on ecosystems 84–6
identifying impact of ecosystems on 86–8
limited public role in setting and achieving 351–2
restoring or rehabilitating ecosystems 119
national development planning, usefulness of MA 112–15
national development plans, integrating MA findings into 117–21
national development strategies, MDG-based 179–86
National Ecological Observatory Network (NEON) 339
National Environmental Action Plans (NEAPs) 88, 182
national governments
action agenda for ecosystem services
bring local voices to the table to influence development projects and policies 39–42
decentralize decisions about ecosystem services 37–9
eliminate perverse subsidies and reform taxation policies 58–60
ensure individuals and communities have secure rights to ecosystem services 36–7
establish conditions for cooperation with communities 43–7
hold elected officials accountable 51–3
identify and manage trade-offs 31–2
incorporate ecosystem stewardship in performance objectives 63–5
increase corporate transparency 55–7
perform regular monitoring and assessment 28–31
publicly decide and track investments in ecosystem services for development 53–5
raise priority of working across levels in national institutions 49–50
support markets and payments for ecosystem services 61–3
use co-management practices 48–9
vision 23
ensuring a strong role and voice in scientific assessments 267–8
involvement in Millennium Assessment 258–9
need to lead the way in environmental policy 193
networks 325–6
national implementation mechanisms, lack of, environmental sustainability 173–5
national institutions, raising priority of working across levels 49–50
national level
ecosystem management, fairness issue 306
mainstreaming MA results through SD strategy 74–5, 82
Sri Lanka 83–92
National Oceanic and Atmospheric Administration (NOAA) 339
National Research Council (NRC) 337
natural assets, decline in 1
natural capital 3–7, 11–12, 161, 349
natural disasters, distribution of vulnerabilities 135
natural resource consultants, involvement in MA 258
natural resource management 39, 283
Nature, Wealth and Power 355
needs assessments, development strategies 182–3
Newfoundland cod stocks, collapse of

12, 340
Newton, I. 239, 240
Nigeria 48, 56
Nissanoff, D. 225
nitrogen challenge, responding to 123
nitrogen (synthetic) fertilizer, applied to crops 4–5
non-governmental organizations (NGOs)
distribution of cell phones in exchange for reporting on environmental conditions 222–3
involvement in ecosystem impact assessments 156–7, 158
monitoring implementation and compliance of environmental policies 286
rise in citizen participation 249
support for national MDG-based development strategies 185
non linear responses, to catastrophic events 301–2
Norgaard, R.B. 235–53
norm shift, communicating MA messages 307–8
North America Free Trade Agreement (NAFTA) 218
north-south balance, implementing MA 77
Nuclear Non-Proliferation Treaty 324–5
nutrition, agriculture for 202

Oberman, J. 30, 223
ocean acidification 204–5
oceans, priorities for World Bank policy 204–5
Official Development Assistance 185, 193, 348
oil and gas sector 40, 56
Olson, D.M. 320
operational budgets, lack of 178–9
operational objectives, lack of 175–6
Order from Strength scenario 217, 219, 220
Organization for Economic Co-operation and Development (OECD) 58, 127, 129, 130, 193, 348
Ostrom, E. 38, 39, 49
Our Common Future 352
outcome-oriented environmental strategies 182–3
overconsumption, reduction, as World Bank priority 196
Ozawa, C.P. 238

paddler and canoe analogy 308–9
Panama Canal 62
Parana basin 135, 136
participation, in ecosystem impact assessments 156–7
participatory approach, decision-making 152
Partnership for Principle 10 (PP10) 42, 356
partnerships
with scientific communities 340–41
strengthening rights of local people 37
voluntary 286
see also Poverty-Environment Partnership
Payment for Ecosystem Services (PES) 289–90
payment schemes, to support ecosystem services 61–3, 200
Pemon knowledge, savannah burning 38
performance objectives, incorporating ecosystem stewardship goals in 63–5
Petkova, E. 34, 41
Philippines 47, 81, 301
Planet Earth Ltd 17–18
Pogge, T. 327
policy choices
ability of MA to evaluate 264–6
devoting resources for analysis of potential 270–72
policy frameworks, adaptive management 288–9
policy makers, priorities for 108–24, 190–212
Policy Priorities Report 191
political organizations, facilitating emergence of 146–7
political patterns, ecosystem change

and accumulation 127–33
political reform 166–8
political trigger points, turning ecological trigger points into 301–2
political will, implementation of environmental policy 173, 207–8
politicians, overcoming short time-frame of 52
Pollutant Release and Transfer Registers 55
pollution
cap and trade schemes 220
impact of 3–5
increasing interconnection between ecosystems and 136
priorities for World Bank policy 205–6
taxing 161–2, 202, 205–6
see also water pollution
Pollution Prevention Pays 300
poor, protecting buying power of 226
population distribution, fragile lands 134, 135
population growth 221, 280–82
population stability 196–7, 198–9
post hoc analyses, of policy choices 270–71
Potable Water and Sanitation System Law 47
poverty
and ecosystem degradation 13, 113, 316, 346
implementation of MA 77–8
need for UN harmony 200
undemocratic countries 40
World Bank environmental policy 199
poverty alleviation programmes 84–6, 95
poverty maps 31
poverty reduction
'3 x M' method for promoting 43
by enabling poor to use ecosystem services to improve their livelihoods 347–8
enterprise solutions 335–7
environmental foundations 356
as main goal of development 196
MDG agenda for 346–7
see also Commission on Macroeconomics and Ecosystem Services for Poverty Reduction
Poverty Reduction Strategies 353
Poverty Reduction Strategy Papers (PRSPs) 88, 173, 174, 175, 183, 347
Poverty-Environment Partnership 39, 348, 350, 353, 354–6, 357, 362
power relations, environmental conflicts 143–4
Powers, N. 35
Prahalad, C.K. 336
precautionary principle 199–200, 288
predictive science 239–40
pricing
carbon dioxide emissions 202
development objectives 300
water, in rural areas 89–90
private investment, and poverty reduction 336–7
private sector, participation in scientific assessments 270
proactive decision-making 218, 220
procedural justice 155
producers, tailoring information for 33–5
progressive governance, understandings of science in 239–41
protected areas 26, 36, 38, 121, 141–2
provisioning services 1, 9, 11, 217, 227
public planning tool 35
Publish What You Pay 51
purchasers, tailoring information for 33–5

Quechua communities 45–7

rainfall, and water resources 104–5
Rajan, S.C. 315–28
Ramsar Convention on Wetlands 110
Ranganathan, J. 21–65
Rao, S. 338
rate of globalization, and rate of ecosystem change 137
reactive decision-making 216–17

Reddy, C. 175
Reed, D. 43, 356
reforestation 62
regulating services 1, 9, 11–12
renewable energy 196, 201
research communities
 action agenda for ecosystem services
  bring local voices to the table to influence development projects and policies 39–42
  eliminate perverse subsidies and reform taxation policies 58–60
  establish conditions for cooperation with communities 43–7
  form bridging organizations 47–8
  frame messages that resonate with the public 32–3
  identify and manage trade-offs 31–2
  incorporate ecosystem stewardship in performance objectives 63–5
  perform regular monitoring and assessment 28–31
  tailor information for citizens, producers and purchasers 33–5
  use co-management practices 48–9
  vision 22–3
resource management 39, 274, 275–84, 283
resource user communities, inclusion in ecosystem monitoring 29–30
Revenga, C. 356
Ribot, J. 38
rights
 of local people, strengthening 35–42
 see also water, rights
Rio Declaration 41, 146, 200, 274, 288
Rio institutions 146
River Rehabilitation Councils 47
Rodriguez, I. 38, 44, 45, 125–47
Rodriguez, M. 138
RWSS (Report on the World Social Situation) 21
Sachs, J.D. 300, 351, 357, 358
'safeguard' policies 193
Salzman, J. 47, 63
Samoan Fisheries Division 49
Sandel, M.J. 249
satellite tracking 30, 34, 338
Sathirathai, S. 6
savannah burning, Pemon knowledge 38
Save the Children 57
scenario-building 265–6
scenarios
 scientific assessments 271–2
 see also Millennium Ecosystem Assessment, scenarios
Scenarios Working Group (MA) 151
Schmidt-Traub, G. 169–87
Schneider, S. 266
science, and environmental governance
 need to rethink relationship between 237
 outdated understandings of 239–41
 public funding 248–9
 reconfiguring 247–50
science-based decision-making 288
scientific assessments
 devoting resources to analyse potential 270–72
 ensuring an ongoing process 269
 ensuring private sector participation 270
 ensuring strong role and voice for national governments 267–8
 see also Millennium Ecosystem Assessment
scientific communities, building partnerships with 340–41
scientific knowledge 163–4
scientists
 Millennium Ecosystem Assessment
  achieving collective understanding 236, 242–3
  balance between natural and social 255
  Board of Assessment Panel 255
  conclusions 250–52
  credibility 256–7
  epistemological challenges 243–7
  geographical distribution 255

invitations for participation 256
mandate 171–2, 173
need to rethink relationship of science and governance 237
review of ecosystem policy and management literature 237
review of the results of various chapter teams 255–6
selection of 3
shift from disciplinary to interdisciplinary perspective 235–6
thinking about sustainability 9
Scott, L. 334
SD *see* sustainable development
sector-based recommendations 191
sectoral planning, integrating MA findings in 118–19
Senegal 38
service areas, determining those relevant to SD 83
service markets 62–3
sewerage/treatment, user charges 91
Seymour, F. 32, 296–310
shame-and-blame, ecosystem degradation 190
Shell 48, 158, 159, 336
short-term decision making 229–30
short-term MDH-based development strategies 184
shrimp aquaculture 5–7, 57
Sikkink, K. 307
Slaughter, A.M. 326
smart globalization 329–44
Snel, M. 356
social impact assessment, West-End Gas Pipeline 48, 158
social impacts, globalization, Latin America 44, 140
social insurance, ecosystem maintenance 32–3, 303–5
social interconnectivity, correspondence to interconnectivity in ecosystems 136–7
social security debate, United States 303–4
social systems, inclusion in hierarchy of ecosystems 244–5
Socialist Republic of Vietnam 183
socially responsible community investment 61, 62
socially responsible corporate funds 209
socioeconomic vulnerabilities, correspondence with vulnerabilities created by global ecosystem change 133–5
solar electricity, as World Bank priority 202
South Africa 31–2, 35, 45, 56–7, 133
south-north balance, implementing the MA 77
Southeast Asia 300–301
Southey, S. 35
Special Report on Land-Cover, Land-Use and Land-Use Change (IPCC) 271
special reports, of MA 266
Speth, G. 209
Sri Lanka, sample application of AIM 83–92
stakeholders
diversity of, environmental conflicts 139–41
improving engagement in ecosystem security 284–7
selling ecosystem stewardship as an insurance policy 304–5
standards, ecosystem monitoring and assessment 30–31
Status Quo Projection 216, 219–20, 228
Stern, P. 249
stock exchanges 56, 230–31
Stockholm Conference on the Human Environment (1972) 165, 288
Stockholm International Water Institute 202
Strategy Group (MDG) 54
Stratospheric Ozone Assessments 258, 260
'streamlining safeguard' policies 193
structural frameworks, influence on balance of power 144
Sub-Global Assessments (SGAs)
BSC review of 322
Caribbean Sea 47

catalyzing more 123
China 259, 262–3, 265
locations 2
models of bridging organizations 47
reframing and collaborative thinking 245–6
relevance for ecosystem stakeholders 151–2
selection and conducting of 259
South Africa 31–2
Vilcanota 45–6
subsidies, elimination of perverse 58–60, 161, 199, 226
Summary of the Intergovernmental Preparatory Meeting of the 13th Commission on sustainable development (SD) 287
supporting services 9
sustainability, environmental *see* environmental sustainability
Sustainable Development Issues Network (SDIN) 286
sustainable development (SD)
Brundtland Commission definition 9
changing the discourse on 123–4
evolution of concept 9
Sri Lanka 83–92
strategy, mainstreaming the MA results through 74–5
triangle 80, 101
sustainable enterprise 35
sustainable tourism, policies on 121
sustainomics 81, 101
Swiss Agency for Development and Cooperation 290

Tanzania 37
tax distribution, ecosystem-based 227–8
taxation policies, reform of 58–60
taxes
consumption-based 59, 224–7
on pollution and waste 161–2, 202, 205–6
technical advice 185
TechnoGarden scenario 217, 220, 221, 222, 223, 315
Ten-Point Plan 209
tenure reform 36–7, 39
Thailand 5, 37
'ticking clock' analogy 33, 309
time dimension, MA policy choices 265
time horizons, lengthening 229–30
tipping points, understanding and anticipating 340–41
Torra Conservancy 39
Toulmin, C. 348
tourism, policies on sustainable 121
trade-offs
framing inappropriate, as wasteful and inefficient 298–301
identifying and managing 31–2
need for careful consideration of 114
traditional beliefs, accommodation of 45–6
training programmes, teaching applied knowledge 164
trans-boundary approach (AIM) 101
transformative dialogue 147
transport sector, priorities for World Bank policy 200
Treaty of Westphalia 216
tree plantations, acceleration, as World Bank priority 203
trust-building 249–50, 283
Tumushabe, G.W. 52

Uganda 37, 42, 52, 356
Ukraine 42, 356
United Kingdom 133, 356
concentration of economic activity 129
Department for International Development 355
improvement of access rights 42
United Nations
Commission on Sustainable Development (CSD) 276–80, 284, 285, 286, 287
Conference on Environment and Development (UNCED) 276
Convention on Climate Change 146, 192
Convention to Combat Desertification 110, 146
Development Programme (UNDP) 39, 40, 75–6, 83, 150, 160, 178,

186, 187, 257, 331, 348, 350, 355, 360, 361
Educational, Scientific and Cultural Organization (UNESCO) 164, 165
Environment Programme (UNEP) 29, 58, 96, 99, 133, 172, 187, 192, 257, 285, 286, 331, 348, 353, 355, 360, 361
Food and Agriculture Organization (FAO) 4, 37, 257, 331
General Assembly Resolution (1992) 288
General Assembly Resolution (1993) 276
Interagency Task Force on Gender and Water 37, 287
Millennium Development Goals *see* Millennium Development Goals
Millennium Project *see* Millennium Project
Population Fund (UNFPA) 281
Rio Declaration (1992) 41, 146, 200, 274, 288
support for national MDG-based development strategies 184, 185, 186
United States
agency responsibility and development of trust 250
anti-environmentalism 192, 193, 207, 209
business schools and dissemination of information 35
concentration of economic activity 129
Controlled Substance Act (1972) 224–5
debate over social security 303–4
divided-world format of scholarly disciplines 239
framing inappropriate trade-offs as wasteful and inefficient 298–301
Internal Revenue Service (USIRS) 58
low level of visibility of MA process and findings 260–61
National Assessment of climate impacts 269
need for incentives to internalize coal pollution costs 201
norm shifts 307
number of dams 133
response to 9/11 301–2
urban sector
environmental conflicts in 142–3
priorities for World Bank policy 206
user charges, pollution 91

Valle de Lurin 143
Venezuela 38, 142
Vilcanota sub-global assessment 45–6
Vithanage, J.C. 105
voluntary partnerships 286

Walmart 333, 334
waste, taxes 161–2
water
adaptive management 288–9
amount impounded by dams 5, 132
Andean traditional view of 46
conservation, as World Bank priority 202–3
intergovernmental cooperation, Africa 50
markets 90
pricing/fees, in rural areas 62, 89–90
resources, and food security 88–90, 103–5
rights, tradable 90
scarcity 103
Water Conservation Fund (FONAG) 61, 289, 290
water pollution 50–51, 90–91, 96
Water War 46–7, 143
waterborne disease 96
Watson, Bob 194
wealth creation, and ecosystem damage 316
web-based calculator 310
Wellington, F. 334
West-End Gas Pipeline 48, 157, 158
wetlands, employment and preservation of 86
White, R. 356
Wickramagamage, P. 104
wild harvesting, limitations on 203

Wildlife Agency 35
Winning the Oil End Game 200
Winston, A. 334, 335
women, as guardians of ecosystem services 37
Women's Association of Water Ministers 50, 283
Wood, S. 356
World Bank 41, 53, 55, 90, 134, 150, 160, 175, 178, 182, 185, 186, 331, 348, 352, 355, 357, 360
  contribution to environmental degradation 194–5
  environmental policy priorities 190–212
    anti-environment movement 192–3
    general policy fundamentals 196–7
    implications of MA 194–6
    political will to implement 207–8
    signs of hope 208–10
    specific policies 198–207
    targeting decision makers 190–92
  opposition to log export bans 197
World Business Council for Sustainable Development 29
World Commission on Dams 5, 40, 132, 191, 194, 202
World Commission on Environment and Development (WCED) 9, 352
World Conservation Monitoring Centre (UNEP) 29
World Conservation Union (formerly IUCN) 9
world development reports 150
World Health Organization (WHO) 160, 178, 186, 361
World Resources Institute (WRI) 2, 24, 29, 30, 31, 34, 35, 37, 38, 39, 40, 45, 49, 55, 60, 272, 296, 331, 337, 348, 354, 356, 360
World Resources Report (2005) 331
World Summit (New York) (2005) 169
World Summit on Sustainable Development 280, 285
World Trade Organization (WTO) 160, 200, 285, 286, 325, 331
World Wildlife Fund (WWF) 30, 43, 338
WorldWatch 203
Worldwide Fund for Nature 355–6

Zucchetti, A. 143